Environmental Organizations in Modern Germany

Monographs in German History

Volume 1
Osthandel and Ostpolitik: German Foreign Trade Policies in Eastern Europe from Bismarck to Adenauer
Mark Spaulding

Volume 2
A Question of Priorities: Democratic Reform and Economic Recovery in Postwar Germany
Rebecca Boehling

Volume 3
From Recovery to Catastrophe: Municipal Stabilization and Political Crisis in Weimar Germany
Ben Lieberman

Volume 4
Nazism in Central Germany: The Brownshirts in 'Red' Saxony
Christian W. Szejnmann

Volume 5
Citizens and Aliens: Foreigners and the Law in Britain and the German States, 1789–1870
Andreas Fahrmeir

Volume 6
Poems in Steel: National Socialism and the Politics of Inventing from Weimar to Bonn
Kees Gispen

Volume 7
"Aryanisation" in Hamburg
Frank Bajohr

Volume 8
The Politics of Education: Teachers and School Reform in Weimar Germany
Marjorie Lamberti

Volume 9
The Ambivalent Alliance: Konrad Adenauer, the CDU/CSU, and the West, 1949–1966
Ronald J. Granieri

Volume 10
The Price of Exclusion: Ethnicity, National Identity, and the Decline of German Liberalism, 1898–1933
E. Kurlander

Volume 11
Recasting West German Elites: Higher Civil Servants, Business Leaders, and Physicians in Hesse between Nazism and Democracy, 1945–1955
Michael R. Hayse

Volume 12
The Creation of the Modern German Army: General Walther Reinhardt and the Weimar Republic, 1914–1930
William Mulligan

Volume 13
The Crisis of the German Left: The PDS, Stalinism and the Global Economy
Peter Thompson

Volume 14
"Conservative Revolutionaries": Protestant and Catholic Churches in Germany After Radical Political Change in the 1990s
Barbara Thériault

Volume 15
Modernizing Bavaria: The Politics of Franz Josef Strauss and the CSU, 1949–1969
Mark Milosch

Volume 16
Sex, Thugs and Rock 'N' Roll. Teenage Rebels in Cold-War East Germany
Mark Fenemore

Volume 17
A Single Communal Faith? The German Right from Conservatism to National Socialism
Thomas Rohrämer

Volume 18
Selling the Economic Miracle: Economic Reconstruction and Politics In West Germany, 1949–1957
Mark E. Spicka

Volume 19
Between Tradition and Modernity: Aby Warburg and Art in Hamburg's Public Realm 1896-1918
Mark A. Russell

Volume 20
Cultures of Abortion in Weimar Germany
Cornelie Usborne

Volume 21
Environmental Organizations in Modern Germany: Hardy Survivors in the Twentieth Century and Beyond
William T. Markham

ENVIRONMENTAL ORGANIZATIONS IN MODERN GERMANY

Hardy Survivors in the Twentieth Century and Beyond

William T. Markham

Berghahn Books
New York • Oxford

Published in 2008 by

Berghahn Books

www.berghahnbooks.com

Library of Congress Cataloging-in-Publication Data

Markham, William T., 1946–
Environmental organizations in modern Germany : hardy survivors in the twentieth century and beyond / William T. Markham.
p. cm. — (Monographs in German history ; v 21)
Includes bibliographical references and index.
ISBN 978-1-84545-447-0 (hardback : alk. paper)
1. Environmental protection—Germany—History—Societies, etc. I. Title.

TD171.5.G3M36 2008
333.7206'043—dc22

2008017885

British Library Cataloguing in Publication Data

A catalogue record for this book is available from the British Library

Printed in the United States on acid-free paper

ISBN: 978-1-84545-447-0 hardback

This book is dedicated with great affection to Charles M. Tolbert,
mentor and friend of over forty years.

Contents

List of Abbreviations

BBU Bundesverband Bürgerinitiativen Umweltschutz
Federal Alliance of Citizens' Initiatives for Environmental Protection

BfV Bund für Vogelschutz
League for Bird Protection

BH Bund Heimatschutz
League for Homeland Protection

BIs Bürgerinitiativen
Citizens' initiatives

BN Bund Naturschutz in Bayern
Bavarian League for Nature Protection

BUND Bund für Umwelt und Naturschutz Deutschland
German League for Environment and Nature Protection

CDU Christlich Demokratische Union
Christian Democratic Union

CSU Christlich – Soziale Union
Christian Social Union

DGB Deutscher Gewerkschaftsbund
German Federation of Labor

DNR Deutscher Naturschutzring
German Nature Protection Ring

EEB European Environmental Bureau

FDP Freie Demokratische Partei
Free Democratic Party

GNU Gesellschaft für Natur und Umwelt
Society for Nature and Environment

NABU Naturschutzbund Deutschland
German Nature Protection League

NSM New Social Movement

ÖDP Ökologisch — Demokratische Partei
Ecological Democratic Party

PDS Partei des Demokratischen Sozialismus
Party of Democratic Socialism

SMO Social Movement Organization

SPD Sozialdemokratische Partei Deutschlands
German Social Democratic Party

WWF Worldwide Fund for Nature

ACKNOWLEDGEMENTS

As a person who spent most of his career writing journal articles, I used to glance through the long lists of acknowledgements at the beginning of scholarly books with amazement. How could the authors run up so many intellectual and personal debts in the course of a single research project, I wondered? And why did they seem to feel so deeply about them? Now, at the end of a nine-year project of my own, I know the answers to both those questions, for I too now have a long list of people and organizations to thank and a deep sense of gratitude to all those who helped along the way.

First, my thanks to Helmut Wiesenthal, Emeritus Professor of Political Science at Humboldt University. In 1999, he responded positively to a request for a visiting appointment from a stranger who wanted to begin work in a new research area in a new country. Without the nine-month research appointment he arranged for me at Humboldt's Institute for the Social Sciences, there could have been no book. Professor Wiesenthal also used his many connections to the German environmental movement and Green Party to help me arrange many of the interviews for this book. I hope the book represents a good return on his investment.

My stay in Berlin was made possible by a paid research leave from my university, the University of North Carolina at Greensboro (UNCG). A grant from the German Academic Exchange Service helped me improve my German by attending a two-month Goethe Institute course the summer before my visit. I am grateful, too, for the support of three Berlin friends and colleagues who made my visit a more pleasant one and have provided continued support for my project over all these years, Christine Hannemann, Doris Felbinger, and Petra Langheinrich. Thanks too to Peter Vagts and the members of the Humboldt University choir, who so warmly welcomed a rather lonely visitor to their ranks.

A second grant from the German Academic Exchange Service made it possible for me to return to Germany during the summer of 2001 to work at the Museum and Archive for the History of Nature Protection in Germany and at several other Bonn area libraries. I am grateful to Hans-Werner Frohn and the staff at the archive for their hospitality and assistance, to Anna-Katharina Wöbse and Helge May for insights into the history of the German League for Bird Protection (now NABU), and to Herr May for access to its archive.

In 2004, Professor Angelika Wolf of the University of Duisburg-Essen became the second German professor to gamble on inviting an unknown American to her university. With financial support from a Fulbright Fellowship and supplemental funding from UNCG, I was able to spend half a year on the Essen campus, teach a course, and make substantial progress on my research. I am also grateful to Professor Wolf for her insights about the German environmental movement and her former employer, the German League for Environment and Nature Protection. Angelika, her husband Gerd, Gerd Mahler, and the members of the University Choir made my and my wife's stay in Essen a pleasant one. Thanks also to my good colleague Dr. Ingo Bode and to Professor Jürgen Krüger for arranging for me to teach a course on German and US environmental organizations at the Duisburg campus of the university.

I also benefited as a Fulbright fellow from two weeks of research at the Research Archive for Environmental History at the Technical University in Neubrandenburg, which contains a treasure trove of documents about environmentalism in the GDR. I am grateful to Professor Hermann Behrens, the Director, and to Dr. Jens Hoffmann and Doris Thiel, the archive's librarian, for their assistance with using the archive and for their warm welcome during a very cold North German February.

In 2005, Arthur Mol and Kris van Koppen of the Environmental Policy Group at Wageningen University invited me to spend a semester as a visiting scholar in Wageningen. Financial support from my visit came from the Wageningen Institute for Environment and Climate Research. The uninterrupted research time, excellent facilities, good colleagueship of the group's faculty, and pleasant environment all helped to move this book, as well as a co-edited book on nature protection in nine countries, several long steps toward completion. Working with my co-editor and valued colleague Kris van Koppen, along with the colleagues from seven other nations who wrote chapters about individual countries for the edited book, proved to be of enormous value in putting the German case in comparative context. I am especially grateful to Kris and to Magnus Boström and Chris Rootes for helpful comments on my drafts, and to Kris and his wife Stella for their hospitality to me and my wife. I am also indebted to Steve Kroll-Smith, former Head of the Sociology Department at UNCG, for supporting my second leave in two years, and to my departmental colleagues at UNCG for their forbearance during my absence.

In addition to the persons mentioned above, a number of German colleagues have offered me encouragement, helpful comments, and invitations to deliver guest lectures. These include, among others, Hellmuth Lange, Tine Stein, Matthias Groß, Fritz Reusswig, Dieter Rink, Hans-Dieter Hellige, Winfried Osthorst, Harald Heinrichs, and Martin Jänicke.

Several research assistants have contributed to this book over the years. A series of exchange students from Germany, most notably Marga Avci and Caroline Kanter, patiently tutored me in spoken German. Almost a decade ago, Sabrina Broselow Moser collaborated with me on the content analysis reported in Chapter 7 and helped me improve my German as well. Brandi McCullough proofread rough drafts and cross-checked references, and Brian Boylston repeated this onerous task for the final drafts, patiently and accurately working through chapter after chapter for both my book projects. Nicole Schwindeler, a graduate student at Humboldt University, performed yeoman service proofreading and verifying the German references, and Liz Wilkinson gave the entire manuscript a final careful proofreading. Gaylor Callahan and rest of the UNCG interlibrary loan department cheerfully and patiently obtained one obscure German book after the other for me over the course of a half decade. My thanks also to Richard Koritz for his comments on several chapters.

Many of my friends and relatives have had to put up with my frequent unavailability and occasional grumpiness, especially during the seemingly endless months when I was "wrapping things up." My thanks go especially to Lori, Priya, Paul, David, Marga, Jan, Jerry, Anne, and Fred. My deepest gratitude, however, is reserved for Peggy, who watched me climb the stairs to my upstairs study night after night and weekend after weekend to "work on the book" and bore it all (usually!) with amazingly good humor. She also took time away from her own work to take care of some tasks I didn't have time to do, gave the bibliography its final proofreading, and was a well of emotional support that never ran dry.

Chapter 1

INTRODUCTION

This book combines a sociological analysis of the history of German environmental organizations in the twentieth century with an analysis of the dilemmas and strategy decisions that confronted them as they entered the twenty-first. The history is a fascinating and sometimes dramatic story of organizations that have doggedly pursued environmental protection through difficult times—times of hyperinflation and war, National Socialist rule, postwar devastation, state socialism, and confrontations with the authorities during the 1970s and 1980s. It is also a history punctuated by the organizations' encounters with powerful social movements from across the political spectrum—homeland protection and socialism in the early years of the twentieth century, the Nazi movement, the antinuclear and new social movements of the 1970s and 1980s, and the movement that brought down the German Democratic Republic and led to reunification.

But turbulent times and confrontational social movements are far from the whole story here. Of almost equal interest are the environmental organizations' struggles to remain relevant and continue their work in periods, like the current one, when environmental concerns were not at the top of the national agenda. For whether in turbulent times or quiescent, environmental organizations must obtain the resources they need, structure themselves effectively, and adapt their programs and goals to the world around them. This is seldom easy. Their organizational environments are complex, and the social actors on which they depend for their survival and legitimacy are rarely in agreement about what they should be doing and how they should be doing it. Strategy formulation in such circumstances becomes a balancing act, interesting in its own right, impressive when executed well, and worthy of scholarly attention to understand how it works.

Four organizations are the focus of my analysis of present-day German environmental organizations. They are the German chapters of Greenpeace and the Worldwide Fund for Nature (WWF), the German League for Environment and Nature Protection (*Bund für Umwelt und Naturschutz Deutschland:* BUND), and the German Nature Protection League (*Naturschutzbund Deutschland:* NABU)—formerly the League for Bird Protection (*Bund für Vogelschutz:* BfV). The last two are the German counterparts of well known British and US organizations, such as the Sierra Club, Audubon Society, Royal Society for the Protection of Birds, and Friends of the Earth. I selected these organizations because they share several specific characteristics. All pursue multiple environmental goals. And except for Greenpeace's opposition to military conflict—a borderline case in any event—they do not have non-environmental goals. All are mass membership organizations, whose supporters are individuals, not other organizations, and their memberships far exceed those of other organizations with similar characteristics. All operate on the national level and are formally organized, with bylaws, formally elected or appointed officers, and formalized procedures.

I have also used these characteristics as a guideline for selecting organizations to include in the historical analysis, although it was sometimes necessary to bend the rules a bit to fit historical circumstances. Environmentalism, conceptualized by both contemporary scholars and environmental organizations as a combination of efforts to protect nature, human health and well-being, and natural resources (Markham and van Koppen, 2007), is not yet four decades old. Before that, these objectives were typically pursued by different groups, and only nature protection advocates showed much propensity to form large formalized organizations of national scope.

Consequently, the history of German "environmental" organizations before the 1970s is perforce mainly a history of nature protection organizations. A few of these, such as the League for Bird Protection, pursued nature protection in relatively "pure" form, but I have also included some that gave nature protection a high priority but combined it with other goals. Also included is the Bavarian League for Nature Protection (*Bund Naturschutz in Bayern:* BN), a regional organization that occupies a key role in the history of German environmentalism and was the predecessor organization to BUND.

Environmental organizations like the ones described in this book are among the most persistent, adaptable, and influential forms of environmental action in Western democracies (Bammerlin, 1998; Rawcliffe, 1998; Bosso, 2005). All three of the largest environmental organizations operating in Germany on the eve of World War I still exist today, and organizations like the Sierra Club and Royal Society for the Protection of Birds have equally long pedigrees. The number of supporters of the four largest German environmental organizations ranges today from over a quarter million to more than a half million, many times the Green Party's 44,000 members, and their combined membership far exceeds that of either of Germany's two largest political parties (Deutscher Bundestag, 2006). Indeed, in a survey conducted for the German Ministry for Families, Seniors,

Women, and Youth, 82 percent of those who reported doing volunteer work for the environment or animal protection said they did so in the context of an environmental or animal protection organization (Gensicke, Picot, and Geiss, 2005). This situation is not unique to Germany. Elsewhere in Europe (Dalton, 2005; Bozonnet, 2004, 2005) and in the US (Bosso, 2005) environmental groups attract large memberships.

Unlike political parties, environmental organizations are free to focus all of their attention on environmental issues. Their professional staffs and financial resources equip them especially well to educate the public about environmental problems, develop proposals for policy change, monitor legislation and administrative actions, offer expert testimony, and lobby national and regional governments. Indeed, they play key roles in each of these areas (Oswald von Nell Breuning Insitut, 1996; Rat von Sachverständigen für Umweltfragen, 1996; Bammerlin, 1998). Their large memberships and budgets also make them "keystone organizations" (Bosso, 2005) in the overall structure of environmental activism. That is, they play a disproportionate role in setting the environmental agenda, and other environmental groups must take them into account. Rucht and Roose (2001a, 2001b, 2001c), for example, found that BUND, NABU, Greenpeace, and the WWF generally ranked at the top when other environmental groups, both large and small, were asked to name the groups with which they exchanged information or formed partnerships for joint campaigns.

The four organizations' activities have earned them a positive reputation and a great deal of public trust. A 1989 Allensbacher Institute poll showed that environmental organizations enjoyed more public trust than government environmental offices (Voss, 1990), and a 1995 survey (INRA (Europe) – ECO, 1995) showed that the German public placed far more faith in them to tell the truth about the environment than political parties, trade unions, the government, industry, scientists and teachers, or the media. More recent surveys (European Opinion Research Group, 2002; Bundesministerium für Umwelt, 2002, 2004; Gruneberg and Kuckartz, 2003; Directorate General Environment, 2005; Kuckartz and Rheingans-Heintze, 2006) show that the public both trusts them more and ranks their capacity to solve environmental problems higher than government bureaus, churches, labor unions, scientists, business, and all political parties—including the Green Party—and that confidence in the environmental organizations has remained high even as public confidence in the other actors declined.

Large, national environmental organizations have, nevertheless, come in for criticism. The critics, in Germany and elsewhere (e.g., Jordan and Maloney, 1997; Brulle, 2000; Blühdorn, 2000; Bergstedt, 2002), typically charge that the organizations have proved unable to induce societies to solve environmental problems and are unwilling to take the lead in social movements or confront core environmental problems and powerful social actors. They are also accused of propagating "checkbook environmentalism," which merely soothes the conscience of their middle-class supporters, whose lifestyles are responsible for a disproportionate amount of the continuing environmental damage. Finally, critics accuse them of ignoring social movements of the poor, who suffer most from environmental degradation. Yet for all of this, it is hard to imagine the environmental scene without them and hard to imagine that things would be better without them (Bosso, 2005; Markham, 2005a).

With a few exceptions (e.g., Rucht, 1989; Mitchell, Mertig, and Dunlap, 1992; Ingram, Colnic, and Mann, 1995), social scientists before the late 1990s seemed content to leave the investigation of environmental organizations to activists and journalists (e.g., Sale, 1993; Dowie, 1995; Bergstedt, 1998). This pattern ended in the late 1990s, when researchers in the US (e.g., Shaiko, 1999; Bosso, 2005) and the UK (e.g., Jordan and Maloney, 1997; Rawcliffe, 1998) began to produce detailed analyses of environmental organizations; however, very little information is available in English about environmental organizations elsewhere. The only recent information about German environmental organizations comes from three short articles (Rucht, 1989; Blühdorn, 1995; Foljanty-Jost, 2004), two of them rather dated. In the German literature, only a little-known book by Bammerlin (1998), now almost a decade old, offers an extended analysis of the German environmental organizations.

The German case is especially interesting for several reasons. Germany has the largest population and the largest economy in Europe, and it has exerted considerable influence on European Union environmental policy (Sbragia, 2000; Schreuers, 2002). Numerous studies have shown a relatively high level of environmental consciousness and willingness to act to protect the environment in Germany (e.g., Commission of the European Communities, 1992; Worcester, 1993; INRA (Europe) – ECO, 1995; Brand, 1997; European Commission, 1999; Wurzel, 2002; Bozonnet, 2005), although several recent studies suggest that Germany may be losing its lead (Bozonnet, 2004; European Commission, 2005; Directorate General Environment, 2005). The German environmental movement in the 1990s achieved what, by some estimates, was the highest level of mobilization in Europe (Blühdorn, 1995; Rucht and Roose, 1999), and German rates of membership, financial contributions, and active participation in environmental organizations today are well above the European average (Bozonnet, 2004, 2005).

Germany also has the most successful Green Party in Europe, and German environmentalists have enjoyed enviable success in institutionalizing their agenda in government regulations, industrial practice, and public education (Rucht and Roose, 1999; Brand, 1999a; Schreuers, 2002; Foljanty-Jost, 2004). Germany is a leader in the percentage of its GNP spent on environmental protection, in energy efficiency, in reductions of greenhouse gases, and in exports of pollution reduction technologies (Jänicke and Weidner, 1996; Brand, Eder, and Poferl, 1997).

Although Germany's reputation as an environmental leader is well deserved, environmentalists and environmental experts (e.g., Jänicke and Weidner, 1996; Pehle, 1997; Jänicke, Kunig, and Stitzel, 1999; Bick and Obermann, 2000; DNR Deutschland Rundbrief, 2001g; Schreuers, 2002; Berliner Zeitung, 2004; Bundesministerium für Umwelt, 2004) can also cite a long list of unresolved problems. Among them are lack of follow-through in enforcing environmental laws, the high and increasing number of endangered species, slow progress in identifying nature reserves, weak provisions for protecting soil from contamination through pesticides or toxic wastes, slow progress in cleaning up waste dumps, rapid loss of scarce open space to development, and the country's strong support of its auto industry and growing reliance on automobile and airline transportation. Failure to impose autobahn speed limits to save energy and reduce pollution is a long-running embarrassment for Germany in international environmental politics, and it recently temporarily blocked an EU effort to require recycling of automobiles. Germany's commitment to environmental protection has historically varied with the degree of economic prosperity, and the country's focus on the problems of reunification and on maintaining economic competitiveness in the face of a weak economy appear to have moved it away from its earlier role as a leader in pressing for higher EU environmental standards (Pehle, 1997, 1998; Sbragia, 2000; Wurzel, 2002).

Nevertheless, the successes of German environmentalism have received a great deal of press attention (Rucht and Roose, 1999), and Germany is sometimes held up as a model for other nations (e.g., Weidner, 1991b; Dowie, 1995). The Greens, in particular, have attracted enormous attention from the popular press and scholars (e.g., O'Neill, 1997; Blühdorn, 2002; Hoffman, 2002), and a good bit has been written in English about German environmental politics and policy (e.g., Blühdorn, Krause, and Scharf, 1995; Jänicke and Weidner, 1996; Pehle, 1997; Wurzel, 2002). Yet despite their high membership and key role in environmental action, Germany's large national environmental organizations have been all but ignored. This book undertakes to fill that gap.

Theoretical Perspective and Research Questions

I have purposefully written this book to be accessible to readers whose primary interest is not social science theory but the environmental organizations themselves. Nevertheless, productive sociological analysis is almost invariably guided by theory, and this book is no exception. Although it remains, for the most part, in the background, I have been guided at every turn by theory—and particularly by theories about the behavior of organizations.

My recourse to organization theory represents a departure from most previous analyses of environmental organizations, which have generally viewed them as components of social movements (e.g., Rucht, 1994; Koopmans, 1995), as interest groups attempting to influence the political system (e.g., Leonhard, 1986; Jordan and Maloney, 1997; Bosso, 2005), or—somewhat less often—as actors of civil society (e.g., Brulle, 2000). Such approaches provide useful insights, and I draw on them frequently; however, as tools for analyzing environmental organizations, they have two serious drawbacks.

First, these approaches are primarily theories about social movements, political systems, or civil society, not about environmental organizations. Consequently, using them tends to divert attention away from analyzing environmental organizations themselves and focus it instead on understanding the organizations' roles in the movement, a system of interest groups, or civil society.

Second, none of these theories are adequate for analyzing environmental organizations in the full range of social contexts in which they operate. Viewing environmental organizations as social movement organizations is not very useful for understanding how they function when under tight state supervision, as German environmental organizations were under National Socialism and in the GDR, or in periods like the decade following World War II, when nature protection organizations continued their work but an environmental movement hardly existed. Interest group models and theories of civil society are also poorly suited to understanding environmental organizations under dictatorship, and they have little utility for analyzing environmental organizations so embedded in confrontational protest movements that their participation in the political system is neither welcomed by other participants nor desired by the organizations themselves—a situation that characterized several German environmental organizations during the period of confrontations with the state and business during the 1970s and 1980s.

Rather than viewing environmental organizations primarily as components of larger structures, which then become the main focus of interest, this book places the accent on the organizations themselves. It asks why they adopt the programs and goals, internal structures, and strategies for coping with the outside world that they do and how these programs and goals, internal structures, and strategies influence one another. The inquiry is guided by general theories of organizations that are typically applied to analyze business firms, government agencies, and other NGOs.

I rely primarily on four strands from this literature. First, I use the open system model (e.g., Thompson, 1967; Katz and Kahn, 1978) to examine how environmental organizations are constrained to choose specific objectives and strategies that allow them to procure the financial resources, labor, and legitimacy they need to survive and pursue their general goals. This model also highlights the dilemmas organizations face when the social actors that can provide them with resources and legitimacy differ in their opinions about what they should be doing. Second, the book draws on literature (e.g., Wilson, 1995; Jordan and Maloney, 1997) about how volunteer organizations manage to solve Mancur Olson's (1965) "free rider problem." That is, how do environmental organizations persuade individuals to support them, even though no single person's contributions have a decisive impact on whether the organization succeeds and even though those who contribute nothing will benefit equally from its successes? Third, the book draws on a line of research and theorizing that can be traced back to Robert Michels' (1949) "iron law of oligarchy." The focus here is on identifying factors that promote or retard professionalization and centralization in NGOs and the consequences of these trends for organizations (e.g., Rucht, Blattert, and Rink, 1997). Finally, the analysis draws on recent theoretical efforts (e.g., DiMaggio and Powell, 1991a;

Perrow, 1993) to analyze organizations as institutions, that is, as social actors that are expected to conform to standard practice and whose goals and strategies are valued and imbued with symbolic meanings by the members.

The institutional perspective, in particular, insists that organizational analysts must take culture and history seriously, and organizational history and traditions are especially important factors in the behavior of NGOs supported by volunteers and donors. Consequently, a historical perspective is key for understanding the present-day problems and strategies of environmental organizations. Moreover, a historical analysis allows us to compare the strategies used by environmental organizations in different historical situations and avoid the myopia that can accompany an exclusive focus on recent history. The German case is especially useful in this respect because of the exceptionally diverse social and political contexts within which environmental organizations there have operated during the twentieth century.

The research questions this theoretical approach suggests are, in the first instance, the classic questions of all organizational analysis. How do environmental organizations acquire the resources they need to survive and move toward their goals, how do they structure themselves internally, and how do they select their goals and the strategies for implementing them? Environmental organizations are, however, not just generic organizations. Like any type of organization, they have their own specific characteristics, and these characteristics raise more specific questions. As NGOs, environmental organizations must acquire financial support from business or government, which are their frequent opponents, or from individuals, who face the classic temptation of the free rider to let others pay the bills and do the work. As agents for social change, they must decide how radical an agenda to embrace and what issues to emphasize. Moreover, they must frequently negotiate relationships with social movements that share their objectives, including segments of movements whose ideologies, style of operation, and objectives differ from their own. Finally, as voluntary associations, they face tradeoffs between designing internal structures to maximize efficiency and quick decisions and to meet prevailing standards for professional operation, versus living up to democratic ideals and activists' desire for a say in decision making.

Data Sources

The information used in this book comes from a variety of sources. First, I conducted a comprehensive review of existing literature about German environmental organizations and German environmentalism from the beginning of the twentieth century to the beginning of the twenty-first. For the historical part of the book, I supplemented my reading of secondary sources with five weeks of archival research at the Archive and Museum for the History of Nature Protection in Germany, the headquarters of the German Nature Protection League (formerly the League for Bird Protection), and the library of the Federal Nature Protection

Office, all in or near Bonn. I also worked for two weeks at the Study Archive for Environmental History at the *Fachhochschule* in Neubrandenburg, which houses a large collection of publications and documents about environmentalism in the German Democratic Republic. I focused attention especially on annual reports and member magazines of the most important organizations, as well as on the organizational histories they themselves have produced.

Chapters 7 to 12 present an analysis of the four largest national organizations as they entered the twenty-first century. Information for this section comes from a wider variety of sources. First, I reviewed all articles about the organizations published between 1997 and 1999 in *Der Spiegel,* Germany's leading news magazine, *Die Zeit,* a major national, weekly newspaper, and the *Frankfurter Allgemeine Zeitung,* Germany's leading newspaper of record. The first two are generally identified as moderate to left in political orientation, while the *Frankfurter Allgemeine* is business-oriented and conservative (Brand, Eder, and Poferl, 1997). I also draw on articles collected less systematically from other newspapers and from periods before and after 1997–1999. Second, I read all issues of the member magazines of three of the four organizations published between 1998 and 2000 page-by-page. I also read all issues of Greenpeace's newsletter for its supporters for this period. I also reviewed all issues of the same four publications for the years 2001 to 2004 for relevant materials. I read the annual reports of all four organizations between 1997 and 2004 page-by-page and reviewed several anniversary issues of member magazines and similar publications, along with a few internal documents that were made available to me.

In Chapters 11 and 12, I also draw on results of a detailed content analysis of the January 1998 Internet pages of the four major organizations conducted by me and Sabrina Broselow. An overview of the research methodology has been published elsewhere (Markham, 1999). I have often reviewed the organizations' web sites since then, and my analysis is informed by this information; however, due to the ephemeral nature of Internet sites, I have cited them only when equivalent information is not available elsewhere.

The analysis also draws on twenty-four semistructured interviews I conducted in 2000, 2001, and 2005. Interviewees included two to four elected leaders or top-level staff members in each of the major organizations except Greenpeace, former leaders of the organizations, and leaders of *Länder* chapters of the organizations that have them. Also included were officials of environmentally oriented foundations and of the umbrella association of German environmental organizations, journalists and press spokespersons who regularly dealt with environmental issues, members of the Bundestag (national parliament) from the three largest political parties who were assigned to coordinate their legislative delegations' activities on environmental issues, several elected officials and staff members of the Greens, and environmental activists, including several from the former East Germany. A number of interviewees could be classified in more than one of these categories. The face-to-face interviews were based on semi-structured interview schedules custom designed for each interview and lasted between thirty minutes and two hours. Greenpeace leaders declined to participate in personal interviews but agreed to respond in writing to the interview questions. I supplemented the written responses with the organization's own 300-page examination of its philosophy and goals, operating procedures, and problems (Greenpeace Deutschland, 1996), which was often referenced in the written responses.

The analysis presented in Chapters 7 to 12 thus rests on information collected during the years immediately before and after 2000. In most respects, the situation that existed then resembles the situation at the time this book went to press, and many of the findings can be generalized to the present. But careful research takes time, and, inevitably, some things, such as the exit of the Greens from the governing coalition in November 2005 and a resumption of slow economic growth in Germany after several years of stagnation, have changed. I speculate about what these changes might mean for the future in the final chapter.

Overview of the Book

To make this book accessible to readers whose primary interest is the German environmental organizations and German environmentalism, I have concentrated detailed discussions of theory in Chapter 2 and parts of Chapter 13. Nontechnical readers may wish to skip over these sections. I have avoided technical terminology and detailed discussions of theoretical issues in Chapters 3 to 12, except in the "Summary and Conclusions" sections, which relate the content of the chapters to the theoretical discussion in Chapter 2; however, even in these summaries I have minimized the use of technical language, so that nontechnical readers can read them without undue distraction.

Chapter 2 begins by looking at three widely used approaches to studying environmental organizations: the social movement, interest group, and civil society approaches. An examination of the strengths and limitations of these approaches suggests that more general theories of organizations can actually provide a more generally applicable approach to analyzing environmental organizations, and the final section of the chapter describes the relevant organization theories in some detail. I rely, in particular, on open systems theory and its various extensions—supplemented by theories of organizations as institutions.

Chapters 3 to 6 look at the most important German environmental organizations in several historical periods throughout the twentieth century. For each period, I first review relevant developments in the condition of the physical environment, social thought, economics, and politics. I then examine how the most important environmental organizations of the period reacted to these trends.

Chapter 3 covers the period from the founding of the first national environmental organizations to the early 1930s. It emphasizes that the roots of most of the organizations were in the conservative reaction to industrialization and urbanism, that environmentalism was closely linked to concerns about national identity, and that the national environmental organizations of the time viewed their mission primarily as protecting endangered species or treasured landscapes, not as fighting industrial pollution. It also examines the place of environmental thought in ideologies of the left and its embodiment in working-class groups with nature protection goals.

Chapter 4 shows how the Nazis used a combination of progressive environmental legislation and anti-modernist ideology to win the support of many nature protection advocates but then turned away from environmental objectives in favor of war preparations. It also describes the absorption of environmental organizations into government-controlled umbrella organizations. The chapter also looks at the postwar struggles of environmental organizations, weakened by the war and tarnished by their association with Nazi ideology, to reestablish themselves in a new West German state preoccupied with recovering from wartime devastation and, later, with enjoying the postwar "economic miracle."

Chapter 5 describes the adaptations of the large, national environmental organizations in West Germany to the tumultuous period from the late 1960s through the early 1980s—a period dominated by ideologies of the countercultural left, mass demonstrations against nuclear power plants, and a bitter struggle to place environmental concerns higher on the national agenda. I emphasize the struggle of older organizations to adapt to this new situation and describe the founding of new organizations, such as Greenpeace, that proved to be better adapted to the changing social context.

Chapter 6 offers a detailed treatment of the little-known history of East German environmentalism, focusing especially on the struggles of the large, state-sponsored nature protection organization to balance nature protection with ensuring its own survival in a one-party state. It also examines the rise of the East German environmental movement centered in less structured and more marginal groups and the rapidly changing landscape of East German environmentalism in the years immediately following reunification.

Chapters 7 to 12 provide a detailed examination of the situation of the four most important German environmental organizations as they entered the twenty-first century. The first three of these chapters provide the necessary background for this analysis. Chapter 7 describes the rapidly changing social and political conditions in the closing years of the twentieth century and beginning of the twenty-first, as well as the challenges the changes posed for environmental organizations. These include the fading of the environmentalist counterculture, the institutionalization of environmentalism in government policy, industrial practice, and public opinion, the problems of reuniting with East Germany, and the stagnation of Germany's economy. Chapter 8 describes the organizational environment facing German environmental groups at the beginning of the new century. It focuses on the German political and economic systems, public opinion, environmental groups other than the large national ones, and the media. Chapter 9 contains detailed profiles of the four most important environmental organizations, including their goals and activities, formal structure, internal organization, publications, memberships, and budgets.

Chapters 10 to 12 are organized around seven strategic dilemmas that the four organizations must deal with in order to survive and prosper. These dilemmas are suggested by the theoretical model developed in Chapter 2, but the specific forms in which they appear are conditioned by both the history and traditions of the organizations and the social context in which they operate. They fall into three categories—internal structure, resource acquisition, and external relations—each of which is the topic of a separate chapter.

Chapter 10 examines professionalization and centralization. Taking its departure from social science literature about persistent trends toward oligarchy and professionalization in voluntary associations, it outlines the advantages and disadvantages of professionalization versus reliance on volunteers and of centralized versus decentralized decision making and administration. Although trends toward increased centralization and professionalization are found in all four organizations, the organizations continue to struggle to find a workable balance between voluntarism and professionalization and between decentralization and centralization. How they resolve these tradeoffs is shaped not only by considerations of efficiency and the demands of the outside world, but also by traditions of volunteerism and grassroots involvement from their history.

Chapter 11 focuses on problems of resource acquisition. As professionalized organizations with a diverse agenda and well-organized opposition, environmental organizations need both a large base of supporters and substantial amounts of cash. Financial support is potentially available from government and business; however, such support is laden with pitfalls, causing the organizations to accept it cautiously or not at all. Instead, all have chosen to seek a mass base of donors, which they can also cite to enhance their credibility. This strategy "solves" the problem of obtaining financial support, but only at the expense of producing new issues. These include criticism that the organizations must mute their message in order to attract the largest possible number of supporters, resistance to their more aggressive marketing strategies, and the possibility that this approach might drive away their most committed supporters and alienate them from their traditions of grassroots activism.

Chapter 12 examines three strategy dilemmas that the organizations face as they seek to get their message across and influence public policy: cooperation vs. confrontation with business and government, cooperation vs. competition with other environmental organizations, and focusing on many issues or only a few. All the organizations face an external environment in which there is less to gain from confrontation with business and industry than in the past, yet too much compromise runs the risk of alienating some of their most committed supporters and accomplishing nothing. All must also choose between cooperation and competition with other environmental organizations. Both competition and cooperation have advantages and disadvantages, so the organizations must cautiously navigate between losing the advantages of cooperation and losing their independent identities.

The final chapter briefly summarizes the major findings, relates them to the theoretical models discussed in Chapter 2, and discusses prospects for the future.

Chapter 2

Environmental Organizations

Theoretical Considerations

This book examines both the history and present-day status of large, national environmental organizations in Germany. It focuses on why they have chosen the goals, strategies, and structures that they have, emphasizing, in particular, those decisions that constitute "dilemmas" for the organizations. I examine the factors in the organizations' political and social contexts that influence their decisions, how the choices made to deal with one dilemma influence the remaining ones, and the consequences of their choices for the organizations and their roles in society.

I believe that these issues can be understood without bombarding readers with the sometimes arcane terminology of social science theory, so I have written the bodies of the ten substantive chapters that follow with minimum use of esoteric terminology and few explicit references to social science theory. I have relaxed these strictures somewhat in the summary and conclusions sections at the end of each chapter, but I have segregated "heavier" discussions of theory to this chapter and the final chapter. Nontechnical readers should, therefore, find the book understandable even if they choose to skip or skim this chapter and some sections of the last.

This book is, however, far from atheoretical. I have been guided at every step by ideas and insights from social science theories: in the questions I pose, in the explanatory factors I examine, and in how I combine these factors into explanations. This chapter summarizes the theoretical perspectives that have guided my inquiry, but the review is selective, as none of the theories were developed primarily for environmental organizations and each contains much that is of only marginal relevance.

Most previous research about environmental organizations has been either primarily descriptive (e.g., Mitchell, Mertig and Dunlap, 1992; Blühdorn, 1995) or firmly anchored in a single theoretical framework, usually either interest group theory (e.g., Rawcliffe, 1998; Shaiko, 1999; Bosso, 2005) or theories of social movements (e.g., Rucht, 1994; Koopmans, 1995; Brulle, 2000). Recently, some authors (e.g., Warren, 2001; Skocpol, 2003) have suggested that theories of civil society can also be used to understand environmental organizations; however, no full-fledged analysis using this approach has yet appeared. In general, political scientists have tended to favor interest group theories; sociologists, social movement approaches. Differences in the emphases of different national or regional sociologies may also play a role in the theoretical approach preferred (Mol, 2006). European scholars, for example, have been especially likely to analyze environmentalism under the rubric of new social movements.

Unfortunately, working exclusively within any one of these traditions has problematic consequences. This is true for three reasons. First, each perspective poses only a limited number of questions about environmental organizations and examines only a limited range of explanatory variables. Moreover, the research questions and explanatory factors characteristic of one approach can be difficult to assimilate into the others. Second, none of the three approaches makes explaining what environmental organizations do and why they do it its primary task. If addressed at all, these questions are treated only as secondary concerns. Third, none of the three perspectives are broad enough to apply to the full range of environmental organizations. Environmental organizations have pursued an exceptionally wide variety of goals using very diverse strategies (Lake, 1998). Some have been central actors in environmental movements. Yet others have eschewed participation in movements, and some have continued to function during periods when movements were, for all practical purposes, nonexistent. Some environmental organizations have functioned like interest groups, working within the system to influence public opinion and public policy. Others, however, focus their attention on practical nature protection or land acquisition and avoid political activity, while still others decline to participate in the political system—or are denied participation—because of their radical ideologies or confrontational strategies. Radical ideologies and confrontational strategies also make these organizations difficult to fit under the rubric of civil society.

My approach to resolving these problems is twofold. First, I draw rather eclectically on concepts and theoretical propositions from theories about interest groups, social movements, and civil society. Second, I have built the analysis not around these theories but instead around general theories of organizations. These theories have three important advantages. First, they were developed specifically to analyze organizations, not social movements, political systems, or civil society. Second, they are abstract enough to apply to all the environmental organizations, both contemporary and historical, described in this book. Third, relevant insights from the other theoretical approaches are, in general, easy to fit within the framework that organizational theory provides.

The primary goal of this book is to better understand the environmental organizations, and most of the book is devoted to that task. Nevertheless, improvements in theory inevitably develop out of encounters with concrete data. Therefore, in the final chapter I develop a comprehensive theoretical model of how German environmental organizations were adapting to their environments as they entered the twenty-first century. I also reflect there on how a more general theory of environmental organizations grounded in general theories of organizations might appear.

Interest Groups

Many political scientists and some sociologists classify environmental organizations—along with business lobby groups, labor unions, professional associations, and citizens' groups working for other causes—as organized interest groups.[1] Interest group theories view such organizations as working within the political system to influence political decisions and their implementation. They accomplish this by using some combination of the following strategies: 1) lobbying or testifying before legislators and government agencies; 2) mobilizing citizens to sign petitions or contact their representatives and relevant administrative agencies directly; 3) conducting public information campaigns and—occasionally—protests to influence public opinion; 4) attempting to influence election outcomes through campaign contributions, endorsements, and voter mobilization; and 5) monitoring the implementation of public policy. Many interest groups push for decisions and policies that will benefit primarily their own supporters, but public interest groups, including environmental organizations, work for what they see as the interests of citizens in general; i.e., they attempt to influence government to provide public benefits that are available to all.

Theories of Interest Groups

Interest group theorists have developed two major models of how interest groups function within political systems. Pluralist theory[2] was developed originally by American political scientists to describe US politics, but it has been applied in other Western democracies as well. It sees environmental organizations and other interest group organizations as independent representatives of various group interests; that is, they are neither state-sanctioned nor state-supported. Interest groups can be formed at will to represent a wide variety of interests, including industries, professions, labor, hobbyists, and groups with moral or cultural agendas.

Interest groups compete with one another for influence in the political arena, where each group's influence is proportional to its ability to mobilize supporters

---

[1] For reviews of interest group theory see Berry (1984), G. Wilson (1990), Walker (1991), Petracca (1992), Mundo (1992), and Burstein (1998).

[2] The seminal statement of the theory is Dahl (1961). Useful overviews include Ball and Millard (1987), Knoke (1990), and Petracca (1992).

and funds and use them skillfully. Some groups have more such resources—and hence more influence—than others, but ordinarily no interest group has enough power to reach all of its political objectives. This occurs because proposals advocated by one group typically work to the detriment of one or more other groups. The more extreme the proposal, the more threatening it is, and the greater the resources oppositional groups can mobilize against it. Groups on both sides of an issue can attempt to increase their influence by forming coalitions with other groups, but groups with nothing at stake in a controversy generally avoid taking sides for fear of alienating some of their supporters. Most decisions are compromises fought out within the political system, with the state acting as referee. The state is also usually responsible for implementing the resulting decisions.

A second model of interest group functioning, neocorporatism,[3] applies best to societies such as Sweden, Austria, and the Netherlands; however, elements of this system are also visible in Germany. Under neocorporatism, governments encourage the formation of national-level associations to represent the interests of broad sectors of society, such as business, agriculture, charities, and labor. In contrast to corporatism under authoritarian rule, neither the formation of such associations nor participation in them is compulsory. These umbrella associations become deeply intertwined with government, which recognizes them as speaking for their sectors and gives them privileged access to behind-the-scenes discussions of political decisions and policy administration. The state acts as a coparticipant and mediator in negotiating compromises among the groups, which are reached with an eye to allowing all of the interest groups to reach at least some of their objectives and generally involve incremental change. The key actors in neocorporatist systems have traditionally been economic. Environmental interest groups were historically not represented, and their incorporation into the system has often been only partial.

Neocorporatism simplifies the task of government by reducing the number of interest groups with which it must deal and giving the groups incentives for cooperative participation within the system. In return for their privileged access, they are expected to accept the overall legitimacy of the system and the legitimacy of the other groups, be willing to compromise, and work to persuade the individuals and organizations they represent to accept compromise decisions, even when they are not fully satisfied with them. Because recognized interest groups enjoy a virtual monopoly, their supporters have little choice but to support them. Interest groups that are accepted within the system rarely need to engage in protest tactics, such as strikes and demonstrations. Being accepted in the system legitimates them, and government may support them financially and allocate responsibility for carrying out important tasks to them. For example, industry may be charged with determining the best practices for pollution control.

---

[3] Basic statements of neocorporatist theory include Schmitter (1979, 1982) and Streek and Schmitter (1985). For overviews, see F. Wilson (1990) and Oswald von Nell Breuning Insitut (1996).

Relevant Topics in Interest Group Theory

The literature about interest groups includes discussions of several topics useful for understanding environmental organizations. The best known of these is the free rider problem. Although frequently mentioned in studies of social movements and voluntary associations, discussion of free riding originated in the interest group literature, and the topic has received its most complete treatments there.

According to the well known formalization by Mancur Olson (1965), the free rider problem occurs because rational actors will inevitably choose not to contribute effort or money to mass-membership public interest groups working for "collective benefits" like clean air or national parks. Such benefits, Olson argued, can be enjoyed equally by "free riders," who have contributed nothing, and no individual's contribution has much impact on whether the organization reaches its goals. Hence it is irrational to contribute. Large national environmental organizations should therefore have serious problems attracting and retaining supporters.

Olson's argument is supported by the fact that the vast majority of citizens do not, in fact, join or donate to environmental organizations or other public interest organizations, even though many citizens support their objectives. However, this leaves unexplained the behavior of the hundreds of thousands of supporters that such organizations do attract. Using the rational actor model, Olson speculated that people might support public interest groups because they are coerced to do so—clearly not the case for environmental groups—or because the groups provide them with "selective benefits," i.e., benefits available only to members. However, while benefits such as magazines with interesting content, information services for members, or reduced prices on merchandise or services available only to members might have some effect on potential members of environmental organizations, it is hard to imagine that such benefits are decisive for many, and when surveyed, members do not rank such benefits high on the list of reasons to participate (Godwin, 1988; Jordan and Maloney, 1997).

Commentaries on Olson's work and reviews of related literature and related theoretical literature on interest groups[4] identify many additional reasons why citizens might join, volunteer for, or donate to environmental organizations and other public interest groups. Supporters may assess their contributions as more important to the success of the organization than is realistic, and the larger groups often use expensive advertising campaigns to persuade them that this is the case. Their advertising often stresses the group's role in preventing terrible outcomes that would occur without the group's efforts, which appears to be especially motivational. For many citizens, signing a petition, attending an occasional meeting,

---

[4] See, for example, the following: Salisbury (1969); Moe (1980); Knoke (1990); King and Walker (1991); Smith (1994); Udehn (1993); Wilson (1995); Zimmer (1996);Jordan and Maloney (1997); Shaiko (1999); and Felbinger (2005).

or writing a check are low-cost actions (Diekmann and Preisendörfer, 1988), hardly worthy of elaborate cost-benefit analysis. Particularly in groups with local chapters, participants may enjoy social contacts with other members, and being known as a hard-working volunteer or significant donor may confer prestige, both within a group and in the broader community. Consequently, organizations and networks often go out of their way to publicize the contributions of their most active and generous supporters. Volunteers may also enjoy the tasks they undertake for the organization or see them as an opportunity to learn new skills or qualify for a new career. If opportunities to rise to leadership roles exist, the chance to gain personal influence may also attract some supporters. Individuals who identify with other supporters of environmental goals, a common characteristic of environmentalists, can reinforce that identity through their participation, and they may develop emotional ties to other members or the organization itself. In addition, humans' ability to reward themselves for following norms or doing what they perceive to be the "right thing," can provide supporters with a "warm glow," allowing them, in effect, to reward themselves for their contributions.

The interest group literature also contains insights about other topics relevant to analyzing environmental organizations. For example, interest group theorists have looked at how public interest groups come to be established and how they acquire the support they require to continue their operations.[5] Earlier theorists sometimes assumed that interest groups form spontaneously in response to needs that are not being satisfied by business or government or a desire for political change held by some segment of the population; however, this does not seem very realistic for large, national environmental organizations, whose supporters are typically widely dispersed. More recent work emphasizes, instead, two other factors: the importance of interest group entrepreneurs, who play a key role in assembling the needed resources and finding ways to attract supporters, and the key role of initial financial support, often from foundations or individual donors, in getting new interest groups off the ground.

Interest group theorists have also examined trends toward professionalization in public interest groups and their consequences.[6] Many such groups, including environmental organizations, find that policy discussions are becoming increasingly technical and complex, that interest groups in other sectors are professionalizing, and that volunteers with the skills needed to cope with these developments are in short supply. They therefore come under pressure to professionalize their own operations. Professionalization leads, in turn, to growing budget pressures, which transform fundraising into a major organizational preoccupation. In some cases, the needed funds may be solicited from foundations, major donors, or government contracts or grants, but for many public interest lobbies, fundraising from a mass of small individual donors through methods like direct mail and telephone solicitation has become the primary funding source.

---

[5] For reviews see Salisbury (1969), Berry (1984), Walker (1991), and Sabatier (1992).

[6] See Berry (1984), Godwin (1988), Jordan and Maloney (1997), and Shaiko (1999).

Interest group theorists have also examined relationships among interest groups.[7] They note that although public interest groups with similar goals do often cooperate, their relationships are also marked by competition because they must compete for supporters and financial support, access to and influence over politics, and media attention. Hence, there is always a tension between cooperation and competition among groups with similar objectives. Some groups respond by purposefully differentiating their goals and strategies so that they can cultivate a distinctive identity and niche. Differentiation of goals and strategies can also result from schisms within existing organizations. While organizations can sometimes successfully secure their position by becoming "niche players," such organizations find it difficult to attract as much media attention or as many supporters as well-established organizations with a broader agenda.

Other literature explores the relationship between the structure of a nation's political system and the goals and strategies of its interest groups.[8] For example, interest groups in federal systems frequently develop strong regional branches with boundaries corresponding to the political subunits, while those in centralized states more often emphasize work at the national level. Interest groups might also be expected to key their efforts to influence the legislative, executive, and judicial branches of government to the relative influence of the three branches, and legal systems that are more open to litigation from groups representing the public interest might prompt interest groups to focus their attention on litigation.

Nations with weak political parties, nations with pluralist political systems, and nations with two-party systems, where forming small parties to represent specific interests is not an attractive option, are likely to have more interest group activity than other societies. Neocorporatist systems, on the other hand, often create tensions because some interest groups, especially new ones such as environmental groups, are excluded from participation. Excluded groups may challenge the legitimacy of the system—especially if they believe that it serves the interests of other groups better than theirs—and turn to protest tactics to gain attention and influence. Interest groups that benefit from existing neocorporatist arrangements, on the other hand, have good reason to guard their prerogatives. They and their allies in government typically perceive aggressive groups representing new interests and new demands as a threat to the system.

Finally, interest group researchers have speculated about the contribution that public interest groups make to the functioning of democracy.[9] Some authors

---

[7] Useful discussions of the relationships among similar interest groups include Salisbury (1969), Moe (1980), Berry (1984), Godwin (1988), and Shaiko (1999).

[8] The following sources offer general treatments of the effects of the political system on interest groups: Berry (1984); Ball and Millard (1987); Smith (1989); G. Wilson (1990); Walker (1991); Oswald von Nell Breuning Institute (1996); Salamon and Anheier (1998); and Wurzel (2002). Stein (2003) focuses on the effects of the legal system on interest groups, while Schmitter (1979, 1982), Streek and Schmitter (1985), and F. Wilson (1990) emphasize neocorporatist systems.

[9] Relevant discussion include Hayes (1983), Berry (1984), Godwin (1988), F. Wilson (1990), G. Wilson (1990), Walker (1991), Rucht (1993b), and Jordan and Maloney (1997).

emphasize their role as intermediaries between citizens and the state, pointing out that such organizations clarify, focus, and bundle citizen needs and discontent and express them effectively to the authorities. Other theorists, however, argue that interest groups, even in pluralist systems, tend to do a far better job of representing business, well-organized constituencies such as labor, and the politically conscious middle class than the poor and dispossessed. Finally, some argue that professionalized public interest groups that rely on mass fundraising operations undermine democracy by teaching citizens that representation of their interests is simply a service to be purchased from professionalized and centralized organizations that neither expect nor want the democratic participation of their supporters.

Summary and Conclusions

Interest group theories offer well-developed models for thinking about the role of environmental organizations in society and about people's reasons for supporting them. They also offer some insights into why environmental organizations behave as they do. Nevertheless, their chief interest is not how environmental organizations and other public interest groups function as organizations, but the political activities of these organizations and their effects on the political system. This strong emphasis on the organizations' efforts to influence politics through relatively conventional means leaves many of their activities unexamined. These include raising money for nature protection projects, mobilizing volunteers for practical nature protection work at the local level, public education campaigns aimed at motivating citizens to choose environmentally friendly lifestyles, and acquisition and maintenance of nature reserves. Finally, although interest group theories do include protest among the strategy repertoire of public interest groups, they provide little leverage for understanding environmental organizations whose ideologies represent a radical challenge to the existing system or organizations that are so embedded in confrontational protest movements that participation in the political system is neither desired by the organizations nor acceptable to other actors in it. In short, although it offers many insights into the actions of some environmental organizations in some times and places, interest group theory is far from a comprehensive theory of environmental organizations.

Social Movements

Most sociologists, as well as some political scientists, analyze environmental organizations using theories of social movements, and social movement theories provide many useful insights into the behavior of environmental organizations.

Although there is fairly broad agreement that social movements are comprised of networks of persons and organizations attempting to influence the direction and nature of social change, the details of the definitions offered by social movement scholars vary widely along numerous dimensions.[10] These include whether the definition of social movements encompasses: 1) only movements that press for political change, versus movements that also emphasize personal lifestyle change, development of group identity, or withdrawal into utopian communities; 2) only movements that press for change, as opposed to countermovements; 3) only movements that press for radical system change, versus reform movements; 4) only movements that represent class interests, which are themselves defined in various ways, versus movements not associated with a particular class; 5) only movements that emphasize protest tactics, versus those that do not; and 6) only movements that involve mass mobilization of the public, as opposed to efforts to promote or resist change that continue even when large-scale mobilization is absent.

Whether an environmental organization is classified as part of a social movement thus depends on the definition of social movement one employs, and it is clearly inappropriate for researchers to unreflectively characterize all environmental organizations in all historical periods as components of social movements without very careful qualification of what definition of a social movement the researcher has in mind. Relatively apolitical environmental organizations that focus on the acquisition and maintenance of nature reserves, organizations that eschew confrontation in favor of lobbying within the system, and organizations that continue to pursue their objectives between waves of mobilization have little in common with very widely used conceptualizations of social movements (e.g., Melucci, 1980; Touraine, 1985; Tarrow, 1991; Rucht, 1993b, 1999a; McAdam, Tarrow, and Tilly, 1996; della Porta and Diani, 1999; Klandermans, 2001) as relatively militant challenges to the system by the less powerful.

Theories of Social Movements

Theoretical perspectives on social movements are almost as diverse as definitions. Here I describe four of the most prominent.

Resource Mobilization Theory

Resource mobilization theory was developed in the United States during the 1970s and 1980s.[11] Among the theories of social movements, it is the one best suited to analyzing environmental organizations because it focuses attention on the central role of social movement organizations (SMOs) in social movements. SMOs are relatively large, formalized, and permanent organizations that func-

---

[10] For reviews of the various definitions see Raschke (1987), McAdam, McCarthy, and Zald (1988), Diani (1992), and Rootes (2004b).

[11] The seminal statement of resource mobilization theory is McCarthy and Zald (1977). The following sources offer good reviews of later contributions to the theory: Jenkins (1983); Klandermans (1986); McAdam, McCarthy, and Zald (1988); Oliver and Marwell (1992); Buechler (1993); Burstein (1998); Edwards and McCarthy (2004); Caniglia and Carmin (2005); Davis et al. (2005); and McAdam and Scott (2005).

tion as key actors in many successful movements. Some of them might also be described as interest groups, but they are often less formalized than interest groups or political parties, and they can be quite confrontational. Research mobilization theory focuses primarily on the strategies and structures of SMOs and on the factors that affect them; it has less to say about the motivations of social movement or SMO supporters or about how environmental organizations fit into larger social and political structures. Many research mobilization theory concepts were adapted from general theories of organizations, so there is some overlap with these theories.

According to resource mobilization theorists, grievances and the desire for social change are ubiquitous in society. Therefore, the rise of SMOs and social movements is best explained not by the grievances themselves but by the activities of social movement entrepreneurs, who manage to bring together the necessary financial resources and public support to build SMOs and social movements. According to resource mobilization theory, environmental organizations thus appear not just because many citizens are worried about the environment but because some persons or groups, operating in a basically rational and planful way, have succeeded in creating and sustaining them.

According to the theory, SMOs are most likely to appear when two conditions are met. First, social movement leaders must develop an interpretation of a social problem and set forth goals and strategies for solving it that attract widespread support without completely alienating powerful actors in the society, who could block their formation. Indeed, in the best case, social movements might even attract support from some powerful actors. Second, the organization must obtain the resources it needs to operate. These include knowledge and technology, access to relevant networks, and volunteers with needed skills. Third, resource mobilization theory places great emphasis on the need to acquire financial resources. Money can come from individual donors, but resource mobilization researchers also emphasize the importance of financial support from foundations and wealthy individuals who support movements as "conscience constituents." In some instances, financial support may even come from government. While obtaining funds from these latter sources can be advantageous, organizations or individuals who provide key financial support may require SMOs to conform to their standards and expectations, robbing them of their independence.

SMOs typically seek to recruit both individual activists and broad citizen support. Research in the research mobilization tradition demonstrates that the activists SMOs and movements attract are generally active and engaged citizens recruited through preexisting social networks, not the alienated and anomic individuals portrayed in some older social movement theories. SMOs that seek to recruit a broad base of less active supporters typically employ the strategies described above for interest groups; however, such supporters are often less committed to the movement and can be recruited only via major expenditures of time and funds.

Many social movements contain multiple SMOs. Taken together, these constitute a "social movement industry." To survive, SMOs must compete successfully with other SMOs in their industry, and research mobilization theorists' discussions of their strategies for accomplishing this largely parallel those offered by interest group theorists. All of the social movement industries that are active in a particular nation and time period make up the "social movement sector," and there may also be competition or cooperation between SMOs in one industry and those in others. For example, environmental SMOs might cooperate or compete with SMOs in the animal rights sector. Some organizations, such as Greenpeace, may bridge two social movements industries. Finally, because successful movements often call forth countermovements, SMOs frequently have to contend with organized opposition.

New Social Movements

New social movement (NSM) theory was developed primarily in Europe as an effort to understand the emergence of the student, peace, feminist, environmental, and related movements.[12] NSM theorists viewed these movements as differing from earlier movements, which they interpreted as based on economic classes, in several respects: 1) new social movements were not as directly focused on gaining economic advantage or political power; indeed, many drew on a countercultural ideology that eschewed economic success as it had been defined in the cultural mainstream and focused on lifestyle and quality of life instead; 2) many NSMs sought to bring about social change not only by means of political action, but also by efforts to transform the dominant culture and lifestyle changes; 3) active participation in NSMs became the basis for important personal and group identities, such as feminist or environmentalist, which were often defined in opposition to the dominant culture. Participants frequently affiliated with such movements to reinforce these identities and set themselves apart from others, and symbolic practices, such as vegetarianism or use of public transit, came to symbolize these identities; 4) NSMs claimed to speak for the interests of society in general, not for specific interest groups; 5) insofar as they attempted to effect political change, NSMs emphasized grassroots mobilization, grassroots democracy, informal networks of local groups, and protest, not conventional political participation or formalized organizations. Indeed, the latter strategies were often regarded quite skeptically. Consequently, NSM theories are better suited to explaining the less formally organized aspects of environmental movements than to analyzing large, formalized environmental organizations.

The most prominent theme in the NSM literature has been efforts to identify reasons for the rise of NSMs. The most commonly cited explanations have been 1) citizen resistance to increasing intrusions into everyday life by the economic sphere and bureaucratized and inefficient government, 2) the growth of postmaterialist values among a postwar generation weaned on prosperity and in search of personal fulfillment, and 3) efforts by the new middle class of professionals and

[12] For early statements of new social movements themes see Eder (1985) and Offe (1985). Useful literature reviews include Klandermans (1986), Brand, Büsser, and Rucht (1986), Brand (1987), Johnson, Laraña, and Gusfield (1994), Buechler (1995), Pichardo (1997), and della Porta and Diani (1999).

intellectuals to gain influence and prestige by asserting the priority of new values. Although these discussions can sound somewhat dated in the social context of the early twenty-first century, they continue to provide insights about the support base for environmental movements and organizations today.

Social Constructionism

Social constructionism explores the role of cognitive and emotional factors in the rise of social movements.[13] It rests on the premise that it is often far from self-evident that social or environmental conditions are social problems in need of solution. They become problems to be addressed by a social movement only when people can be persuaded to replace old ways of looking at things with new ones that define the conditions as problems. Draining a wetland for farming, for example, might be defined either as progress or as loss of an important ecosystem and a threat to endangered species.

Social constructionists place heavy emphasis on factors that help to establish the credibility of movement claims that particular environmental conditions constitute a social problem and that social movement activity holds promise for solving it. These factors include the persuasive use of science and scientific experts, the credibility of those making the claims, the claims-makers' skill in presenting their arguments, and successful use of the media. The last topic in particular has received a great deal of attention by social constructionists. Researchers in this tradition have examined questions such as how environmentalists tailor their strategies to attract favorable media coverage and how the media filters and interprets environmental news.

Probably the best-known contribution of the social constructionist perspective is the concept of framing.[14] Framing refers to attempts by social movement ideologists to place conditions and events that they view as problems in cognitive "frames" that will resonate with the public's preexisting ideas and convince them that a problem indeed exists and that a solution is possible. Framing is important because many environmental issues can be framed in several ways. Rainforest destruction, for example, might be framed biocentrically, as the loss of intrinsically valuable biodiversity, or anthropocentrically, as the loss of species that might someday be sources of valuable medicines. Framings are likely to be attractive when they are seen as credible, internally coherent, and consistent with accepted facts. Successful framings also usually fit comfortably into a culture's core values and narratives. It is also helpful if they are consistent with "master frames" that already exist in the culture. For example, framing environmental problems in terms of environmental injustice fits into the widely used master frame that people at the bottom are not treated equitably.

---

[13] For reviews of this perspective see Klandermans (1988), Gamson (1995), Hannigan (1995), and Williams (2004). Gamson and Wolfsfeld (1993) provide an especially useful treatment of the role of media in social constructionism.

[14] For a pioneering statement of framing theory see Snow et al. (1986). Relevant literature reviews include Gamson (1992), della Porta and Diani (1999), Benford and Snow (2000), and Boström (2004).

Frames that are attractive to one population group may fail to resonate with others, so social movement advocates face the difficult task of either tailoring their appeals to different groups or finding a framing that appeals simultaneously to activists, potentially sympathetic bystanders, and the press. In either case, it is also highly desirable that frames not hopelessly alienate powerful opponents. Sometimes this challenge is met by the emergence of specialized SMOs with framings and identities tailored to fit the tastes of various constituencies. However, social movement ideologies that become too extreme can drive away supporters. Frames are often borrowed or adapted from other groups, and framings produced by social movements are frequently rebutted by counterframes from other organizations.

Social constructionism was not developed to be a complete theory of social movements, much less as a theory of social movement organizations. Nevertheless, it does provide important insights about the interaction between environmental movements and their supporters, helping to explain why environmental organizations choose the goals and strategies that they do.

Political Opportunity Structure

Social movement scholars using what is sometimes called the "political process approach" have focused attention on the influence of the "political opportunity structure" on social movement goals and strategies.[15] They argue that public perceptions that a social problem exists are a necessary but not sufficient condition for the formation and continuation of social movements. Potential participants must also believe that change is possible and that the costs of bringing it about will not be too high. This depends, in part, on the ability of the movement to produce results.

According to political opportunity structure researchers, whether a movement appears and whether it succeeds depend, to a significant extent, on the characteristics of the political context in which it operates. Researchers have thus sought to identify key dimensions of political systems that explain the emergence, strength, ideologies, goals, and strategies of social movements. Some researchers working in this tradition have suggested, for example, that both extreme openness to new interests and ideas and extreme repression on the part of the political system discourage social movement formation, while intermediate openness encourages it. Other studies have examined how the political opportunity structure influences movements' recourse to confrontational, violent tactics.

Researchers using this approach have identified a wide variety of characteristics of political structures that might affect social movements. Some of these are features of the political system's formal structure, such as whether it is federal or centralized and how much influence various branches of government possess. Federal systems and systems where the executive, judiciary, and legislature all have influence and operate relatively independently are thought to create multiple

---

[15] For reviews of this approach see the following sources: Tarrow (1991, 1996); Rucht (1996); Rootes (1999c); Kriesi (2004); Meyer (2004); Caniglia and Carmin (2005); and van der Heijden (2006).

points of access for social movements, which then tailor their strategies accordingly. Also important is the formal openness of the political system to challengers. For example, political systems differ in the ease with which new parties can gain representation in the legislature. Other potentially important characteristics of the political opportunity structure are based on its informal modes of operation. For example, political elites vary in the strategies they use to deal with challengers, including their readiness to use repression. Moreover, some states have the capability to act decisively and effectively in response to new demands, while others do not. Finally, several characteristics of the more general political environment have been identified as important. These include the significance of traditional political cleavages based on religion or class, changes in economic structure or other events that put the power structure in a state of flux, the availability of alliance partners for social movements, the existence of divisions within elites, and the stability of ruling electoral coalitions. Unfortunately, the very large number of independent variables employed, the unclear, inconsistent, and overlapping definitions of these variables, and the problems with measuring them have worked against the cumulation of knowledge about the effects of political opportunity structure.

Researchers using the political opportunity structure approach have typically focused on explaining when social movements appear, what goals they pursue, and what strategies they use. Such research emphasizes social movements that attempt to influence the political system and largely ignores those that engage in identity cultivation. The approach also pays little attention to the internal structures and decision-making processes of SMOs. And except for the insight that people affiliate with social movements when they believe there is a realistic prospect for change, it has little to say about the motives of individual movement participants or about how they are recruited to membership in SMOs. Nevertheless, the political opportunity structure approach does complement resource mobilization theory by helping to identify external conditions to which environmental organizations respond in choosing their goals and strategies.

Relevant Topics in Social Movement Theory

Social movement researchers have explored several issues that are especially relevant to the study of environmental organizations. Among the most prominent are tendencies toward centralization of power in SMOs, life cycles and institutionalization of social movements, waves of social movement protest and mobilization, modes of organizing social movements, the strategy choices facing social movements, and the relationship between social movements and the identities of their supporters.

Probably the best known of these topics is tendencies toward centralization of power and democracy in SMOs.[16] Research and theorizing about this topic

---

[16] The following sources contain useful commentary on tendencies toward centralization of power and professionalization in social movement organizations: Zald and Ash (1966); Knoke and Wood (1981); Wiesenthal (1993); Kriesi and Giugni (1996); Rucht, Blattert, and Rink (1997); Staggenborg (1988); Lahusen (1998); Rucht (1999a, 1999b); Fromson (2003); and Bosso (2005).

originated with Robert Michels' famous iron law of oligarchy (Michels, 1949), which he derived from his study of turn-of-the-century European labor unions and political parties. Michels argued that—despite democratic ideologies and forms—such organizations almost invariably evolve toward centralization, bureaucratization, and rule by an entrenched elite. This occurs because movement leaders typically desire to remain in office in order to retain the power, prestige, and perquisites of leadership. Incumbent leaders therefore work to make themselves irreplaceable by mastering the technical aspects of leadership and cultivating networks of useful contacts, and they regularly tout their accomplishments to their supporters. They also use every mechanism available to control the selection of new leaders, manipulating ostensibly democratic procedures to perpetuate their rule. Rank-and-file members, on the other hand, frequently lack the interest, knowledge, skills, and time to participate actively in decision-making, so they are prone to abdicate participation in democratic governance in favor of oligarchical leaders. Challenges to entrenched leaders are consequently rare and are usually beaten back successfully. The upshot of this is that influence in SMOs typically becomes centralized in the hands of a self-perpetuating clique, which prefers cautious strategies over confrontational protest because its claims to power and advantages rest on working within the system.

More recent work predicts a similar evolution for SMOs, but traces it to a different source: the tendency toward professionalization. The reasons social movement researchers give for increasing professionalization of SMOs are similar to those cited by interest group theorists, but they have paid more attention to the effects of professionalization on organizational structure and strategies. Specifically, they suggest that professionals seek to take over functions formerly handled by volunteer activists, push for formalization of procedures, and prefer to concentrate power in their own hands. They do this because they desire to secure their own positions, because formalized procedures and increased staff influence are more conducive to putting their professional skills to work, and because government agencies and other professionalized organizations may prefer working with professionals. The resulting gains in efficiency in certain activities, such as fundraising, member service, and lobbying, make professionalization an attractive option, but the result is centralization of power and an emphasis on cautious strategies and working within the system.

Some SMOs do manage to remain democratic; however, this evidently requires specific conditions, such as high member commitment to consensual decision-making, federated structures that keep the central headquarters weak, rules that discourage careerism in leadership, homogeneity and high organizational commitment among members, and low reliance on specialized skills. SMOs can also alternate between periods of centralized and democratic control, and their desire to present themselves as citizens' movements may prompt them to present a democratic facade that conceals the underlying centralization of power.

Professionalization and centralization of power can lead to serious problems for SMOs. To the extent that their members or the public expect them to adhere to democratic norms, control by entrenched cliques or professionals can open them to criticism and undermine their support. Professionalization can also lead to power struggles between paid staff and activists or elected leaders. All of these issues are quite visible in the histories of environmental organizations (e.g., Dowie, 1995; Bergstedt, 2002; Diani and Donati, 1999).

A way of thinking about patterns of change in social movements that is somewhat more comprehensive than merely noting tendencies toward centralization and professionalization is efforts to describe their life cycles.[17] Noting that new movements form regularly and that few movements manage to sustain themselves indefinitely, early models of movement life cycles described a rather rigid pattern of movement evolution. Movements begin with unfocused public discontent, which is followed by successful mobilization of the public, some gains for the movement, and finally, institutionalization. Institutionalization, which includes professionalization, bureaucratization, and a focus on maintenance and self-preservation instead of new goals, occurs for numerous reasons. The movement's successes reduce the motivation of its supporters, resulting in reduced levels of mobilization and protest. Formalized SMOs come to dominate the movement, processes of professionalization and centralization get under way, and SMOs become an accepted part of society's institutional landscape with rather stable resource flows. SMOs or political parties that originated during periods of high movement mobilization often live on as institutionalized remnants long after the movement has faded, focusing on protecting or marginally extending past gains. During this phase, they resemble interest groups more than activist participants in social movements

This model provides many useful insights, and it applies fairly well to the recent history of environmental movements in many Western societies (van Koppen and Markham, 2007). Nevertheless, it has been justifiably criticized as oversimplified. Critics point out that real-world movements can be hard to distinguish from one another and often lack readily distinguishable beginnings or ends. Activists are frequently involved in multiple movements, and they often transfer their attention from one movement to another. Moreover, new framings or the rise of new SMOs can unite previously separate movements or divide old ones. The intertwined history of the environmental and anti-nuclear movements provides a good example of this pattern (Joppke, 1993). Many nascent movements fail to get off the ground at all, and some suffer repeated and devastating defeats. Rather than becoming institutionalized, they may devolve into sects, isolated countercultures, or relatively small and inactive networks of former activists. Nor is successful institutionalization necessarily the final curtain. Changed circumstances or new framings sometimes allow movements to resurrect themselves, building on institutionalized SMOs or residual activist networks that have kept movement goals and ideology alive between waves of public support and mobilization.

---

[17] For a pioneering statement of the theory of social movement life cycles see Blumer (1969). Important reviews of later literature include Raschke (1987), McAdam, McCarthy, and Zald (1988), Snow and Benford (1992), Meyer and Whittier (1994), Brand (1995), Rucht, Blattert, and Rink (1997), Rucht and Roose (2001c), Minkoff (2001), and Caniglia and Carmin (2005).

A further insight from recent work is that overall levels of social movement activity and protest tend to follow a cyclical pattern that extends across social movement industries and nations.[18] Advocates of this model argue that changes in the political opportunity structure or the invention of new framings or tactics can lead to striking successes for new or previously moribund movements. This can set off an intense wave of social movement and protest activity, as new supporters are attracted by the movement's successes and mobilization spreads to related social movements or to other nations.

These waves of mobilization and protest are impressive sociological phenomena, but they tend to be self-limiting. Competition among SMOs and movements can lead to a spiral of radicalism. At times this leads to revolution or civil war, but more frequently it results in repression of the movement or drives away moderate supporters. Factional conflicts often break out, and activists in intense waves of movement mobilization can easily become burned out. At the end of the cycle, a new equilibrium is reached and the movements fade away or go into hibernation. This model provides a good approximation of the course of NSMs in many Western nations during the 1970s and 1980s.

Waves of mobilization and protest have complex effects on environmental organizations. New SMOs, such as Greenpeace, often form to take advantage of the movement's energy, and existing organizations may come under strong pressure to modify their goals and strategies and function as SMOs for the movement. By doing so, they can maintain or increase their support and forge alliances that help them to attain their goals. On the other hand, becoming active participants in confrontational social movements can bring about many difficulties for such organizations. The new members they attract may include movement activists who press for new goals and strategies that existing supporters find unpalatable. The result can be internal conflict and the withdrawal of either activists or existing supporters. The more radical and confrontational posture required to be part of the movement can also threaten SMOs' acceptability to their funding sources and to negotiating partners in industry or government, or invite repression.

Social movement theorists have also explored in some detail the relative advantages and disadvantages of various social movement structures.[19] The most commonly made distinction is between loose national or regional networks of smaller groups versus more formalized associations or political parties. Networks can facilitate information exchange, provide mutual assistance, coordinate the efforts of their members, and convey the concerns of their constituent groups to politicians and media at the national or regional levels. Member groups ordinarily retain considerable autonomy, and the loose network structure provides for grassroots democracy and flexibility. On the other hand, local groups focused

---

[18] For a key statement concerning cycles of movement activity see Tarrow (1991). Useful overviews include Snow and Benford (1992), Rucht (1993b), Lofland (1996), McAdam, Tarrow, and Tilly (1996), and Koopmans (2004).

[19] Useful discussion of movement structures include the following: Jenkins (1983); Mayer-Tasch (1985); Rucht (1996); Diani and McAdam (2003); and McCarthy (2005).

on local affairs often lack the money and the desire to fund larger networks adequately, and they may guard their autonomy and democratic traditions so jealously that coordinated action becomes difficult.

At the other extreme lie centrally controlled organizations, such as the WWF. They are governed by self-perpetuating boards, and their supporters have no opportunity to participate directly in decisions. These organizations enjoy all the advantages associated with bureaucratic administration: rationality, predictability, and efficiency. Nevertheless, their ability to involve and mobilize their members for action is minimal, and they are subject to criticism for being undemocratic. There are also a variety of hybrid organizational forms that hope to realize the major advantages of each approach. Typically, these are national organizations with a federal structure and semiautonomous local chapters.

In recent years, globalization of the world economy, growing consciousness of the international dimensions of social problems, and the growth of international and regional governance organizations have introduced another level of complexity for the environmental and other social movements—the need to operate on the international level.[20] The result has been a rapid increase in the number and importance of international SMOs. Organizational forms found at the international level echo those found at the national level. They include relatively loose networks, such as Friends of the Earth and the Rainforest Action Network, as well as tightly structured and relatively hierarchical organizations like Greenpeace. These international organizations and networks are of interest not only in themselves, but also because of their potential impact on their national affiliates. Moreover, when they are successful in influencing the policies of international governmental organizations, they have indirect effects on national politics and policy, which, in turn, influence national-level social movements and SMOs. Finally, in some cases, international movements and SMOs may also undertake to influence national policy directly through protests, boycotts, and the like.

Social movement theorists have also paid considerable attention to the advantages and disadvantages of various social movement strategies.[21] The most frequently discussed dilemma concerns the advantages and disadvantages of confrontational strategies as opposed to working within the system. Confrontational strategies can be highly motivational for committed activists, they are often effective in drawing media attention, and they can influence politicians. On the other hand, too much confrontation can repel casual supporters and the public and lead to a hardening of battle lines, ridicule, or outright repression. Less confrontational strategies focused on working within the system can provide SMOs with access to decision-making circles and some influence, but the gains realized may

---

[20] Recent years have seen the publication of numerous articles and books about international nongovernmental and social movement organizations. See, for example, Smith (1997, 2005), Wapner (1996), della Porta and Kriesi (1999), Rucht (1999a), and van der Hiedjen (2002, 2006).

[21] For discussions of movement strategies see Lipsky (1968), Turner (1970), Jenkins (1983), Tarrow (1991), McAdam (1994), Lofland (1996), della Porta and Diani (1999), Diani and Donati (1999), and della Porta and Rucht (2002).

be modest, and SMOs may come under pressure to compromise key objectives in order to retain their seat at the table. The result can be disillusioned activists and diminished support or factional splits. New protest groups that adopt more extreme tactics can pose competition for the established organizations, but they could also strengthen the other organizations' position by making the latter appear to be the more approachable negotiating partners.

A final important contribution of the social movement literature to understanding environmental organizations comes from discussions of participation in social movements and in SMOs as a source of personal identity or personal transformation.[22] Theorists and researchers who work in this tradition point out that many social movements seek not only to transform the political or economic system, but also to alter the viewpoints and lifestyles of their members. One of their strategies for doing so is to create a collective identity among their members, who then feel bound to one another and the movement.

Movements as diverse as the labor, feminist, and environmental movements have at various times become important sources of personal identity for their core supporters, who may adopt distinctive styles of dress, language, and behavior to symbolize their attachment to the movement and their opposition to mainstream norms. Environmental organizations, for example, frequently seek not only to persuade government to control power plant emissions, but also to encourage their supporters to buy fuel-efficient cars, travel by bicycle, and use mass transit. Persons deeply involved in social movements may also spend time interacting with other participants in the movement "scenes," such as union halls, coffee houses, or nature reserves. At the extreme, they may even eschew political activity altogether and withdraw into utopian communities, where they can live in accordance with their own values; however, this approach is more typical of smaller, less formalized groups than of large, formalized organizations.

Summary and Conclusions

Social movement theory offers a very useful set of tools for analysis of environmental organizations. It provides insights about supporters' motivations for participation in environmental organizations, organizational structure and strategy, the impact of larger social and political structures on environmental organizations, and, in at least a limited way, the role of environmental organizations in society.

There are, however, serious drawbacks to automatically classifying every environmental organization as a social movement organization. The problem arises because influencing political decisions and public policy rank low on the agendas of some environmental organizations, and some are decidedly nonconfrontational and rarely use protest tactics. For example, even during the wave of mobilization and protest in the 1970s, organizations like the WWF and the Nature

---

[22] See, for example, Curtis and Zurcher (1974), Melucci (1985), Gamson (1991, 1992), Friedman and McAdam (1992), and della Porta and Diani (1999).

Conservancy in the US remained largely apolitical and nonconfrontational. Nonconfrontational environmental organizations were a fixture of the environmental scene in the post-World War II years in most Western nations (van Koppen and Markham, 2007), and some researchers argue that this style of operation is gaining renewed prominence today. They give two reasons for this. Some (e.g., Mol, 2000) argue that the growth of ecological modernization and the increasing involvement of environmental organizations in cooperative decision-making with business and industry have made cooperation a more appealing strategy. Others (e.g., Blühdorn, 2000) claim that radical ecology has simply lost its appeal.

In situations like these, environmental organizations can be defined as SMOs only by adopting the broadest possible definition of social movements. Moreover, classifying them as SMOs frequently leads to major problems in analyzing them because the authors of many social movement theories intended their theories to fit *only* organizations that challenge the power structure and use confrontational tactics. Their theories explicitly exclude apolitical environmental organizations with members more identified with protecting nature reserves or cooperating with established interests.

These problems can be reduced by employing only those social movement theories that are appropriate to the particular environmental organizations under study or by excluding from the research those organizations that fail to fit the particular social movement theory being used. However, the first alternative often leads to the exclusion of large segments of social movement theory, while the second often excludes many important environmental organizations. These considerations make it clear that social movement theory does not and cannot provide a theory suitable for analyzing all environmental organizations in all time periods. Social movement theories do, however, provide a wealth of insights about many environmental organizations in many times and places, and I draw on them frequently in the remainder of this book.

Theories of Civil Society

A more recent theoretical approach with potential applicability to environmental organizations is built around the concept of civil society.[23] Most definitions of civil society describe it as the sphere of social life that is structured by private, voluntary arrangements among individuals and groups pursuing their own goals—which may, however, include working for what they perceive as the common good—rather than by the family, the state, or the market economy. That is, participation in civil society is not determined by birth, occupational requirements, or state coercion, but by free choice. Although some authors (e.g., Put-

[23] This section is based on the following expositions of theories of civil society and their limitations: Held (1995); Foley and Edwards (1996); Minkoff (1997); Smith (1998); Brulle (2000); Putnam (2000); Deakin (2001); Warren (2001); Skocpol (2003); van Tatenhove and Leroy (2003); Fung (2003); Walzer (2004); Edwards (2004); and Felbinger (2005).

nam, 2000) include informal cliques and neighboring in civil society, the core of the concept centers on organized associations. Concretely, this includes groups as diverse as amateur sports leagues, hobby clubs, self-help groups, neighborhood associations, charitable organizations, public interest lobby groups, and social movement organizations. Environmental organizations are frequently classified as components of civil society; however, for reasons described below, some writers have reservations about including SMOs in civil society, so environmental organizations that function as SMOs occupy an ambiguous status in the theory.

Theorists of civil society typically argue that neoliberal democracies with market economies, whether in mature or newly emerging democracies, function best when accompanied by a well-developed civil society. Yet there is considerable disagreement about whether civil society is growing stronger or weaker in mature democracies. European theorists (e.g., Beck, 1986; Marks, Hooghe, and Blank, 1996; Jordan, 2001; van Kersbergen and van Waarden, 2004) are more likely to argue the former position. They cite three major reasons for the growth of civil society: 1) declining public trust in the capability of nation states to cope with increasing levels of complexity and risk; 2) a growing realization that efforts to bring about change through command and control regulation by government lead only to resistance and polarization; and 3) increasing reliance on complex systems of governance in which influence is shared among international, national, and local government, as well as among business and nongovernmental organizations. These governance systems are viewed as incorporating civil society associations in decision making by inviting them to participate fully in every stage of decision-making processes. The involvement of environmental groups and organizations in such governance processes is cited as a prime example of this trend (Mol, 2000; van Tatenhove and Leroy, 2003; Buttel, 2003).

Scholars from across the Atlantic (e.g., Putnam, 2000; Skocpol, 2003), on the other hand, are more likely to identify decreasing participation in local groups and civic associations and the rise of single-issue interest groups as signaling a decline in civil society. They argue that groups like Greenpeace mobilize masses of isolated and passive supporters with media campaigns and extreme positions, creating a politics of bitter contests among interest groups rather than rational deliberation.

There is agreement on both sides of the Atlantic, however, that the neoliberal state's retreat from providing public services is opening new opportunities and posing new challenges for civil society, which is increasingly expected to take up the slack in providing various types of services, such as provision and management of nature protection areas. There is also broad agreement that globalization undermines the political and economic power of the nation state and that civil society organizations and networks are growing in number and coming to play an increasingly important role in world politics. According to some, new international organizations and networks, such as Greenpeace and WWF, are the harbingers of a global civil society that unites people who face similar problems around the world and puts global social problems on the international political agenda.

Most discussions of civil society claim that the associations of civil society also serve important functions for society. First, they are credited with building social capital—the network of overlapping memberships and mutual trust that binds citizens to one another and to society and reduces social isolation. Second, the market economy and the state often fail to deliver many specialized services and collective goods (Weisbrod, 1988; Zimmer, 1996). Civil society organizations may step in to provide them on their own or on behalf of the state. Third, civil society supplements the formal democratic structures of the state by bringing important problems to light, providing additional mechanisms for public participation in setting societal goals, representing the interests of groups of citizens, and acting as a check on government and business power. This is especially important in societies where democracy is under threat, but civil society organizations can provide an alternative to immobile state bureaucracies in any society. Fourth, through participation in the work and governance of associations, citizens learn skills in self-government, along with a sense of efficacy and appreciation of the necessity of carrying their share of the load. Finally, a strong civil society is thought to contribute to the development of a deliberative politics. In deliberative politics, knowledgeable citizen groups take part in reasoned deliberations among all the parties involved in an issue in order to reach a rational decision. This tends to eliminate destructive struggles among interest groups.

Despite the celebratory character of much writing about civil society, notes of caution have also been sounded. First, persons from higher levels of the social class ladder are strongly overrepresented in most civil society associations, including environmental organizations, so giving such organizations a greater role in governance may do little to reduce existing power inequities. Second, civil society theorists may have placed too much faith in the efficacy of civil society institutions to solve social problems. Provision of public goods by the nonprofit sector, for example, can be spotty and underfunded. Third, mutual respect and rational deliberation do not necessarily make conflicting interests disappear. Powerful interest groups may prove unwilling to compromise, and some organized groups may even promote intolerance. Finally, fitting social movements into theories of civil society has proved difficult because of their inclinations toward protest and working outside the system, their tendencies toward oligarchy, and their occasional propensity to stubbornly embrace extreme positions and combative strategies. Civil society theorists have, therefore, been reluctant to include SMOs in their models. Realistically, however, social movements are unlikely to disappear from the scene, and they may prove unwilling to be "domesticated." Furthermore, even confrontational SMOs can perform many of the functions attributed to civil society organizations.

Summary and Conclusions

The civil society approach provides a limited but valuable perspective on environmental organizations. It is limited because it focuses almost entirely on the place of environmental organizations and other associations in the governance structures and functioning of society. It offers little insight into how such organizations set their goals, organize themselves, and choose their strategies, and it provides only a few hints about why individuals choose to support environmental organizations. On the other hand, the theory's focus on how environmental organizations and other groups contribute to the common welfare fills a significant gap in theories of social movements and organizations and offers an alternative to the pluralist and corporatist models described by interest group theorists.

Organization Theory

The last theoretical perspective, general organization theory, has thus far been used only infrequently in analyzing environmental organizations, although resource mobilization theories of social movement borrow much from it. Nevertheless, organization theory has great potential utility for this purpose. Unlike the three approaches reviewed above, it places the accent on the organizations themselves, not their roles in larger social systems. It asks why organizations adopt the goals, internal structures, and strategies that they do and how these choices influence one another.

Theories of Organizations

In this section, I focus on general theories of organizations, not those specifically designed to analyze business firms. Two branches of organization theory stand out as especially promising in this context: the open systems model of organizations and its derivatives and institutional theory.

Open Systems Theory

Open systems models[24] attempt to explain how an organization's goals, structures, and strategies are influenced by the broader social context in which it functions. They emphasize the following explanatory factors: 1) the general social and cultural milieu in which an organization operates; 2) the preferences and actions of individuals and organizations from which an organization acquires the resources it needs to sustain its operations; 3) the preferences and actions of organizations with which an organization competes or cooperates; 4) the preferences and actions of government agencies and other organizations that regulate, accredit, or certify an organization or otherwise report on its activities; 5) the preferences and actions of an organization's customers or the individuals and organizations that are the target of its actions; and 6) the preferences and actions of groups that oppose an organization or some or all of its actions.

Figure 2.1 summarizes an open systems analysis of an environmental organization diagrammatically. The organization derives the inputs of labor, services, and

---

[24] For seminal statements of open systems theory see Thompson (1967) and Katz and Kahn (1978). For reviews see Miner (1982), Preisendörfer (2005), and Scott and Davis (2007).

Figure 2.1. Open Systems Model of Environmental Organizations

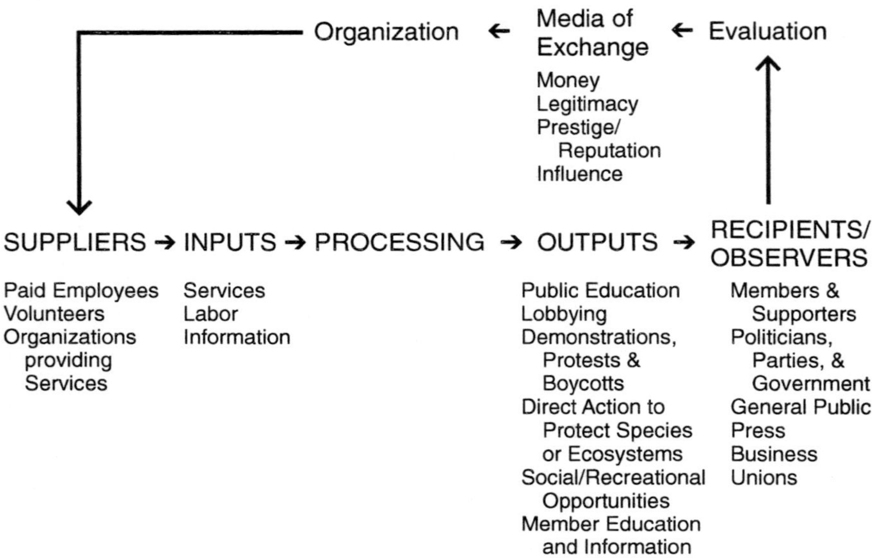

knowledge that it needs from its employees, volunteers, and other organizations. It then processes these inputs in some way to produce one or more outputs. These might include, for example, public education, lobbying, purchase or care of nature reserves, demonstrations, and social opportunities for its supporters. These outputs affect or are observed by relevant individuals and organizations in the organization's environment. These can include ruling and dissident elites, media, government and other regulatory bodies, opponents, and potential sympathizers, donors, and supporters. Depending on the favorableness of their evaluation of the organization and its activities, they decide whether to provide it with "generalized media of exchange" (Parsons, 1970). These include the following: 1) money, in the form of purchase of products or services produced by the organization, membership dues, personal or corporate donations, and government appropriations, subsidies, or contracts; 2) legitimacy, i.e., the perceived right of the organization to exist and pursue its activities; 3) prestige, including especially its reputation for effectiveness; and 4) influence, the likelihood that individuals and other organizations will respond positively to the organization's suggestions or demands. The more of these resources an organization commands, the more easily it can procure additional inputs, remain in operation, and effectively pursue its objectives.

Whether or not an organization is positively evaluated by the various actors in its environment depends on both the mixture of outputs it generates and on the preferences of these actors. This often creates dilemmas for environmental organizations because a striking feature of their social contexts is dissensus among the individuals and organizations that evaluate them about what objectives environmental organizations should pursue and what strategies they should utilize. Goals and strategies that win approval and support from some recipients or observers may therefore provoke withdrawal of support or hostile opposition from others. Strongly worded critiques of government from an environmental organization, for example, may inspire activists but infuriate government officials, and hiring professional staff to prepare expert reports and negotiate with government and business may win acceptance from these groups but leave volunteer activists feeling excluded.

Where two or more organizations with the same general resource needs and objectives operate in the same social context, they must either compete for resources or find a way to increase the total amount of resources available. They also become subject to comparisons by evaluators, potentially affecting the amount of resources each obtains. This means that an organization typically monitors the activities of other organizations in the same "industry" and must carefully manage its relations with them.

Drawing especially on ideas from Thompson (1967), the open system model was extended by resource dependence theorists,[25] who emphasized the effects of organizations' dependence on other organizations in their environments for resources and organizations' strategies for coping with this situation.

Resource dependence theory argues that organizations necessarily assign high priority to ensuring that they continue to obtain the money, prestige, influence, legitimacy, and key inputs that they need to operate and execute their programs. To accomplish this, they must maintain satisfactory relationships with the actors who supply them with these resources, especially if they cannot obtain them elsewhere. An organization's concern about acquiring needed resources will be heightened when there is concern that the flow of these resources might be interrupted. An environmental organization that depends heavily on the support of a broad base of individual donors for funds, for example, would be expected to choose goals that appeal to its donors and allow the organization to demonstrate success. It would also be likely to spend much money and time cultivating its media image, to recruit celebrities or respected government officials to speak for it, and perhaps to seek external accreditation of its effectiveness and fundraising practices. To reduce its dependence on this funding source, it might attempt to diversify its sources of funds by producing consulting services for sale or obtaining grants from government agencies. Long-term grants or contracts that ensured a continuing flow of money would be especially desirable. Should it become heavily dependent on a particular government agency for financial support, an organization might attempt to ensure continued support by giving the agency a

---

[25]The most comprehensive statement of resource dependence theory is Pfeffer and Salancik (1978). For later developments and reviews see Hall (2002), Preisendörfer (2005), and Scott and Davis (2007).

voice in its decisions by placing an official from the agency on its board of directors, even if this cost it some of its autonomy. Dependence on individuals or other organizations for funds and the undesirable effects of fluctuations in funding could be reduced by accumulating cash reserves as a buffer.

No organization can devote itself, however, to pleasing only one of its constituencies, as other actors in its social context might respond by withdrawing their support. An environmental organization that becomes too confrontational, for example, risks losing grant support from government or business, and an organization that has grossly exaggerated its accomplishments to impress donors might come under media scrutiny and find itself deprived of the legitimacy and prestige needed to exert influence or attract supporters. Because loss of legitimacy is especially damaging to nonprofit organizations, they can be expected to vociferously counter threats to their legitimacy, such as claims that environmental reforms contribute to job loss. For most environmental organizations, then, management becomes a delicate balancing act in which organizations try to develop goals and strategies that will satisfy the actors in their environment on which they are most dependent without unduly offending other groups on which they also depend.

Resource dependence theorists have also devoted considerable attention to understanding how organizations manage their relationships with competing organizations. For example, to reduce competition with other environmental organizations, organizations might seek out specific niches, and such specialization might be ratified by informal understandings or formal agreements among the organizations. In some situations, competing organizations might also benefit from forming alliances to act together for the welfare of all. For example, environmental organizations might join together in an association to lobby government for increased grant support for environmental projects, or they might hold a joint press conference to criticize a corporation for environmentally damaging actions. Corporatist political structures strongly encourage the formation of umbrella organizations to formalize and promote such cooperation, and dictatorial regimes may even mandate them. Joint efforts can reduce the costs borne by each organization; however, the participating organizations are likely to be very concerned about maintaining their individual visibility in order to reap the donations, legitimacy, and influence that result from successful efforts. Desires to protect their own identities might also be obstacles to another common method of reducing competition, merger.

While resource dependence models emphasize the strategies that organizations use to manage their relationships with the outside world, contingency theory models focus on how organizations might modify their internal structures to better adapt themselves to their environment.[26] Much of the work in this tradition is directed toward business firms and not easily adapted to environmental organizations, but the contingency theory literature does contain some useful insights. For example, it suggests that organizations that are having difficulty obtain-

[26] For reviews see Miner (1982), Preisendörfer (2005), and Scott and Davis (2007).

ing needed resources frequently modify their goals or direct their efforts to new audiences or new geographic areas.

Population ecology models[27] also begin with the assumptions of the open systems model; however, they doubt that organizations' planful efforts to adapt to their environments are the best explanation of changes in the goals, structures, and strategies of organizations in a particular "industry." They argue instead that the capacity of individual organizations to respond planfully to forces in their environments is very limited. The social context in which many organizations operate is complex and difficult to analyze, and the information needed to plan intelligently is scarce and expensive to obtain. Major changes in strategy may prove impossibly expensive because they require writing off previous investments in equipment, procedures, and employee skills, and employees and supporters may resist radical changes in goals or strategies. Consequently, organizations are usually unable to adapt to rapidly changing circumstances.

Since organizations find it so hard to adapt, changes in the typical goals and strategies of organizations of the same type occur primarily because organizations that were poorly adapted to social change fade away and are replaced by new ones better adapted to the new conditions. Organizations that have selected goals and strategies to fit a specialized niche fade away especially quickly if support for their particular goals declines. For example, during the 1970s, nature protection organizations that refused to broaden their goals to include environmental protection ran the risk of being supplanted by new organizations that embraced both kinds of goals (van Koppen and Markham, 2007). Organizations with less specialized goals may be better able to ride out fluctuations in the environment, but they may find it difficult to compete with those adapted to specific niches.

Change in the goals, structures, and strategies of organizations in an "industry" thus results, in part, simply from the survival of organizations best adapted to new circumstances. But once a new pattern has proven successful, it tends to be diffused from organization to organization through such mechanisms as business school instruction, business magazines, consultants, and movements of personnel between organizations, making the newly founded organizations likely to adopt it.

The Institutional Approach

Viewing environmental organizations as institutionalized structures provides additional insights about how they behave. Organizations are said to be institutionalized when they are governed by shared assumptions and normative standards that prescribe specific roles, goals, and activities as appropriate for them. Operating within these parameters has two major advantages. First, it increases an organization's legitimacy with government, other organizations, the public, and its supporters. Second, institutionalized goals and norms provide taken-for-

---

[27] The seminal statement of population ecology is Hannan and Freeman (1977). Useful recent reviews include Pfeffer (1997) and Hall (2002).

granted solutions to vexing problems concerning organizational goals, strategies, and structure.

Early writers in the institutional school[28] stressed that organizations, their goals, and their practices, can become infused with symbolic meanings for employees and supporters who are strongly vested in the organization. Organizational practices can then take on the characteristics of myth and ritual, a point also emphasized by the organizational culture perspective (e.g., Trice and Beyer, 1993). The story of Greenpeace's brave founders sailing their tiny boat into a nuclear test zone, for example, is recited repeatedly in Greenpeace documents, and Greenpeace supporters usually believe that confrontations like these constitute the core of what Greenpeace is all about.

Because of the taken-for-granted nature of institutionalized goals, strategies, and patterns of action—as well as their symbolic significance—organizations often follow them even when they are not optimally efficient. Moreover, organizations frequently encounter resistance when they seek to alter institutionalized patterns. Environmental organization activists and donors who identify strongly with a specific goal or program may thus complain or reduce their support if resources are diverted to other goals and programs, even when the case for doing so is well argued. In such situations, opponents of new goals or strategies frequently invoke the organization's traditions in their counterarguments.

More recent statements of the institutional perspective, often referred to as neo-institutional theory,[29] place more emphasis on how institutionalized beliefs and practices from an organization's environment influence its goals and operating procedures. In this view, organizations innovate not so much to improve their efficiency as to maintain their legitimacy and their reputations as trendsetters by responding to the legal requirements and normative expectations of the outside world. Innovations are thus most likely to spread from organization to organization when they come to be defined as normal practice by business schools and management consultants, peer organizations, regulatory or accrediting agencies, and professional associations. For example, adoption of the term "sustainable development" by a number of prominent, prestigious environmental organizations and experts soon won this concept a place in the goals and vocabulary of practically every environmental organization.

Applying Organization Theory to Environmental Organizations

Unlike students of interest groups, social movements, and civil society, organizational theory scholars have displayed relatively little interest in environmental organizations, focusing instead on private-sector organizations and public administration. Environmental organizations have, however, sometimes been inves-

---

[28]A key early statement is Selznik (1957), and Perrow (1993) offers an overview of early work. Reviews of recent developments include Aldrich (1999) and Scott and Davis (2007).

[29]Key statements of neo-institutional theory include Meyer and Rowan (1977) and DiMaggio and Powell (1991b). The following offer useful overviews: Zucker (1983, 1987); DiMaggio and Powell (1991a); Aldrich (1999), and Scott and Davis (2007).

tigated using models derived from the open systems and resource dependence approach, usually under the rubric of voluntary association research (e.g., Knoke and Wood, 1981; Knoke, 1990). The major topics taken up in this line of research, including obtaining needed resources, motivating members to participate, and maintaining democratic governance, have paralleled those examined by scholars using the interest group and social movement approaches. In recent years, however, research about voluntary associations has been largely assimilated into the study of what are variously called nonprofit, nongovernmental, or third sector organizations (e.g., van Til, 1988; Strachwitz, 1998; Bode, 2003). This line of research emphasizes charitable organizations and nonprofit organizations that deliver services on government contracts, diverting attention away from environmental organizations as a subject of study.

Despite the marginal standing of environmental organizations in organizational research, organization theory models offer great promise for analyzing them. Because they offer a *general* theory of organizations, many organization theory concepts and models apply equally well to all environmental organizations, whether they are nonconfrontational nature protection organizations or activist social movement organizations. This distinguishes organization theory from theories of interest groups, which fit only environmental organizations whose major goal is to influence the political system, and from social movement theories, most of which are designed for relatively confrontational environmental organizations embedded in social movements.

A second advantage of organizational theory, especially open systems models, is that virtually all the useful insights from interest group theory and social movement theory reviewed above can be relatively easily incorporated into it. Discussions of the free rider problem and of how social movement entrepreneurs gather the resources needed to get social movements off the ground, for example, can be treated as special cases of the need to obtain needed inputs. Environmental organizations' decisions about how to frame environmental problems in their presentations to media and their supporters can be interpreted as strategies for acquiring money, prestige, legitimacy, and influence and viewed as conditioned by institutionalized goals and norms to which the organization's various constituencies are committed. Choices about whether to adopt goals and strategies that emphasize cooperation or confrontation can be interpreted within the open systems framework as problems faced by organizations trying to find a strategy that will simultaneously allow them to acquire needed resources, avoid offending powerful organizations in their environment, and move toward their goals. The political opportunity structure approach fits into organization theory by helping to identify factors that influence which strategies are most likely to work to acquire needed resources and influence the political system. And new social movement theory's hypotheses about the importance of environmentalist identities among supporters dovetail nicely with the institutional school of organizational analysis.

Because organizational theory is general enough to fit essentially all environmental organizations, and because most insights from other theoretical approaches can be incorporated into it, I have used it as the primary framework for the analysis in the following chapters; however, I have freely incorporated insights from the other theoretical perspectives where they can enrich the analysis. To make the book accessible to nonspecialists, I have kept theory in the background, except in the chapter conclusions and parts of the last chapter. Nevertheless, careful readers of this chapter will find it easy to discern the links between the theoretical perspectives described here and the analysis that follows.

Chapter 3

THE ORIGINS OF NATURE PROTECTION ORGANIZATIONS IN GERMANY

Conservative Reactionaries, Protectors of Nature, and Social Democracy at the Beginning of the Century

National-level organizations concerned with protecting the environment first arose in Germany—as in the US and most other European countries (Dalton, 1994; Kline, 2000)—near the end of the nineteenth century.[1] They appeared in the context of a nation state then only a quarter of a century old, at a time when national-level interest groups in business, labor, and social welfare were also coalescing (Reutter, 2001), and when political debate and legislation about environmental damage and the establishment of government agencies to protect nature were just getting underway.

The turn of the twentieth century was also marked by disruptive changes associated with rapid industrialization, population growth, and urbanization. This transformation produced not only obvious air pollution, water pollution, and damage to nature, but also sweeping changes in the social fabric. Rural peasants became factory workers, toiling long hours for low pay and lodged in massive tenements in unhealthy cities, and rural folkways and traditional verities came

[1] To avoid repetitive referencing throughout the chapter, sources drawn on throughout the chapter are listed here. The following sources contain brief summaries of this period: Rat von Sachverständigen für Umweltfragen (1996); Oswald von Nell Breuning Institut (1996); Williams (1996); Brand, Eder, and Poferl (1997); Gröning and Wolschke-Bulmahn (1998); Brand (in press). More comprehensive treatments include: Bergmann (1970); Wey (1982); Linse (1986); Dominick (1986, 1992); Hermand (1991); Knaut (1993); Chaney (1996); Riordan (1997); Geden (1999), and Olsen (1999).

under threat. Long-established patterns of privilege were disrupted by the growing wealth and power of capitalists and the rise of an organized working class. These disruptions contributed to a flowering of social movements of every conceivable stripe. Consequently both discourse about nature and organized efforts to protect it were strongly influenced by ideologies and social movements ranging from the reactionary right to the socialist left.

Early History of Environmental Protection in Germany

Concerns about pollution, natural resource exhaustion, and threats to nature, along with efforts to combat them, arose well before national-level environmental organizations.[2] Laws to protect useful birds and distinct landscape features, such as caves or ancient trees, in the principalities that later became Germany go back several centuries; however, the first nature protection area was not established until 1833. Its purpose was to protect the *Drachenfels,* the scenic ruins of an ancient castle perched on high rock outcroppings over the Rhine, from quarrying. Beginning at mid century there were also attempts to protect green areas near Berlin and scenic areas, such as Alpine lakes, the Harz mountains, and heathlands, as well as efforts by amateurs to inventory threatened landscape features and species. Bird protection laws in various regions date back to the early years of the nineteenth century, and the first national bird protection law, which focused on protection of songbirds and economically useful birds, was passed in 1888. By the second half of the nineteenth century, writers and foresters were decrying accelerating deforestation brought on by population growth and increasing demand for wood, leading Prussia to pass a forest protection law in 1875.

Environmental Problems in Early Twentieth Century Germany

Earlier threats to nature and the environment were, however, overshadowed by the scale of the new problems that accompanied rapid industrialization and urbanization.[3] With the support of a national government that viewed industrialization as synonymous with progress, Germany was transformed during the years between its unification in 1871 and World War I from an industrial backwater to the world's second largest industrial power. The war and the subsequent hyperinflation and depression slowed the pace of growth, but postwar governments,

---

[2]In addition to sources cited in note 1, discussions of developments before the late nineteenth century include: Hoplitschek (1984); Erz (1987); Hermand (1993); Bick and Obermann (2000); Louis (2003); and Kneitz and Kley (1986).

[3]In addition to sources cited in note 1, especially Wey (1982), see the following: Schoenichen (1930); Sieferle (1984); Brüggemeier (1990); Hermand (1993); Fischer (1994); Rollins (1997); Lorenz (1996, 2002); Wöbse (2003b); Louis (2003), and Uekötter (2003c).

the labor unions, and the labor-dominated Social Democratic Party all remained committed to prosperity through industrialization.

As in the US and Britain, German industrialization was accompanied by rapid population growth and urbanization. In 1871, only a third of Germans lived in cities. By the eve of World War I, two-thirds were urban, including many housed in crowded tenements plagued by poor sanitation and disease. Population in rural areas, by contrast, stagnated as a generation of young people chose industrial work in the cities over field labor and rural poverty. Heavily industrialized areas like the Ruhr Valley and Saar were transformed from quiet agricultural areas into densely populated, and industrialized urban agglomerations.

Industrialization and urbanization brought numerous environmental problems in their wake. Smoke and soot from smelters, industrial plants, and home heating—along with sulfur dioxide other pollutants—threatened human health, crops, and forests. Waterways were polluted by industrial discharges and untreated sewage from newly constructed sewage systems, and drinking water was often unsafe. These problems were sometimes defined as the price of progress, but they were troubling nonetheless, and they became the focus of scientific research, growing citizen concern, and occasional protest.

The deleterious effects of industrialization extended beyond cities. Air pollution spread over large areas, especially after taller smokestacks were built to reduce pollution in the immediate area, and rivers carried pollution downstream. Industrialization also led to construction of railways, power lines, roads, and canals, to damming and channelization of rivers and streams for electric power, barge traffic, and flood control, and to loss of green space. Rapid urban growth and the beginnings of suburbanization converted agricultural lands and open space near cities to urban use, while further afield, damage came from throngs of urban tourists, who used the new rail system to escape to the countryside for recreation. With them came overused trails, litter, tourist hotels, observation towers, and scenic railways.

Germany's growing population greatly increased the demand for agricultural products, and the Treaty of Versailles at the end of World War I cost the country much valuable agricultural land. The result was the extension of agriculture to previously forested land and pressures to use new technologies to increase farm productivity. Small fields separated by woodlots, streams, paths, and hedgerows gave way to larger fields more suited to mechanized agriculture. Fields were joined, roads and streams straightened, and woodlots and hedgerows removed. Rationalized forestry replaced mixed-growth forests with monocultures, and trees were planted in neat rows with underbrush removed. Even forest clearings were planted. These changes led, in turn, to declines of birds and other species due to habitat loss.

The Political Response to Environmental Problems

The national and *Länder* (regional) governments did attempt to respond to the growing environmental problems, but the political structure of twentieth century

Germany placed many obstacles in the way of successful action.[4] Consequently, the government response was frequently halting and ineffectual. Between unification and World War I, German politics was dominated by "iron and rye," a powerful conservative alliance of industrialists from the West and owners of large agricultural estates in the East. Both the power of these interests and the dominant ideology, which emphasized progress through technology, industry, and rationalization of agriculture, stood in the way of effective action.

In early twentieth-century Germany, industrial pollution and urban waste were framed and addressed as localized property damage or public health issues, not as part of an overarching concept of environmental degradation. Most *Länder* had at least some laws that could be used to regulate emissions; however, decisions about issuing discharge permits and enforcement gave considerable weight to the economic benefits of factories. Enforcement was the responsibility of local governments, which often lacked the legal powers, staff, and financial resources to do more than respond to complaints and deal with the most egregious problems, often by compromises, such as requiring taller smokestacks to disperse air pollution. Responsibility for combating the problems was dispersed among numerous offices, and bureaucratic procedures slowed enforcement. The national and *Länder* governments had small offices concerned with water quality and public health. They could offer technical advice based on the limited knowledge available but had no enforcement powers. Suits for damages were possible only when a property owner could show direct damage from specific sources, which became increasingly difficult as sources of emissions multiplied.

Although industrial firms sometimes acknowledged pollution problems, displayed willingness to compromise, and paid damages, they also frequently resisted controls, raising objections about overly bureaucratic procedures, the difficulty of assigning blame and measuring damage, and the costs of cleanup. Settlements usually involved payments for damages rather than shutting down or cleaning up. Cities were also reluctant to acknowledge the effects of dumping urban waste or discharging untreated sewage into rivers and to undertake the costs of cleanup. The overall result was a system of half-measures adopted in response to complaints or crises, which succeeded only in limiting the most egregious damage

Government action was somewhat more effective in the area of nature protection, which fell under the purview of the *Länder.* The most important initiatives in this area originated with biologist Hugo Conwentz, the serious-minded and hard-driving director of a regional museum and author of a landmark report for the Prussian government on the need for protection of "natural monuments." The term natural monument, popularized by Alexander von Humboldt, was used to describe unusual, striking, or beautiful geologic formations, natural springs,

[4] In addition to sources cited in note 1, especially Wey (1982), information about government activity in this period appears in: Schoenichen (1930, 1954); Sieferle (1984); Erz (1987); Rösler, Schwab, and Lambrecht (1990); Brüggemeier (1990); Wettengel (1993); Uekötter (2003c); Oberkrome (2005). For developments in Prussia, see Milnik (2003), and for Bavaria, Hoplitschek (1984) and Wolf (1996).

stands of ancient or rare trees, and the like, which were seen as the natural equivalents of manmade monuments.

In 1906, Conwentz was appointed director of the newly established Office for the Protection of Natural Monuments in Prussia, which was by far Germany's largest *Land.* The authorizing legislation included not only natural monuments as traditionally defined, but also remnant populations of endangered species and the micoenvironments, such as moors and heathlands, where they lived. The establishment of national parks had been debated in the Prussian legislature, but Conwentz did not believe large nature protection areas or parks were a practical objective for Germany, a country with little remaining wilderness and a limited nature conservation budget. He advocated instead establishing smaller reserves and parks.

Despite its tiny staff, Conwentz's new office launched an ambitious nature protection program, including staging and attending domestic and international conferences, publication of periodicals and other documents, promotion of instruction about nature protection in the schools, nature study trips, and inventories of endangered species and natural monuments. Conwentz also set out to establish volunteer local nature protection committees in each Prussian province to publicize and promote nature protection, inventory natural monuments, and work for the designation of natural monuments. World War I and subsequent economic problems hindered this effort; however, these committees set the precedent for using volunteer committees to help discharge local government's responsibility for nature protection, which remained standard practice until after World War II—and in the GDR until reunification. The local committees typically included government officials, scientists, museum directors, teachers, and representatives of local nature protection groups, and their membership overlapped heavily with these groups. Generally no public funds were provided to support the committees' work.

In 1902 and 1907, Prussia put in place laws that prohibited defacing scenic areas and villages with billboards and the like; however, there were no penalties until the 1920s, and neither the local committees nor Conwentz's office possessed the staff, financial resources, or enforcement authority to vigorously enforce the laws. A 1912 effort to allow compulsory protection of natural monuments on private land failed to pass, so the nature protection offices had to content themselves with inventorying important sites and persuading cooperative landowners to preserve them. There was no money to purchase significant tracts of land. Despite these handicaps, Conwentz's system achieved significant successes in designating natural monuments and other biologically sensitive areas as protected areas. His efforts to prevent loss of bird habitats to agricultural rationalization projects were less successful, and he normally eschewed direct criticism of industrial development and did not focus his attention on industrial emissions.

Hesse, one of the smaller *Länder,* also enacted a provision for the protection of natural monuments as part of a 1902 law aimed at protecting manmade monuments, but it was the Prussian example that was imitated by other German *Län-*

der, including Saxony, Württemberg, and Bavaria. The Bavarian approach was especially noteworthy. It established an all-volunteer *Land* nature preservation committee with a small budget in 1906, a system of volunteer local committees, and an interpretation of nature protection that was considerably broader than the Prussian emphasis on protection of natural monuments. In 1908, it passed legislation protecting prehistoric sites, and a 1912 law authorized local authorities to protect endangered species. Nevertheless, like its counterparts in Prussia and elsewhere, the Bavarian nature protection effort was understaffed, underfunded, and heavily reliant on the support of volunteers.

World War I led to increased damage to nature through intensified agriculture, interrupted government nature protection, and weakened volunteer nature protection groups. Worse still, prewar patterns and problems in protecting nature and controlling pollution resurfaced in the postwar Weimar Republic. The 1919 constitution did make nature protection an official government responsibility, and almost every German *Land* passed a nature protection law during the Weimar period; however, landowners' demands for financial compensation for private lands set aside for nature protection made it impossible to pass truly effective legislation, and the laws that did pass were far from uniform. Nature protection efforts remained understaffed and underfunded, and there was no money for land purchases. The Weimar Republic's struggles with war reparations, inflation, depression, and political turmoil kept nature protection low on the agenda of the national government, and the *Länder* resisted passage of a national nature protection law. In industrialized areas, prewar patterns of dispersed responsibilities, understaffing, relatively weak enforcement of laws regulating pollution and waste disposal, and compromise solutions continued.

The Social and Ideological Context of Early Environmental Protection Efforts

National environmental organizations appeared not only in the context of growing environmental problems only partly countered by government efforts to cope, but also as part of a larger set of reactions to industrialization and urbanization. These included influential right- and left-wing ideologies and social movements, as well as a variety of "life reform" movements. Each of these ideologies and movements had concerns that overlapped with environmental protection; therefore, environmental issues could easily be incorporated into their agendas, and advocates of environmental protection had to take them into account.

Nature Protection and the Reactionary Critique of Modernization

The social movement most relevant to environmental protection—which it defined primarily as protection of nature—was based on an ideology of reactionary resistance to technology and industrialization, the rising influence of capitalists, rationalization of agriculture, rapid urbanization, and the rise of the urban working class.[5] These developments threatened established patterns of life based on hereditary social status, village life, small business, and craft production as well as the prestige and influence of intellectuals, small business owners, and the hereditary aristocracy. The result was a reactionary movement centered around efforts to restore traditional verities and ways of life—including, quite prominently, the protection of nature—which reached maturity in the first third of the twentieth century.

This reactionary critique of modernization had its roots in much earlier intellectual critiques of the Enlightenment's emphasis on rationality, individualism, materialism, and the leveling of social distinctions and cultural differences. These critiques, articulated in German Romanticism and, more explicitly, in the late eighteenth century writings of Herder, saw society as a unique, organic whole based on traditions evolved and tested over generations. It viewed nature sentimentally and anthropocentrically, emphasizing its beauty and soul-restoring power.

These ideas found sharper definition in the mid–nineteenth century writings of Wilhelm Riehl, a cultural historian, professor, and museum director best known as the originator of the systematic study of folklore. Reacting to early manifestations of the social changes described above, he denounced industrialization, the destruction of nature, and the growth of large, cosmopolitan cities as the epitome of all that was wrong with society. Neither economic liberalism nor socialism offered adequate solutions because they substituted individualism or class solidarity across national borders respectively for the unity of the German *Volk* and the traditional order. He advocated instead a harmonious, ordered, and stratified society rooted in the German soil and peasantry. Indeed, according to Riehl, the corrupting influence of socialism could only be countered by stable rural communities with citizens embedded in the particularities of the local situation. Nature protection was never his central concern, but he did argue for preserving wilderness as a counterpoint to industrialized and urbanized areas.

During the same period, biologist Ernst Haekel developed monism, a second precursor to the reactionary resistance to change. Monism was an effort to unite scientific ecology—Haekel was the first to use the term ecology and a pioneer in the field—with Goethe's pantheism to produce a biocentric philosophy. It incorporated elements of Riehl's emphasis on protecting the German *Volk* from liberalism, an argument that societies were founded on natural laws, and an interpretation of Darwinism that focused on the survival of cultures or races rather than of individuals. Haeckel himself belonged to the right-wing Pan-German League and later to the Thule Society, one of the precursors of National Socialism.

---

[5] Especially thorough treatments of the reactionary critique of modernization appear in several of the sources cited in note 1 (Bergmann, 1970; Hermand, 1991; Knaut, 1993; Geden, 1999; Olsen, 1999), as well as in the following sources: Muthesium (1981); Sieferle (1984); Andersen (1989); Knaut (1991); Rollins (1993, 1997); Fischer (1994); Lorenz (1996); Jeffries (1997); Kerbs and Reulecke (1998); and Radkau and Uekötter (2003). Other useful articles treat the reactionary ideologies and movements of this period as the forerunners of National Socialist nature protection (Dominick, 1987; Speitkamp, 1988; Biehl, 1995; Statham, 1997; Bensch, 1999) or the 1970s counterculture (Renn, 1985).

As the transformation of German society and accompanying environmental destruction gathered speed, a new generation of late nineteenth century writers began to forge a stronger link between reactionary social criticism and advocacy of nature protection. The most prominent of these was Ernst Rudorff, a German musician turned social crusader. Rudorff, the son of a privileged family, spent much of his youth and free time as an adult hiking the family's rural estate and the surrounding area. He was appalled by the damage industrialization, rationalization of agriculture, and urbanization were doing to the German landscape and national character, and he set out to stop it.

Like Herder, Riehl, and many German intellectuals of his day, Rudorff believed that a society's structure and well-being were based on the biological characteristics and culture of its citizens, which evolved in the context of the society's natural environment and in turn transformed the environment through agricultural practices and settlement patterns. The outcome of this process of reciprocal adaptation was a "*Volk,*" a people with an organic, cultural, and spiritual unity rooted in its native soil. Rudorff thus viewed the German national character as rooted in its intimate connections to an organically evolved, harmonious, and interconnected folk culture and the German landscape. Materialism, urbanism, and the rationalization of life threatened not only nature and rural landscapes, but also archetypical Germanic virtues—hard work, simplicity, peacefulness, joy, and reverence—and the social order associated with them.

Rudorff's indictment of the threats to the health of German society included a long bill of particulars. Industrialization despoiled the countryside with unsightly railways and power lines, scenic rivers were dammed and channelized, and industrial emissions fouled water and air. Mechanization and rationalization of agriculture replaced bucolic landscapes of small fields, woodlots, hedgerows, winding paths, villages, and cottages with vast and monotonous fields with geometrically regular boundaries. Forests were cleared and wetlands drained, and forest clearings were planted with trees. Rather than living in harmony with nature, people came to see it as resource to exploit or destroy. Nature was further despoiled by a flood of urban tourists.

Ignoring the exploitation, harsh living conditions, and high rates of infant mortality and tuberculosis among the peasantry, Rudorff idealized the rural village and vilified cities. Urbanization, he argued, depopulated rural communities and replaced rural *Gemeinschaften* with a dehumanized, alienated industrial proletariat susceptible to the appeals of socialist agitators, who advocated what Rudorff contemptuously termed "*Gleichmacherei*"—the foolish idea that all should be equal—and urged them to identify with the international working class. Urbanization was also accompanied by superficial urbanity and cosmopolitanism, moral decay, loss of respect for authority, and an unfulfilled longing for traditional verities and ways of life.

Rudorff's writings contain numerous blanket condemnations of modernism, technology, industrialization and urbanism; however, in his more cautious moments, he expressed a more qualified and nuanced view that acknowledged some of their benefits. He rarely attacked capitalism directly, focusing his critique instead on growing materialism, the evils of cities, damage to rural landscapes and scenic areas, and loss of traditional architecture and customs. To combat these problems, he prescribed not revolution, but public education and persuasion coupled with various meliorist measures. These included educating the public about local customs, traditional architecture, and nature, promoting government action to protect scenic landscape features, limiting tourism, and designing dams, railways, and power lines to be compatible with the landscape.

Rudorff's nature protection agenda emphasized not so much saving wilderness, which he recognized had already all but disappeared from Germany, but protecting traditional rural landscapes and the species endangered by their destruction. He also condemned deforestation and declines in bird populations. His concerns were grounded mainly in sentiment and aesthetics, but arguments more reflective of scientific ecology do appear intermittently in his writings.

The writings of Rudorff and his intellectual allies resonated with aristocrats, well-educated professionals and artists, government employees, and small business owners—all groups that enjoyed comfortable lives but found their social position and lifestyles threatened by the growing influence of industrialists and the organized working class. They were less attractive to capitalists, farmers, and the urban working class.

The German Left and Environmental Protection

The second major critique of social change in late nineteenth- and early twentieth-century Germany was the criticism of unregulated capitalism and industrialization from the left.[6] The basic outlines of this Marxist-based critique are too familiar to warrant rehearsal here. Less well-known is the fact that intellectuals of the left and the organizations of the working class at times raised their voices against destruction of nature. The problems they cited differed little from those listed by the reactionary right, but their critique laid the blame on capitalist greed and class privilege rather than industrialization and urbanization.

Early worker protests, such as those conducted by the Luddites, sometimes manifested an anti-technology bias not too different from Rudorff's, but after Marx the main thrust of socialist thought shifted toward seeking material prosperity for all through worker ownership of the means of production, not the rollback of industrial progress. Nevertheless, Marx, Engels, and the German socialist theorist August Bebel sometimes included deforestation, soil erosion and degraded fields, consumption of nonrenewable resources, and industrial emissions in the catalog of sins that capitalism was visiting upon the world, and some passages contain surprisingly modern ecological analyses. Socialist deputies in the national

---

[6] In addition to the sources in note 1 see the following: Sieferle (1984); Würth (1985); Hannigan (1995); Foster (1999); and Groß (2001).

parliament and *Länder* legislatures sometimes proposed legislation to protect worker health from dangerous substances, preserve natural areas for recreation, and limit industrial pollution. Yet compared to gaining a bigger slice of the pie for the working class, environmental issues were never a top priority for parties of the left, and their successes were few.

Organized action on behalf of the working class, including efforts to preserve nature, had to be carried out by independent working-class organizations,[7] for class divisions at the turn of the twentieth century were sharp, leaving little room for the proletariat in the associations of the middle and upper classes. Socialist leaders often promoted separate workers' organizations to educate and uplift workers, pull them out of deadening routines of long work hours, crowded tenements, and taverns, build class consciousness, and unite them in the struggle for a better life. Such organizations were most active in Germany between the lifting of the government's ban on the Social Democratic Party in 1890 and the rancorous split between Social Democrats and Communists that developed in the mid 1920s. There were sports, chorale, theater, language study, women's, amateur radio, and even abstinence and freethinker associations, many of them paralleling corresponding middle-class organizations. Within this social and ideological context, workers' groups focusing on experiencing and protecting nature were a natural development.

The working-class intellectuals who led these organizations believed that promoting the welfare of the working class required not only unions and labor parties to fight for workers' economic and political interests, but also organizations to promote a distinct working-class culture and consciousness and contest the dominance of bourgeois culture. These workers' organizations encouraged their members to be active in union and party affairs, and party and union leaders recognized the contributions such organizations made to supporting and recruiting for the movement. Nevertheless, they also worried that workers' organizations might divert attention from party and union activities or that their members might be infected by ideas from parallel middle-class organizations.

Other Social Movements

The social ferment at the turn of the twentieth century spawned not only the reactionary and socialist movements, but also many other movements and organizations with concerns relevant to the environment.[8] One prominent set of organizations was middle-class hiking clubs for young people, such as the Wandering Birds (*Wandervögel*), which romanticized experience in nature as an escape from

---

[7] In addition to sources cited in notes 1 and 6, useful discussions of workers' organizations of the period appear in Winterer (1955), Wunderer (1977, 1980), Kramer (1984), Upmann and Rennspieß (1984), Zimmer (1984, 1989), Gröning and Wolschke (1986), and Erdmann (1991).

[8] For descriptions and analyses of these movements see the sources cited in note 1 and the following: Muthesium (1981); Sieferle (1984); Renn (1985); Gröning and Wolschke (1986); Biehl (1995); Kerbs and Reulecke (1998); and Lorenz (2002). Garden cities and other city beautification efforts are also discussed in Bergmann (1970) and Rollins (1993, 1997).

city life. A second set of movements included a wide array of "life reform" movements, which emphasized bringing about social change, not so much through direct challenges to the system as through the cumulation of individual changes. These included vegetarianism, abstinence, natural healing, rural communes, life simplification, clothing reform for women, nudism, and mystical belief systems like theosophy and anthroposophy. A third set of relevant movements and organizations devoted itself to the beautification of cities and the landscape, expansions of urban parks and gardens, and the development of garden cities, which were promoted as a solution to urban problems and a way to recreate the folk *Gemeinschaft*.

Notably absent, however, were nationally organized movements to combat the air and water pollution generated by industry and urban agglomerations. Although there was no shortage of short-lived local protests, no national networks emerged. The fact that the authorities were able to respond successfully to some of the worst cases, and some complaints were defused by financial settlements probably hindered the development of such organizations.[9]

National Environmental Organizations in the Early Twentieth Century

With the exception of the weak and relatively short-lived Association for the Protection of Bird Life, founded in 1875 (Wöbse, 2003b), national-level organizations addressing issues we now classify as environmental did not appear in Germany until the end of the nineteenth century. When they did appear, all emphasized nature protection rather than pollution or resource consumption, and most combined nature protection with other goals. Organizations that pursued nature protection in combination with other environmental goals but did *not* have non-environmental goals—the focus of Chapters 7 to 12 of this volume—did not appear in Germany until after World War II (Bammerlin, 1998).

The four organizations I describe in the remainder of this chapter are those that historical research suggests had the strongest and most sustained impact on environmental problems and set the precedent for today's environmental organizations. Two of them are national organizations that pursued nature protection in combination with other types of goals. One had only a single goal, bird protection, and one was a regional nature protection organization. The last two are of special interest because they are the precursors of two of the four most important environmental organizations in Germany today. Other national and regional organizations, such as the People's Alliance for Homeland Protection, the League for Nature Protection, the Association for Nature Protection Parks, the Association for the Protection of Alpine Plants, the Berlin-Brandenburg Nature Protection Ring, and the Isar Valley Association, were also active during this

[9] See note 1 and Bammerlin (1998) and Uekötter (2003c).

period. However, because of narrow foci, regional scope, or short life spans, their long-term impact was not as significant.[10]

The League for Homeland Protection

The League for Homeland Protection (Bund Heimatschutz: BH) grew out of critiques of effects of industrialization and urbanization on rural communities, cultural landscapes, and nature.[11,12] This critique manifested itself not only in the writings of Rudorff and his compatriots, but also "homeland" art and literature, which glorified the simple life, and "homeland" architecture, which emphasized traditional building styles. It manifested itself in the formation of new associations to protect local communities, architecture, and customs. These groups supplemented and often worked with existing organizations devoted to the study of local history and local beautification.

Rudorff, the most visible proponent of homeland protection, was not only a theorist but also a propagandist and activist with a wide network of connections to government and Germany's intellectual elite. During the closing years of the nineteenth century, he concluded that Germany needed a national homeland protection organization like those in Britain and France. After several unsuccessful attempts to interest existing organizations in his plans, he determined to push for a new organization. Several years of preparatory work on a constitution and recruitment of prominent supporters ensued before Rudorff and 150 cofounders established the BH in 1904. A board of directors, including Rudorff, elected by an annual assembly of members served as its central governing body. Daily business was to be conducted by an executive secretary. Conwentz agreed to join the leadership group as leader of the working group for protection of natural monuments. The BH's first president was Paul Schultze-Naumburg, a well-connected artist, successful architect, and prolific author, whose ideas paralleled Rudorff's in many respects. He believed that educating the public about the BH's concerns was the key to effecting change. His writings emphasized the need for architecture and landscape design that harmonized with the natural world, rather than decimating it in the name of progress and profit, but his approach was less backward-

[10] In addition to sources cited in note 1, see Schoenichen (1954) and Rollins (1997) for information about these organizations.

[11] Among the works cited in note 1, Knaut (1993) offers the most complete discussion of the League for Homeland Protection. Other useful secondary sources include the following: Sieferle (1984); Andersen (1989); Knaut (1991); Rollins (1993. 1997); Lorenz (1996); and Jeffries (1997). Schmoll (2003) provides a useful biography of the organization's first president, while Koshar (1998) and Speitkamp (1988) document its work on non-environmental issues; Wettengel (1993) emphasizes its relationships with government. This section also draws on three histories published by the organization itself: Hempel (1930); Zuhorn (1954); and Fischer (1994).

[12] The BH modified its name in minor ways several times over the course of the twentieth century, but for simplicity I have referred to it as the BH throughout. The German word *Heimat* is conventionally translated as "homeland," but it also implies strong emotional attachment based on familiarity, family ties, and tradition.

looking, less hostile to technology and progress, and more compromise-oriented than Rudorff's. Over time it came to be the dominant perspective.

The BH's objectives included not only the preservation of natural monuments, native species, and the German rural landscape, but also protection of historic ruins, monuments, and buildings, as well as the preservation of local customs, dialects, traditional architectural styles, styles of dress, arts and crafts, and festivals. Indeed, the uniqueness of its vision lay in its effort to preserve the whole ensemble of features that it believed constituted the core of Germany's national heritage and in its focus on preservation of the landscape as a whole, not just specific landscape features or species. It thus viewed natural monuments not just as geological curiosities or scientifically interesting plants, but also as important symbols.

The BH aspired to become the central organization for a nationwide homeland protection movement; however, numerous societies for homeland protection, local beautification, folklore, historical, and natural history, such as the highly influential Saxon Homeland League, already existed at the local and *Land* levels. Therefore, the new organization set out not just to attract individual members and donors, but also to recruit these groups as members. It also cultivated close ties to the anti-modernist Dürerbund and to other organizations devoted to preserving rural communities and sought out alliances with homeland protection groups abroad.

The BH attracted numerous individual members during its early years, including many luminaries of the homeland and nature protection movements. Nevertheless, it also encountered many of the typical problems of federations of local organizations. Local and *Länder* homeland protection groups were heavily focused on local affairs because planning, building regulation, and nature protection were responsibilities of the *Länder*—except in Prussia, the largest *Land*, where provinces were responsible for these duties. Moreover, the core task of homeland protection, the protection of local communities and distinctive landscapes, was inherently local and regional, so supporting a national organization seemed of questionable utility. *Länder* groups also valued their autonomy and feared that members' loyalties might be split between them and the national organization or that their regional focus might be subordinated to a centralized national bureaucracy. Finally, local and *Länder* groups worried that a strong national organization might place heavy demands on the very limited financial resources that their members' dues and government subventions made available for homeland protection. These difficulties were to plague the BH for many years, but their immediate effect was that many preexisting and newly formed homeland protection groups declined to affiliate with the new national organization.

In an effort to reduce these difficulties, the BH reconstituted itself in 1908 as an umbrella organization, with regional and local homeland protection organizations rather than individuals as its primary members. Individual members from areas where there were established local affiliates were transferred to these affiliates, and individuals could now be members of the national organization only in special circumstances, such as living in an area without a regional affiliate. Nevertheless, due to the lack of enthusiasm from *Länder* and local groups for supporting the national organization financially, individual supporters continued to furnish a significant fraction of its financial support. Local homeland and nature protection work became the exclusive responsibility of the local affiliates; the national organization focused on its functions as an information exchange, public relations, and lobbying group. The board of directors was now elected by an assembly of representatives from the regional units.

These changes reduced the power of the national organization and placed it in a very precarious financial situation. The result was a sustained and sometimes bitter conflict that pitted the *Länder* organizations, especially the powerful branch in Saxony, against the president, executive secretary, and their allies on the national board in a struggle over revenues and influence. The conflict was reduced on the eve of World War I by the resignations of Schultze-Naumburg and the executive secretary and a new constitution that weakened the power of their offices. These changes made affiliation with the national organization more attractive, and the BH grew rapidly until World War I, reaching almost 30,000 members. The war led to a pause in growth and the suspension of key activities, such as publication of the magazine; however, growth later resumed.

In addition to member dues, the BH received substantial financial support from government and private donors, including several board members; private donations also supported a foundation associated with the BH that proved crucial to its survival during the war years. In 1929, the BH was strengthened by emergence of a sister organization, the Association of Friends of German Homeland Protection, which supported the cause via fundraising and publications.

The BH generally preferred to work through public education and behind-the-scenes persuasion. It used confrontation only as a last resort. It published a magazine to inform its members about homeland protection issues and its own programs, but budget shortfalls led to constant format changes and gaps in publication. The shortfalls resulted not only from external events, such as World War I, but also from the *Länder* organizations' preference for supporting their own periodicals. The BH also produced homeland protection postcards, pamphlets on topics such as how tourists could protect nature, and books on local history and natural monuments. It sent many publications directly to public officials, administrative agencies, and schools, and it strongly favored educating schoolchildren about Germany's folk heritage and the need to preserve nature's beauties. It also sponsored lectures and slide shows, archived slides and relevant publications, funded professorships in natural history, conducted architectural competitions, and worked to set up local homeland museums. In general, its efforts to work with the authorities were more successful than its efforts to influence public opinion. Still, by the 1920s, homeland protection had achieved broad currency and was no longer a concern solely of isolated intellectuals.

The BH cultivated close relationships with the local clergy and other opinion leaders, as well as with government officials and relevant administrative agencies at all levels of government. Conwentz, the director of the Prussian nature protection office, remained on its board until his death, and his successor became its vice president. The BH served as an information exchange for drafts of legislation, and it worked with some success to persuade the authorities to put in place regulations for historic preservation, protection of scenic areas and natural monuments, billboard regulation, bird protection, and the like. It also worked to protect and restore traditional features of the landscape, such as hedges, small fields, and mixed forests, and against observation towers and tourist railways that might deface scenic areas; however, it lacked the financial resources to undertake significant land purchases itself. It participated only occasionally in efforts to fight industrial pollution or untreated sewage, generally leaving such matters to local government or local citizens' initiatives.

Homeland protection organizations were founded in several other European countries around the turn of the century, and the BH participated regularly in international conferences. It maintained particularly close ties with regional homeland protection organizations in German-speaking Austria.[13]

The BH was at best an ambivalent critic of capitalism and technology. Its leadership ranks always contained some strong critics of capitalism and technology. (Rudorff himself often leaned in this direction.) Nevertheless, the majority, led by Schultze-Naumburg, argued for finding a balance between modernization and homeland protection, not a rollback of all change. Still, the BH's activities did at times stir up opposition from industry, billboard advertisers, property owners, and local citizens who expected economic benefits from proposed projects. The BH's first major campaign, a fight against damming scenic rapids on the upper Rhine for hydroelectric power, attracted wide support from diverse quarters but ended in defeat when the BH was forced to settle for merely offering input about beautification of the new facility. Perhaps because of experiences like this, the BH generally chose not to oppose dams, railways, and the like outright, but to negotiate with business and the authorities about their placement and design. It also sometimes accepted donations from industry.

The BH devoted considerable effort to nature protection, but its approach was marked by the inherent limitations of its perspective. It sometimes invoked arguments reminiscent of modern ecology, as when it argued that saving hedgerows and scattered groves of trees prevented wind erosion and provided nesting places for bird species that devoured insect pests, that plowing all the way down to streamside degraded water quality and fish habitats, and that forests regulated water runoff and flooding. Nevertheless, the science of ecology was not yet widely known, and the aesthetic and romantic arguments about the beauty of the German countryside and the rootedness of the German *Volk* in the German forest and soil dominated its argumentation.

There were also disagreements about how nature was to be protected. Some BH leaders leaned toward the American national parks model, which implied

[13] On this point, see also Dalton (1994) and Wolf (1996).

giving priority to protecting the relatively few large areas in Germany that most closely approximated their original state. This was the objective of the Nature Protection Park Association, with which the BH had a relationship that mixed mutual support and competition. Founded in 1909, this association achieved rapid membership growth; however, industrial and agricultural interests viewed it as a threat to their interests, and Conwentz and his allies argued that the virtual lack of wild lands in Germany and high land acquisition costs made the organization's program impractical. After several false starts, the association was able to combine government subsidies and its own fundraising to purchase enough land to set up a park in the Lüneburger Heath, but it failed to persuade the government to initiate a program to purchase land for additional national parks, and a true national park system was not established until much later.

Probably the largest fraction within the BH saw nature protection as part of an effort to protect, beautify, or restore the whole ensemble of features that made up Germany's scenic landscapes, including rural villages and fields, winding paths, and traditional architecture. If power lines, quarries, canals and railroads, factories, or generating stations had to be built, they should be designed to fit into the existing landscape as unobtrusively as possible. The most prominent exponent of this view was Schultze-Naumburg.

A third group, centered around Conwentz, argued that the most practical course for nature protection was to preserve natural monuments of particular scientific interest, ecological significance, uniqueness, or beauty. They were criticized by the other factions for the modesty of their aims, but Conwentz's reputation guaranteed a prominent place for the preservation of natural monuments on the BH's agenda. Immediately before World War I, the BH's leadership in this area was contested by a new organization, the League for the Protection of Natural Monuments; however, it failed to survive the war.

Nature protection remained an important part of the BH's agenda until after World War II, but its first quarter-century saw a gradual but noticeable drift away from Rudorff's original heavy emphasis on nature protection in favor of its other objectives, especially promoting traditional architecture. Schultze-Naumburg's background in architecture and the fact that the BH often scheduled its meetings simultaneously with the German Day for the Care of Monuments and later absorbed this conference as one of its activities probably contributed to this change. But even as the emphasis on nature protection decreased, some of its *Land* and local groups continued to give it high priority.

The BH's membership was comprised primarily of the well-educated middle class, including pastors, artists, writers, doctors, journalists, scientists, numerous teachers and professors, and small business owners. There were also many state and local employees, especially from offices concerned with nature protection, regulation of buildings, and museums and libraries. The organization's focus on preserving monuments and historical building styles also led to a heavy representation of architects. Finally, there were also some aristocrats and owners of agricultural estates. Capitalists, members of the working class, and farmers were almost completely absent. The organization did not recruit the latter two groups, and it set its dues with the middle class in mind. The BH had very few women members.

The BH remained nonpartisan and claimed to approach the state as an advocate of the general good, not of specific interest groups. The high representation of government officials among its members also restrained any tendency to become strongly confrontational. Although it aspired to a large base of supporters, its leaders did not envision a people's crusade, at least not if that meant alliances with the Social Democrats or other working-class organizations. Some even expressed skepticism that the working class possessed sufficient sensitivity to truly appreciate nature.

The Friends of Nature

The Friends of Nature (*Naturfreunde*) was founded in 1895 in Vienna as a workers' association for the enjoyment and appreciation of nature.[14] It grew out of the same social context that also spawned unions, socialist and communist parties, and other workers' organizations. Its founders were a group of working-class intellectuals who believed that such an organization could uplift, educate, and mobilize the workers and provide an alternative to the taverns.

Tourism had already developed among the middle and upper classes as a way to escape from the hectic pace of city life and learn about nature. Young people in organizations like the Wandering Birds took part in highly romanticized nature hikes, and mountain hiking clubs with primarily middle- and upper-class adult members had existed for several decades; however, hiking as practiced by the middle and upper classes was an expensive hobby, and the hiking clubs were not receptive to proletarians.

The workers who affiliated with the Friends of Nature were occupationally diverse, but skilled workers were better represented than unskilled. There were members of all ages, but adolescents and young adults were so numerous that a separate youth auxiliary was soon formed. Women joined in relatively small numbers, usually as part of a family membership, and were generally limited to sex-stereotyped tasks and absent from leadership roles. The skilled workers who made up the Friends of Nature's core membership were often very mobile geographically. Some moved across national borders from job to job, while some young men followed the tradition of traveling to various countries to learn their trades as apprentices. The organization expanded rapidly as these travelers established many new chapters in nearby countries, including the first German chapter in Munich in 1905. There was even a chapter in New York.

---

[14] General histories of German environmentalism often neglect the Friends of Nature, but the following articles and books provide good coverage: Wunderer (1977, 1980); Upmann and Rennspieß (1984); Kramer (1984); Zimmer (1984, 1989); Erdmann (1991); and Bagger (2002)—as do three older summaries produced by the organization itself: George (1955); Georgi-Valtin (1955a); and Winterer (1955). Hermand (1993) describes the Friends of Nature's efforts to protect the German forests, and Linse (2002) explores the role of women in the organization.

World War I saw a sharp membership decline, but rapid growth resumed after 1918. By the early 1920s there were almost a thousand local groups and well over 100,000 members in Germany, exceeding even the number in Austria. In 1921, a German business office was established in Nuremberg, and in 1925 the Germans were authorized to establish a separate board of directors. Germany was divided into eighteen districts, each with its own board; however, the German branch remained subordinate to the headquarters in Vienna. Local groups enjoyed considerable freedom to set their own agenda until the late 1920s, when a trend toward centralized control from Nuremberg developed.

The Friends of Nature organized hikes and excursions to forests and mountains and built and maintained trails and hiking shelters. It set up libraries about hiking, nature study, and the outdoors, and it explored geologically or ecologically interesting sites and collected specimens of rocks, plants, and prehistoric artifacts. It also held lecture series to inform members about natural history, prepared museum exhibitions, and staged festivals and family outings. As the organization grew, local chapters established interest groups for natural history, folklore, nature photography, skiing, mountain climbing, and canoeing—as well as folk music choirs and gymnastics, theater, and dance groups. Many of the activities centered around the Friends of Nature houses, rural retreats that served as bases for hiking trips and as places to socialize, pursue hobby interests, or browse a small library. By the end of the 1920s there were well over 200 houses, as well as campsites, pools, and youth hostels.

Early versions of the Friend of Nature bylaws emphasized nature study and the aesthetic appreciation of nature, and a 1910 revision elevated nature protection to an official goal. However, as with the BH, nature protection had to share the stage with other objectives—romantic wandering, enjoyment of nature, nature study, and related hobby interests—which were often more prominent. In its publications, the organization criticized the ongoing destruction of nature by capitalist profiteers and complained about monopolization of beautiful lands by rich landowners. At the level of practical politics, the Friends of Nature lobbied government and staged protest actions, focusing on issues like logging of scenic forests, railroad construction in scenic areas, strip mining of coal and peat, river channelization, drainage of wetlands, billboard construction, developments of vacation houses for the rich, destruction of land by military exercises, and the closing of hiking paths by private landowners. Opposition to the land closures even included mass "hike-ins." The Friends also argued for creation of national parks and wildlife refuges. Concerned about the effects of mass tourism, the organization tried to educate its own members and other tourists about nature protection.

The Friends of Nature justified its goals in terms of the ideology of the left and urged its members to be active unionists and participate in working-class political parties. Some of its members studied the work conditions of downtrodden rural workers and distributed political leaflets to people they encountered while hiking. Nevertheless, labor union and socialist party leaders worried that the Friends might evolve into a politically uninvolved hiking club, draw workers away from union and party work and events, and distract members from the more important goal of obtaining a greater share of industry's bounty.

Because of the organization's obvious effectiveness as a recruiting tool for the workers' movement, such criticism generally remained muted, but it was not altogether unjustified. Only a minority of members were fully committed to leftist ideology, and the organization was generally content to leave political work to the political parties and unions. Like the predominantly middle-class Wandering Birds, the Friends of Nature saw romantic treks in nature as providing physical and mental respite from unpleasant conditions in the cities and a chance to counter the increasing dominance of technology. Like their middle-class counterparts, Friends of Nature leaders and publicists rhapsodized about the joys of wandering in nature, and the Friends also sang folk songs and studied folklore. As the organization grew, it attracted a growing minority of members with orientations more like those of members of the Wandering Birds and BH, and some primarily middle-class hiking clubs joined its ranks. Adherents of vegetarianism, nudism, and other life reform movements were also present. Nevertheless, the Friends of Nature did give nature protection a less elitist and more confrontational orientation, arguing that firsthand experiences in nature would sensitize working people to their exploitation in the sweatshops of the cities and to efforts by the rich to exclude working people from free access to the land or to rape the land for profit. Moreover, excursions in neighboring countries helped to unite workers across national borders.

When sharp conflict did erupt within the ranks of the Friends of Nature, the fault line was not between middle-class romantics and working-class activists, but between Social Democrats and Communists. This confrontation, which developed in the second half of the 1920s and reached its height in the early 1930s, mirrored the growing split between the Communist and Social Democratic Parties. The Friends of Nature was officially nonpartisan, but Social Democratic Party activists were prominent among its leaders. Consequently, the increasingly aggressive efforts of Communist members to instrumentalize the organization to recruit for their cause, the Communists' desire to affiliate with international communist sport associations, and their commitment to bringing down the Weimar Republic, where the Social Democrats participated in the governing coalition, inevitably led to a split within the organization. The leadership reacted at first by trying to maintain its official nonpartisan stance, even as it remained informally allied with the Social Democrats; however, during the early 1930s the organization moved toward more open support of the Social Democrats, whom its leaders saw as the only hope for fending off the growing influence of the National Socialists. In the end, over 200 Communist-dominated local groups with about half the organization's total membership were expelled.

The League for Bird Protection

The BH and Friends of Nature were strongly influenced by the major ideological currents and social movements of the late nineteenth and early twentieth centuries, and both combined nature protection with other objectives. Other organizations founded in this era were less strongly influenced by movements of the left and right and focused exclusively on nature protection. The most important of these were the League for Bird Protection and the Bavarian League for Nature Protection. Although their members overlapped with the BH in worldview and social backgrounds, both limited themselves to nature protection.

Bird protection organizations were founded in most northern European counties and the US in the late nineteenth century (Dalton, 1994; Kline, 2000). Germany had an Association for the Protection of Bird Life as early as 1875, and several other bird protection societies were founded by the end of the nineteenth century; however, only the League for Bird Protection[15] (Bund für Vogelschutz: BfV)[16] attained a large membership or much longevity. The BfV was founded in Stuttgart in 1899 by citizens concerned about the decimation of bird populations by habitat destruction and by the hunting and trapping of birds for sport, house pets, and ornamental plumage.

The organization's founders and core supporters were a mixture of teachers, ornithologists, bird watchers, government officials, and titled aristocrats—the latter ranging up to the kings of Baden and Württemberg. Business leaders were also represented, and the BfV had an unusually high proportion of women among its leaders and life members. Although not uninfluenced by the homeland protection movement, it limited its goals to protecting birds. In an unusual but highly successful move, the founders selected Lina Hähnle, mother of six and wife of a successful industrialist, as their leader—a position she was to hold for almost half a century. Hähnle proved to have both charismatic appeal and strong administrative and political skills. She devoted long hours to the organization and proved adept at both working with economic and political leaders and maintaining the common touch.

Generous financial support from the wealthy Hähnle family, along with contributions from other donors and small government subsidies, allowed the BfV to set its dues at a very low level. Hähnle and other board members were tirelessly on the road holding lectures and slide shows, and local groups held lectures and staged exhibitions for the general public and school groups and prepared posters and other educational materials for the schools. Hähnle's son, who assisted in administering the organization, was a pioneer in producing slide shows of bird photographs and recording bird songs. The nominal dues, ambitious program of public education, and widespread public interest allowed the BfV to quickly become the largest and most influential German bird protection organization,

---

[15] Material about the early history of the BfV appears not only in the sources cited in note 1 (see especially Dominick, 1986, 1992), but also in Rauprich (1985) and Wöbse (2003b). The organization has produced two histories: Hanemann and Simon (1987) and May (1999). Its annual reports for the period (e.g. Bund für Vogelschutz, 1914, 1921, 1925) also provide insight into its operations.

[16] The BfV experienced several minor name changes before adopting a completely new name after German reunification to reflect ongoing expansion of its agenda, but for the sake of simplicity I have used the abbreviation BfV for the entire pre-1990 period.

although it suffered a setback in the 1920s, when runaway inflation reduced its membership by about a fourth. It aspired to attract members from a broad spectrum of society and maintained a resolutely nonpartisan public stance

The BfV expressed sympathy for broader nature protection efforts, but it limited its focus rather strictly to protection of birds and the ecosystems that supported them. It was ahead of its time in recognizing the importance of habitat preservation, so it purchased, leased, and maintained bird sanctuaries, but it lacked the funds for extensive land purchases. It also sold nesting boxes and bird feed, conducted amateur ornithological research, and published birding guides and other books about birds and bird protection.

The BfV also lobbied for international treaties and domestic legislation to protect birds from such hazards as unprotected power lines and indiscriminate hunting. Working alongside the BH, it lobbied hard for the 1908 revision of the national bird protection law. Finally, it organized boycotts of feathered fashions, circulated pledges to abstain from such fashions, and tried to enlist dignitaries to support this cause. It did not, however, involve itself in efforts to limit industrial pollution. Several factors led it to prefer public education and behind-the-scenes lobbying over confrontation or mass protests. First, its leadership ranks included significant numbers of well-heeled philanthropists, titled nobility, and government officials. Second, it relied heavily on the financial support of the Hähnle family, whose wealth was based on family-owned factories, and on other business donors such as industrialist Robert Bosch. Third, it received government subsidies and depended on the authorities to help it procure nature reserves.

The Bavarian League for Nature Protection

A somewhat later arrival on the nature protection scene was the Bavarian League for Nature Protection (Bund Naturschutz in Bayern: BN), founded in 1913.[17] For almost six decades it limited its activities to Bavaria, but it warrants consideration here for two reasons. First, it grew to be the largest of the regional nature protection organizations and was located in Bavaria, which—after the breakup of Prussia following World War II—became West Germany's largest *Land.* Second, in 1975, it became the basis for one of Germany's most important national environmental organizations, the German League for Environment and Nature Protection.

The BN originated as an effort to provide support and encouragement to the Bavarian government's Committee for Nature Preservation. This committee, described above, was established by the government in 1905 with support from the Alpine Association and BH and charged with coordinating the activities of the nature protection associations and the Bavarian nature protection authorities and

[17] The following sources are excellent supplements to those cited in note 1: Lense (1973); Hoplitschek (1984); BUND (1995); and Wolf (1996). Useful internal reports from the period include Bund Naturschutz in Bayern (1925), von Reuter (1926), Reuß (1926), and Blätter für Naturschutz und Naturpflege (1931).

providing information and assistance to the government. Most of the members were representatives of nature protection associations or government officials. By the time of the founding of the BN, the committee was suffering from rapid leadership turnover and had experienced a series of failures in its efforts to protect scenic landscapes. Nevertheless, it had spawned numerous subsidiary committees and activists at the local level, and members of these committees became an important part of the membership base of the BN.

In addition to representatives from the state's Committee for Nature Preservation, which hoped that the new organization would support its efforts and raise money for nature protection, the BN's founders included members of the Bavarian Botanical and Ornithological Societies and the Natural History Society. Its founders intended it to educate the public about nature protection and strengthen the state bureaucracy's hand against threats to nature.

Despite its initial small size and lack of financial resources, the BN developed a comprehensive program to advance its objectives. These included efforts to persuade the government to protect scenic areas from damage from road building, quarrying, power stations and construction of second homes or tourist facilities by making them protected areas. The BN also sometimes purchased such areas; however, its limited financial resources ruled out extensive land acquisition. It also worked to educate tourists, the public, and schoolchildren about nature and nature protection by distributing posters and educational materials, and it set up patrols to protect sensitive mountain areas. It also cosponsored the German Nature Protection Day, published a regular magazine for members, and issued other publications such as maps of important natural areas.

Like the BH, the BN's publications sometimes railed against the destruction of nature due to human selfishness and greed, but they rarely named specific offenders. The BN often settled for minor modifications of projects it opposed and was not active in efforts to combat industrial emissions. Its distaste for confrontation was partly a result of its very close working relationship with the Bavarian nature protection offices. Many of its leaders and most active members were state employees, its board of directors overlapped considerably with the Bavarian Committee for Nature Preservation, and it often held its meetings in the offices of the Bavarian interior ministry. It also enjoyed the protection of the royal family of Bavaria.

After 1922 the BN's newly established local groups, ordinarily chaired by a local official, were given quasi-official status. They worked to educate the public about nature protection, inventoried natural monuments and scenic areas and promoted their protection, prepared expert reports when requested by local officials, suggested steps to protect nature, and monitored construction projects and other potential threats to nature. In 1926, the BN merged with the former state Committee for Nature Preservation, giving it a quasi-official status, and government subsidies provided a significant fraction of its budget.

The BN experienced a brief spurt of growth after this merger, but otherwise it grew slowly and failed to realize its aspiration to recruit large segments of the population to the cause. Its membership rolls were dominated by teachers, professors, and state nature protection and forestry officials, and it attempted to recruit opinion leaders, such as clergy and doctors, to membership. There were few farmers, business owners, or laborers. Only about 7 percent of members were women, most of whom were spouses of male members. Its leaders were typically state officials and professors.

Summary and Conclusions

The first national organizations concerned with environmental issues in Germany appeared at the turn of the twentieth century, and the conditions of that period decisively shaped their structures, goals, and strategies. Rapid industrialization and urbanization, already well underway, were creating serious air pollution, water pollution and solid waste disposal problems in the cities. Outside the cities, construction of railroads, power lines, dams, and roads, the extension of agriculture to new areas, the rationalization of agriculture and forestry, and mass tourism all threatened scenic areas and traditional rural landscapes. Government's efforts to cope with these challenges were halting and often ineffectual. Industrialization and urbanization also disrupted traditional ways of life, depopulated the countryside, created an urban proletariat, and produced harsh working and living conditions in the cities. Industrialization also called forth a new elite of industrialists, who accumulated enormous wealth and exerted great power.

The industrialization and urbanization of Germany created multiple grounds for challenges to this new social order. Urban dwellers were troubled by smoke, stench, and polluted water; workers had ample grounds to wish for better wages, work conditions, and housing; aristocrats, small business owners, and the well-educated middle class resented their declining influence and prestige; lovers of scenic vistas and relatively undisturbed nature mourned their disappearance, and citizens from every walk of life were unsettled by the pace of change.

Mobilizing these discontented groups to support national organizations that would address environmental problems was no small challenge, for there were no historical precedents. The organizations that emerged were shaped by the class cleavages, ideological currents, social movement mobilizations, political structures, and government actions of the era, and by the ability of interest group entrepreneurs and social movement entrepreneurs to attract the resources required to get national organizations off the ground and develop the persuasive framings of problems needed to attract support.

National organizations to combat urban and industrial pollution were notable for their absence. Several factors may explain this. German political structures allocated responsibility for combating these problems to the local level, providing more incentive for local organization than national, while powerful political forces and a dominant ideology that associated industrialization with progress limited the prospects for national legislation. The subpopulation most directly affected by pollution problems, the industrial working class, was focused on the immediate problems of securing a larger slice of the pie and isolated from potential middle-class allies by sharp class divisions. Local initiatives to fight urban pollution did spring up episodically, but matters were left mainly to the authorities and the property owners most directly affected. The latter's motivation to mobilize was probably defused by the opportunity to receive compensation for damage, at least in the most egregious cases.

Building national organizations with nature protection goals proved a more feasible project. By the beginning of the twentieth century, Germany had only isolated remnants of wild nature, and efforts to protect them achieved little traction; however, changes that threatened valued birds and animals and drastically altered the appearance of scenic mountain vistas, river valleys, and rural landscapes could mobilize many committed opponents. In the absence of well-developed ecological science, their concerns were based more in aesthetics than science, but they were no less heartfelt. The new organizations founded to address these concerns had to assemble persuasive framings of the problem, funds, and other necessary resources to get the organizations off the ground. In the cases of the BH, BfV, and Friends of Nature the task fell to talented organizational entrepreneurs like Lina Hähnle and Ernst Rudorff.

The new nature protection organizations reflected the class and ideological cleavages of the day in the diversity of their support bases, goals, and strategies. The BfV and the BN appealed to the educated middle class, but they maintained a studied distance from ideologies and movements of both the left and right. The BH attracted a similar demographic group but linked nature protection to a well-developed ideology and an emerging movement that focused on protecting German rural life and the German countryside, which it viewed as the source of the nation's virtues. All three of these organizations could draw on their members' respectability, connections to government, financial resources, and organizational skills. The Friends of Nature made nature protection part of a broader agenda focused on building working-class solidarity and culture and drew its support from alliances with labor unions and political parties of the left.

All four organizations sometimes used the rhetoric of social movements, but only the Friends of Nature actually functioned in this way. The BH had a well-developed movement ideology, to be sure, but this proved as much a burden as an advantage. While its anti-modernist rhetorical salvos helped it to attract some highly motivated supporters, it encountered the dilemma of reconciling ideological purity with practical politics and the deep integration of many of its members into the political establishment. Only after considerable pulling and hauling did practical politics emerge as dominant. The BH then joined the BfV and BN in functioning more as an interest group than a confrontational social movement.

There were good reasons for this. The establishment-oriented leaders of the BH, BfV, and BN never envisioned mass mobilization for protest, but they did aspire to build a broad base of public support, which they believed would help them change citizen behavior and increase their political influence. In fact, however, class and ideological cleavages and the free rider problem stood in the way of achieving even this goal. Direct confrontation with capitalists, the dominant ideology of progress, and government generally proved unrewarding, so the three organizations generally had to settle for working within the system as relatively uninfluential interest groups and accepting compromise solutions. Moreover, the organizations' well-educated middle-class leaders, with their many connections to government (many were government employees themselves), were more comfortable with the classic strategies of interest group politics—public education, petition drives, behind-the-scenes lobbying, and cooperation with government nature protection agencies—than with mass demonstrations and confrontation.

All of the organizations experienced difficulties overcoming the free rider problem. The BH and Friends of Nature relied on the persuasive power of framing nature protection concerns in terms of social movement ideologies. The BfV was able to attract members by combining an ambitious campaign of public education and member recruitment with dues set so low as to make membership essentially cost-free. The BN and BfV also engaged in practical nature protection projects at the local level, providing their members with the chance to build social relationships and see and enjoy the results of their work, while the Friends of Nature offered a wealth of social rewards and enjoyable activities. Using these strategies, all of the organizations but the BN attained substantial memberships, though none approached their ambitious goals.

The cases of the BH and the Friends of Nature also illustrate the perils of combining nature protection goals with other goals. By affiliating with organizations that functioned as social movement organizations for movements in which nature protection was only one goal among many, nature protection advocates hoped to magnify their influence and advance their cause. Yet they ran the concomitant risk that nature protection would be subordinated to other objectives, and after years of internal conflict, the BH did, in fact, evolve in this direction. In the Friends of Nature, nature protection had to share the stage with other objectives, and the whole enterprise was viewed with considerable skepticism by the unions and parties. Indeed, it was ultimately almost destroyed by factional infighting between the Communists and Social Democrats.

In the early twentieth century, government agencies concerned with nature protection were small, as was the base of scientific knowledge about environmental problems. Consequently, it was possible for national nature protection organizations to be staffed almost exclusively by volunteers. Power struggles between professionals and volunteers could thus be avoided, but conflicts over centralization of power could not. Efforts to incorporate preexisting local groups into the newly established BH encountered heavy resistance from local and regional groups concerned about maintaining their autonomy, holding on to scant financial resources for their own use, and maintaining their focus on local concerns. Satisfying these groups required reducing the influence and budget of the national organization at the expense of its regional and local units. Similar problems also cropped up in the Friends of Nature; however, this time the conflict involved the German chapter's desire for more independence from the international headquarters in Vienna.

The experiences of environmental organizations in the first decades of the twentieth century established important precedents for subsequent developments. But before they could enter the recent era, the organizations had to cope with the unique challenges associated with the disintegration of the Weimar Republic, Nazi rule, World War II, and its aftermath.

Chapter 4

NAZISM, THE WAR, AND ITS AFTERMATH

The Causes and Consequences of Right-Wing Ecology

By the 1920s, national organizations with environmental goals, all emphasizing nature protection, had achieved significant memberships, considerable stability, and a modest degree of institutionalization and influence in Germany, occupying differentiated niches with their own objectives and constituencies.[1] All but the Friends of Nature received small government subsidies and had at least occasional input into political decision making, while the Friends of Nature had its own network of connections to parties of the left and other workers' organizations. This arrangement was disrupted in the early 1930s by the rise of National Socialism, a powerful social movement with an ideology that included elements attractive to many of the more conservative nature protection advocates. After the Nazis seized power in 1933, their efforts to protect the environment proved popular with most nature protection organizations, but their subordination of the organizations to state control transformed the social and political context in which the organizations operated and led to the temporary demise of the Friends of Nature.

Germany's devastating defeat in World War II led to yet another radical dislocation. Forced to begin almost from scratch after the war, the West German organizations found themselves in a situation in which rebuilding the economy and then enjoying Germany's newfound prosperity threatened to shoulder aside

[1] Comprehensive treatments of the history of the entire period appear in Dominick (1992) and Chaney (1996, 2005), while Hermand's (1991) history emphasizes cultural and ideological trends. An overview of the experiences of the environmental organizations during this period is available in Rat von Sachverständigen für Umweltfragen (1996).

environmental issues. Although they survived the war, years were required to reach a secure footing and rebuild.

The Nazi Era

The rise of National Socialism confronted environmental organizations with a mixture of uncertainty and hope, challenges and opportunities.[2] The regime's early accomplishments in nature protection provided some cause for celebration and shaped German environmentalism for decades to follow, but the organizations also found themselves under state control or banned. The tragic war and its outcome undermined the Nazis' early accomplishments and left the organizations weakened and fighting to survive.

The Rise of National Socialism

Conservative, nationalistic, and anti-modernist strains in German thought, which were closely associated with the rise of homeland protection in the late nineteenth century, were reinforced by events that preceded Hitler's rise to power.[3] The patriotic wave that swept over Germany during World War I was accentuated rather than quelled by the humiliating defeat and the punitive Versailles Treaty, which stripped Germany of a significant portion of its territory, forced drastic reductions in military strength, and imposed unfavorable trade terms and massive reparations. Conservative intellectuals also interpreted it as a defeat for German culture and the German way of life.

Runaway inflation during the early 1920s, economic depression in the early 1930s, and the weakness and instability of the new Weimar Republic compounded citizens' frustration and resentment and strengthened the longing among more conservative elements of the population for a strong state and a return to traditional values. Plagued by a proliferation of small parties, including far-left parties that would settle for nothing less than communism and far-right parties that longed for restoration of autocratic rule and militarism, Weimar Germany was governed by a series of progressively weaker coalition governments. An entrenched and conservative bureaucracy and military leadership resisted reforms, and some elements of it longed for restoration of the monarchy. The republic managed to survive several coup attempts, but it gradually devolved into chaos—and near its end into rule by decree under the emergency powers of the president.

---

[2] In addition to the sources cited in note 1, the following books and articles provide good accounts of National Socialism and the environment: Bergmann (1970); Sieferle (1984); Gröning and Wolschke-Bulmahn (1987, 1998); Dominick (1987); and Gröning (1996).

[3] Useful histories of the political problems of the Weimar Republic and rise of National Socialism include Halperin (1946) and Broszat (1987). Discussions of ideological developments include Martindale (1981), Bramwell (1985), and Geden (1999). Ditt (2003) and Wöbse (2003a) describe the Weimar Republic's legislative paralysis regarding nature protection.

In this context, passing and enforcing environmental legislation proved next to impossible. This further disillusioned nature protection advocates, who dreamed of a united nation committed to nature protection and strong, uniform national legislation and adequate resources for its enforcement.

The pessimistic tenor of the times was reflected in the popularity of books like Oswald Spengler's *The Decline of the West* (1926), published in the last year of World War I. Echoing many of Riehl's and Rudorff's nineteenth-century themes, Spengler's polemic argued that the organic unity and health of German society, rooted in the land and the virtues of the small farmer, were being undermined by the growth of huge "world cities" like Berlin, with their rootless cosmopolitan population, proletarian rabble, and materialistic, corrupt, and shallow culture. Spengler did not emphasize nature protection, but his ideas, and those of like-minded authors inspired by his work, had great appeal to the homeland protection advocates and other conservative elements, and they stirred a flurry of back-to-the-land movements and plans to rescue the peasantry.

Spengler saw the decline of Germany as inevitable, but other authors believed that a solution could be found, and numerous "*völkisch*" organizations, some with ties to the budding Nazi movement, sprang up to promote a return to an idealized past. It was in this context that the National Socialists offered their own ideology and program of change as the antidote to Germany's political, economic, and spiritual malaise. Their skillful use of political rhetoric that played to people's fears, mass rallies, and propaganda attracted a broad constituency grounded in the lower middle class and farmers. This, in turn, allowed them to transform a marginal political party into a mass movement, achieve a dominant position among the parties of the right, and gain control of the government. Once in control, Hitler moved quickly to eliminate other political parties and power bases and extend government control into all spheres of life.

Nazi Ideology

Nazi ideology was cobbled together from an assortment of preexisting and sometimes contradictory elements. These included nationalism, anti-communism, anti-modernism, anti-urbanism, racism, territorial expansionism, confidence in technology, and quasi-mystical ideas about the rootedness of the German *Volk* in the German soil and forests and the primacy of the *Volk* over individuals.[4] It was propounded in different versions, and it evolved over time in response to political exigencies and power struggles within the Nazi hierarchy as individual Nazi leaders and propagandists emphasized particular themes to suit their purposes or the audience of the moment.

---

[4] In addition to sources cited in notes 1 and 2, the following articles and books describe Nazi ideology as it related to nature and the environment: Kuhn (1971); Broszat (1987); Speitkamp (1988); Gröning and Wolschke-Bulmahn (1993); Staudenmaier (1995); Riordan (1997); Olsen (1999); Ditt (2003); and Blackbourn (2003). Bramwell (1985) and Gerhard (2003) provide biographies of Walther Darré, the leading proponent of Nazi nature protection, but the former is often criticized as unduly favorable. Schmoll (2003) highlights the role of Schultze-Naumburg, the BH's first president.

Hitler, himself a vegetarian and animal protection advocate, had been influenced by the agrarian romanticism of nineteenth-century German nationalists like Riehl and was, in general, favorably disposed toward nature protection. Yet in comparison to racial purification and territorial expansionism, saving nature never numbered among his top priorities. Nature protection occupied a more prominent place in the views and writings of other important Nazi leaders, including Himmler, Hess, and Walther Darrè, and it received some attention in the Nazi press, often in association with racist and nationalist themes.

Many nature and homeland protection advocates were apolitical, but elements of Nazi ideology proved appealing to the more conservative among them, many of whom had been unsettled by Germany's economic and political weakness and the growth of individualism and modernism. Hitler's promise to build a new "*Volksgemeinschaft,*" which would recapture the solidarity of a mythologized German past in the form of a new and powerful *Reich* rooted in the peasantry and free from class conflict and the pernicious influence of communists and Jews appealed to them. Equally seductive were Nazi arguments for the primacy of the *Volk* over individuals and the idea that preserving the German *Volk* and saving nature must go hand in hand.

Some homeland protection and nature protection advocates were also attracted by the phrase "*Blut und Boden*" (blood and soil), popularized by National Socialist theorist Walther Darré. Darré, who later served as minister of agriculture, was an agricultural breeding expert, political activist, and Nazi race expert who advocated breeding a "new nobility" of German farmers. The slogan summarized in capsule form a favorite National Socialist theory. It held that the strength of the German *Volk* was derived from its unique culture and racial superiority, which had evolved out of the successful efforts of the German *Volk* to adapt to and modify a challenging physical environment through permanent settlement and development of distinctive cultural landscapes. Urbanization and industrialization had undermined the nation by breaking this key link between the German people and their native soil. It now needed to be restored. Darré's writings and speeches coupled idealization of the peasant community and love of nature with anti-urbanism, anti-modernism, anti-materialism, and contempt for the Jews, whom he viewed as the epitome of rootless urbanity. He advocated ending rural-to-urban migration, strengthening peasant communities, promoting organic farming, and strengthening nature protection as antidotes to Germany's moral and economic problems. Darré was strongly influenced by Paul Schultze-Naumburg, the first president of the BH, whose ideas had taken on an increasingly racist cast after World War I.

The image of a strong state that would take the lead in nature protection, advocated by Darré and other Nazi leaders, was also very seductive for homeland and nature protection associations. They had a long history of close cooperation with government, but they found themselves continually frustrated by the inability of the Weimar government to pass the nature protection legislation they so badly wanted. With the exception of Schultze-Naumburg's close relationship with Darré, nature and homeland protection activists exerted almost no direct influence on Nazi ideology or the party's policies, and many nature and homeland protection advocates thought the better course was to avoid politics and focus on the work at hand. Nevertheless, even nature protection advocates who had reservations about the Nazis could nurture hopes that a National Socialist government would benefit their cause.

Many officials from state offices charged with nature protection also found things to admire in the Nazis' promises. Some, including Walter Schoenichen, the director of the Prussian nature conservation office, who joined the party in 1933, published articles linking preservation of racial purity and German culture to protection of nature and the countryside. Later, they interpreted the takeover of western Poland as an opportunity to create large new national parks and reduce Germany's population density by resettling the area, and some landscape architects became involved in plans to develop areas as model communities for German resettlement after the native population was removed.

Nazi Environmentalism: Hopes and Realities

The seductive appeal of National Socialism for environmentalists was not limited to rhetoric.[5] In the first years after their rise to power, the Nazis could also point to significant achievements. The new minister of agriculture, Darré, worked to strengthen traditional peasant agriculture, promote large-scale organic farming, and halt erosion. The government also succeeded in concentrating previously dispersed responsibilities in the hands of the national government and produced a flood of new nature protection legislation and regulations. Under the leadership of Hitler's top lieutenant Göring, the regime broke a longstanding bureaucratic logjam and published the detailed regulations required to implement a plant and animal protection law that had been on the books since the 1920s. A 1933 bill outlawed unnecessary cruelty to animals. Progressive hunting, forest management, and homeland protection laws—the latter including a provision for protection of natural monuments—were also put in place during the regime's first two years, and billboards were banned from the countryside. Municipal parks were built, enlarged, and upgraded, many new nature protection areas were officially designated, and reforestation and wetland protection programs begun.

In 1935, to the delight of nature protection advocates, Göring, who also served as minister of forests, used his connections with Hitler to claim broad responsibility for nature protection and push through a long-delayed federal nature protection law. This legislation and the implementing regulations that were put in place in subsequent years replaced uneven *Länder* legislation with a uniform national law. It covered natural monuments and larger areas that warranted pro-

[5]In addition to the sources in notes 1 and 2, numerous sources contain information on Nazi environmental legislation and its implementation: Bramwell (1985); Rösler, Schwab, and Lambrecht (1990); Wettengel (1993); Gröning and Wolschke-Bulmahn (1993); Staudenmaier (1995); Olsen (1999); Ditt (2003); Lekan (2003); Zeller (2003); Klueting (2003); Radkau (2003); and Oberkrome (2003b, 2005).

tection on account of their beauty, rarity, distinctiveness, or scientific significance and authorized regulations for the protection of specific animal and plant species. It also charged the authorities with maintaining lists of protected areas and species. It created zones from which factories, dumps, railroads, power lines, and billboards were excluded, and it required government agencies to consult with officials responsible for nature protection before implementing building plans that could damage nature. Using the power of the authoritarian state to overcome an obstacle that had paralyzed legislation for decades, it also authorized control of land and seizures of property for nature protection without compensation.

The new law also identified government officials responsible for nature protection at the regional and local levels, although nature protection remained only a secondary responsibility for them, and established a national advisory nature protection office based on the older Prussian office founded by Conwentz. The new office was charged with conducting research, issuing advisory reports, and monitoring enforcement of the nature protection law. In addition, it published a monthly magazine and a newsletter. The law also retained the established pattern of assigning the responsibility for advising local authorities about nature protection and promoting nature protection to volunteer officials and committees. These committees were financed and chaired by the administrative official in charge of nature protection for the area. The Nazis also continued the tradition of close cooperation between the nature protection authorities and nature protection groups. Finally, the new regime increased the amount of school instruction about nature protection, albeit often in conjunction with teaching racist ideology.

Still, the Nazi regime's actions could hardly be described as unambiguously pro-environment. Ignoring calls for action by experts, the government did little to improve controls on air and water pollution, a step that would have been much more of a threat than nature protection to German industrial power. Many of Hitler's lieutenants openly advocated rationalism, modernism, and technology, not agrarian romanticism, as the solution to Germany's problems. Even Darré focused greater attention on reforming agricultural marketing policy, racial purification, and the elimination of foreign land ownership than on nature protection, and his ambitious plans to halt the flow of farmers to the city and promote organic agriculture foundered on opposition from other Nazi leaders.

The government's efforts to revive the economy, achieve national self-sufficiency, and increase Germany's military power required draining wetlands and removing hedges to increase land under cultivation, mechanizing agriculture and increasing use of fertilizer, increasing wood production, building new industrial and military facilities, and constructing autobahns. These steps proved hard to square with nature protection, even if the regime did appoint landscape architect Alwin Seifert to ensure the new autobahns would fit into their natural surroundings and avoid damage to sensitive areas.

As war preparations, and then the war effort itself, intensified, nature protection was first subordinated to other concerns, and then all but ignored. As early as 1936, one of Göring's lieutenants in the forest ministry was forced out for placing more emphasis on nature conservation than on wood production. Contrary to Darré's advocacy of organic agriculture on small farmsteads, rationalization of agriculture to increase food production was accelerated. These measures—along with dike construction, river channelization, intensive forestry, and draining of wetlands—led Seifert and other nature protection advocates to warn that parts of Germany could be transformed into dry prairies. Göring also began issuing orders "simplifying" the enforcement of existing laws. Darré became a voice in the wilderness and was retired in 1942.

But even in the mid-1930s, the heyday of Nazi nature protection, reality on the ground had fallen well short of the regime's pretensions or the hopes of the nature protection organizations. The much-vaunted 1935 nature protection law, for example, included exemptions for the army, public transport right-of-way, and "vital" economic programs, and the number of new protected areas created appears to have been exaggerated. Advisory committees at the regional and local levels operated with minimal budgets, no professional staff, and no enforcement authority of their own, and some local nature protection volunteers had little sympathy for or understanding of the new law's more ambitious landscape planning goals. The national protection advisory office had even fewer staff members than its understaffed Prussian predecessor, and it experienced additional cuts during the war.

The government also failed to end the longstanding fragmentation of responsibility for nature protection among government bureaus. Instead, the problems were exacerbated by power struggles among Hitler's lieutenants and competition between nature protection advocates and landscape architects to influence land use planning in Germany and the conquered territories. Requirements that nature protection authorities be consulted about building plans were also frequently ignored.

The Nature and Homeland Protections under State Control

Whether they found Nazi ideology attractive or unattractive, the majority of nature protection groups and their leaders kept their distance from the Nazis in public until after their rise to power.[6] But after 1933, nature protection organizations, like associations in other spheres of German life, had to deal with the regime's policy of *gleichschaltung* (literally, the setting of all switches in the same way). *Gleichschaltung* meant the establishment of direct government control over formerly independent voluntary associations in most fields of German life and the alignment of their programs, internal organization, and goals with the objectives of the regime. Organizations that refused to cooperate could disband or face forced dissolution. Given this stark choice, as well as the affinities of some members for Nazi ideology, it is not surprising that most nature protection organizations chose to cooperate and that many of their leaders joined the Party.

[6] For information about the Nazi state's strategies for dealing with environmental and similar organization see sources cited in notes 1 and 2, Kuhn (1971), Olsen (1999), Geden (1999), Reutter (2001), Lekan (2003), and Brand (in press).

The *gleichschaltung* process and the nature of the resulting government supervision were by no means unambiguous or uniform. The Nazi state was characterized by competition for power and confusion over the rapidly changing jurisdictions of various government and party bureaus and officials. Consequently, nature protection organizations were repeatedly shifted from the oversight of one government organization or official to the other. This provided them with some leverage in their efforts to maintain their independence, and some smaller groups survived with their leaderships more or less untouched. Larger, more visible groups, however, were subjected to closer scrutiny, and the specter of tighter government control or outright dissolution was always present. As a result, the major nature protection organizations ceased to operate as independent organizations and became—albeit to varying extents—arms of the state.

The League for Homeland Protection

Not surprisingly, the BH embraced National Socialism more enthusiastically than the other organizations.[7] The writings of its founders contained many parallels to Nazi ideology, the BH had portrayed World War I as a crusade to save the German homeland, and by the 1920s, veiled racism was becoming increasingly common in the writings of homeland protection advocates. A few of its regional leaders and ideologists had been open racists even before World War I. After the war, others, including its first president, the increasingly reactionary and racist Schultze-Naumburg, began linking the organization's concerns to the emerging Nazi racist ideology.

In addition to his association with Darré, Schultze-Naumburg was active in a variety of right-wing organizations, and his home was a center of right-wing activity. He had rushed to join the Party even before it gained control of the government, and he served the Nazis as minister of education in Thuringia and as a representative in the national Reichstag. Later he became a principal spokesman for the Militant Alliance for German Culture, a Nazi front organization that advocated saving "true" German culture from subversion by rootless urbanism and degenerate art and architecture. After 1933, many other leaders of the BH followed Schultze-Naumburg's lead, including his successor Werner Lindner, who enthusiastically embraced the Nazis and *gleichschaltung.*

Even BH leaders who harbored doubts about the Nazis could be seduced by an ideology that appeared to take their concerns seriously and to promise them a voice in policy making, and in 1933, the BH voted to incorporate itself into a new government-controlled organization concerned with preservation of the homeland and folk traditions. Its top posts were divided between its former leaders and party functionaries, and several of its active members and leaders, including Alwin Seifert, found positions in the government. Yet despite its pro-

---

[7] In addition to the sources cited in notes 1, 2, and 6, the following sources describe the BH's experiences with Nazi rule: Bramwell (1985); Speitkamp (1988); Knaut (1993); Williams (1996); Riordan (1997); and Schmoll (2003). For the BH's own account see Fischer (1994).

Nazi leanings, the BH's encounter with *gleichschaltung* was unsettling, especially for its local groups, which were accustomed to considerable autonomy. After a short period, it was able to extract itself from the Nazi-controlled umbrella organization and negotiate a quasi-independent status, using the argument that the essentially local nature of homeland protection was incompatible with central control; however, in the midst of the war, it was again placed under direct government supervision. Until the war swept other concerns aside, the BH was generously funded and strongly supported in pursuing many of its historic goals, for the Nazis saw in it a tool for building national unity based on patriotism and spreading an ideology compatible with their own.

The League for Bird Protection

Because bird protection was less ideologically loaded than homeland protection, the BfV shared less common ground with the Nazis than the BH.[8] Nevertheless, the BfV too had at times linked its goals to German nationalism. Its annual reports during World War I (e.g., Bund für Vogelschutz, 1914) were filled with florid rhetoric linking bird protection and wartime patriotism.

Officially, the BfV greeted the Nazi takeover with enthusiasm. The organization expressed delight about Nazi nature protection measures in its annual reports, and its longtime leader, Lina Hähnle, voiced similar sentiments in a 1933 speech before the general membership meeting, and again in a speech about Hitler's views on bird protection a year later. How sincere this support was is hard to gauge in retrospect. The evidence suggests that Hähnle and other BfV leaders saw the Nazis' rise to power as an opportunity to advance their cause, even if they had private reservations about the new government and its leaders. Certainly, the organization's leaders were overjoyed when the new regime passed the 1935 nature protection law, an objective it had been pursuing for years.

Almost immediately after the Nazis took power, the BfV was assimilated into the same government-controlled organization as the BH, but it was subsequently shifted from bureau to bureau in response to struggles for influence among the Nazi leaders. Lina Hähnle, a nonmember of the Party who was by then 82 years old, was rather incongruously given the title "Führer." The BfV was allowed to re-elect her until she stepped down in 1938; however, her survival was in part due to the support of Party members within the organization. When she retired, a Party member who had earlier been made second vice president took over her office; however, day-to-day control remained in the hands of Hähnle's son and other former leaders. In 1934, the BfV altered its bylaws almost without comment to exclude members not of German blood, but it did not engage in anti-Semitic rhetoric.

---

[8] In addition to sources cited in notes 1, 2, and 6, the following provide information about the BfV under the Nazis: Cornelsen (1991); Bergstedt (2002); and Wöbse (2003a, 2003b). Additional information is available in the organization's own histories, Hanemann and Simon (1987) and May (1999), and in its publications from the Nazi era (e.g., Reichsbund für Vogelschutz, 1936, 1944).

The BfV continued to pursue most of its former objectives. Its annual reports from the period chronicle these activities and are full of praise for Hitler's triumphs, including many completely unrelated to nature protection. There were also efforts to use various pronouncements by Hitler as a justification for the BfV's goals and activities, such as its program to replant thickets and hedges to shelter birds. On the other hand, articles in the organization's newsletter and magazine did sometimes express regret over loss of bird habitats resulting from intensification of agriculture.

Gleichschaltung cost the BfV its independence, but in other respects the organization profited from its support of the regime. It was renamed the Reichs League for Bird Protection, and in 1938 the Nazis made it the only recognized bird protection organization. Smaller birding clubs had to join it or disband. The BfV was also given control over the magazine of the ornithological society, and the government authorized it to establish chapters in "*Großdeutschland,*" the Nazis' term for the larger German nation they intended to carve out of their conquests in Austria, Sudentenland, and parts of Poland. Membership increased from 32,000 in 1933 to a high of 55,000 in 1943.

The Bavarian League for Nature Protection

The BN also found the Nazi rhetoric and program seductive. Even before 1933, scattered passages in its journal had linked nature protection to ultranationalism and references to the superior German *Volk*.[9] After 1933, its publications welcomed the Nazis' rise to power and dutifully parroted the Nazi line, including occasional anti-Semitic references and justification of German conquests of less populated areas in Eastern Europe as providing new opportunities for nature protection. It also embraced Nazi rhetoric about the connection between preserving the race and saving the natural environment. Like the other nature protection organizations, the BN enthusiastically welcomed the government's new environmental legislation, although as the reality of construction projects, wetlands draining, and land redevelopment for mechanized agriculture became apparent, its journal sometimes expressed disapproval.

The BN's experiences with *gleichschaltung* were similar to those of the BfV, except that it was able to maintain a somewhat greater degree of autonomy. After the National Socialists took control of the government, its constitution was rewritten to conform to Nazi specifications, and its leader was also designated "Führer." After election by the member assembly and confirmation by the National Socialist Party, he was free to ignore the advice of the board of directors and act unilaterally. The close connections between its operations and the Bavarian authorities continued, and it worked hard to have additional areas set aside for nature protection. Like the BfV, it prospered under the wing of the regime, more than doubling its membership during the 1930s.

---

[9] In addition to sources in notes 1, 2, and 6, see Hoplitschek (1984), Wolf (1996), and the BN's own history in Lense (1973). Also useful are articles from BN publications of the time (e.g. Reuß, 1933, 1935, and Blätter für Naturschutz und Naturpflege, 1935).

The Dissolution of the Friends of Nature

The left-leaning Friends of Nature suffered quite a different fate.[10] In the years before the National Socialists seized power, it had been seriously weakened by membership declines resulting from economic depression and the bitter struggles between its Communist and Social Democrat factions; however, it retained over 100,000 members. Its weakened position, coupled with its leaders' desire to avoid dissolution and the loss of the Friends of Nature houses, had mitigated against the organization's playing a prominent role in opposition to the rise of the Nazis. Nonetheless, the new government viewed the group as an ally of the hated socialists and communists, and frantic efforts to save the organization by breaking ties with the international headquarters in Vienna and emphasizing its anti-communist stance and contribution to the health and welfare of the population proved futile. The Friends of Nature was dissolved, its houses, libraries, and financial assets were seized, materials from some of its libraries were burned, and some leading members were imprisoned or executed.

Even after dissolution, some chapters continued to meet informally—including some that functioned as part of the resistance. Other chapters managed to reconstitute themselves as politically acceptable hiking clubs. Some individual members also joined other hiking clubs, and some former Friends of Nature leaders even found places among their leaders.

The Aftermath of War

The devastating defeat in World War II left Germany physically shattered, occupied by conquering armies, and divided into four administrative zones controlled by the Americans, British, French, and the Soviet Union.[11] Postwar chaos was exacerbated by the ten million refugees from the parts of Germany that were given to Poland after the war and from other nations that expelled their German minorities. Moreover, the lost territories included about a fourth of prewar Germany's agricultural and forest lands. Germans had to deal not only with having initiated a devastating war, but also with the exposure of wartime atrocities, genocide, and concentration camps. It is hardly surprising that they came to refer to the day after their surrender as "*Stunde null*"—the zero hour from which they had to begin anew. The priority issues of the immediate postwar period were rebuilding factories, housing, and infrastructure, providing jobs, assimilating refugees, and reestablishing civilian government and institutions and a stable

---

[10] In addition to the sources cited in notes 1, 2, and 6, see Wunderer (1977, 1980), Dulk and Zimmer (1984), and Erdmann (1991), as well as two of the organization's own histories, George (1955) and Georgi-Valtin (1955b).

[11] For overviews of the postwar social, political, and economic contexts see the sources cited in note 1 and Gröning and Wolschke-Bulmahn (1998), Würth (1985), Fulbrook (1995), Moeller (1997), and Olsen (1999). Garner (1995) provides an account of denazification in public service, and Wettengel (1993) discusses how rebuilding Germany's political institutions influenced environmental protection.

social structure. Environmental issues, though not completely forgotten, were of relatively low salience.

In the last months of the war, the remnants of the national nature protection advisory office managed to escape Berlin. It was allowed to continue its operations in the British occupation zone and take up contacts with nature protection offices in other areas, but it could not be formally reestablished until the new West German government emerged. The Allies adopted rigorous, although not always consistently implemented, policies of denazification, barring former Nazis from holding office in government or private associations. Many Nazi sympathizers and officials, including members of the nature protection organizations who served as volunteer nature protection monitors for local government, were removed from office; however, some prominent nature protection officials who had allied themselves with the Nazis, such as Walter Schoenichen, the former head of the nature protection office, escaped unscathed.

The nature protection organizations, like other voluntary associations that wanted to reestablish themselves after the war, had to obtain the approval of the occupying authorities in their zone. Policies in the four zones were not always uniform, and permission to reunite chapters of organizations in the different zones or even to hold national meetings was slow in coming. But even after these restrictions were eased, it proved impossible to reestablish national nature protection organizations that included all of Germany. By the early 1950s, the Soviet occupation zone had been transformed into an authoritarian socialist state allied with the Eastern Bloc. Nature and homeland protection organizations there, like other voluntary associations, were absorbed into state-dominated associations. Meanwhile, the American, French, and British zones were reunited as West Germany under a government dominated by the moderate to conservative Christian Democrats. The remainder of this chapter and Chapter 5 focus on West Germany. I recount the history of East German environmentalism and environmental organizations in Chapter 6.

Nature and Environmental Protection in Postwar West Germany

As immediate postwar rebuilding evolved into the "economic miracle" of the 1950s and 1960s, the West Germans focused on expanding their economy and enjoying their increasing prosperity.[12] Environmental issues were not a high priority for postwar Christian Democratic–led governments (Moeller, 1997). Reports of environmental problems and threats to nature in the press, while far

[12] For overviews of postwar nature and environmental protection efforts see the sources cited in note 1, especially Chaney (1996), as well as Brand, Eder, and Poferl (1997) and Schreuers (2002). More complete treatments are available in the following sources: Wey (1982); Rohkrämer (2002); Engels (2003); Uekötter (2003b, 2003c, 2005); Oberkrome (2005); Körner (2005); and Hünemörder (2005). Sources dealing mainly with environmental politics include Küppers, Lundgreen, and Weigart (1978), Müller (1986), Meroth and von Moltke (1987), Weßels (1989), and Malunat (1994). Right-wing environmental politics is discussed by Sieferle (1984), Jahn and Wehling (1990), Wüst (1993), Ulbricht (1993), and Geden (1999).

from absent, especially in the late 1950s, tended to focus on specific problems in particular regions rather than on national patterns of environmental degradation. The same was true of efforts to combat potentially damaging local projects.

The new constitution gave the national government the right to draft general framework laws governing nature protection and care of the landscape, but more detailed legislation was the prerogative of the *Länder*. The 1935 nature protection law, which—except for its preamble—was largely untainted by Nazi ideology, remained everywhere in force as *Länder* law until the early 1970s, when some *Länder* began passing replacement legislation. The national nature protection office was assigned to the agriculture ministry, and *Länder* nature protection offices were established; however, wartime destruction of offices and materials and the need to replace officials who had retired, died, or been removed through denazification initially limited their effectiveness. Proposals to replace volunteer nature protection advocates with professionals succeeded only in creating a handful of paid positions at the *Land* level. Local-level nature protection commissioners continued to be volunteer positions for over two decades more, and funding remained scant, limiting the effectiveness of the offices.

In an effort to find an alternative to the discredited conservative framings of nature protection in terms of homeland protection, landscape architects and regional planners worked hard to convince the government and public of the need to institute scientifically based, comprehensive programs for the care of the landscape and regional planning. The proposed goals included repairing wartime damage, ensuring orderly development, limiting the increasingly striking losses of open space to new construction, protecting attractive rural vistas, conserving and protecting forests and other natural resources, and providing recreational opportunities for an urbanizing population. One prominent manifestation of this effort was the "Green Charter of Mainau." This document, authored by a coalition of nature protection advocates, landscape architects, and leaders of nature protection agencies, pointed up the growing environmental threats to human welfare and called for better land use planning, more green space and parks, new measures to combat air, water, and soil pollution, protection of nature from unnecessary intrusions, and public education about environmental problems.

Support from nature protection officials and organizations for landscape planning and preservation efforts was, however, often ambivalent. Although they supported its general goals, nature protection advocates feared competition for financial resources and worried that too much emphasis on recreational use might undermine nature protection. A great deal of ink was spilled over efforts to promote the new approach; however, at the end of the day, apart from the establishment of a number of new nature parks intended primarily to provide recreational opportunities, concrete accomplishments were few. Instead, government activity in nature protection and urban and regional planning settled into prewar patterns of passivity, understaffing and underfunding, jurisdictional struggles and fragmentation of responsibilities, and unwillingness to risk steps that might interfere with economic growth.

In the absence of effective countermeasures, Germany's rapid economic recovery, population growth, and urbanization, as well as the growing industrialization of agriculture, led to increasing consumption of energy, water, and other natural resources and to growing water, air, and soil pollution. Water quality in the Rhine and other rivers became increasingly degraded, and clouds of air pollution hung over the Ruhr and other industrial regions, stirring a degree of public concern. In 1951, groups concerned about water supplies and water purity united to form the Alliance for the Protection of Germany's Waters, and there were protests from fishermen and farmers, and from citizens who had to drink polluted water and breathe polluted air. Air pollution problems attracted considerable public concern and press attention, and a 1959 survey showed that two-thirds of the public wanted stronger air pollution control measures (Uekötter, 2003c). Some labor unions also expressed concern, and the government acknowledged the problem by appointing a study commission. By 1961, Willy Brandt, leader of the opposition Social Democrats, was making political capital by calling for blue skies over the highly industrialized Ruhr Valley.

Problems of air and water pollution had, in fact, been under discussion in the German ministries since the 1950s. Beginning in 1952, a small but committed multiparty group of legislators organized themselves as a working group and advanced proposals for environmental legislation. However, except for its concern to secure an adequate supply of water for its own use and reduce the most egregious water pollution, German industry was generally skeptical of pollution control proposals. The *Länder* jealously defended the constitutional provisions that allocated primary responsibility for air and water quality to them, jurisdiction over water and air pollution at the federal level was divided among various offices, and prevailing ideologies equated industrial growth with progress. Consequently, at the federal level, only relatively weak pollution control measures, such as the 1957 federal Water Management Law and the 1959 Clean Air Maintenance Law, were passed. On the other hand, during the 1960s some *Länder,* especially North Rhine-Westphalia, laid the foundation for later national legislation by passing more comprehensive and better-enforced air pollution control measures.

As the problems worsened, public calls for change began to grow louder. A series of apocalyptic essays and novels from the 1950s and 1960s crystallized worries about spiraling production and consumption, worldwide population growth, atomic weapons, water and air pollution, and nuclear power, and the 1964 translation of Rachel Carson's *Silent Spring* received considerable attention. By the end of the 1960s, advocates of a broader environmental agenda were beginning to substitute the term "environmental protection" for the traditional "nature protection" and to call for a broader approach. Nevertheless, until the late 1960s, environmental protection remained well down the list of national priorities, local protests did not generally build networks with similar protests or last beyond the resolution of the issue at hand, no political party made environmental problems a prominent part of its campaign platform, and the courts and administrative agencies typically favored economic growth and business interests over citizen complaints.

Nature protection advocates found themselves not only without widespread public or government support, but also ideologically adrift. Technological development and economic growth seemed to be bringing more problems than solutions, so ideologies that associated nature protection with conservative resistance to change still had an appeal—at least for some; however, the association of the conservative critique of modernity with National Socialism had cost this theme its legitimacy. It was now represented only in the platforms of marginalized right-wing political parties like the National Democratic Party and the writings of isolated, marginal figures, not mainstream advocates of nature or environmental protection. Nature protection advocates wanted very much to preserve the legislative and nature protection accomplishments of the Nazi era, but invoking past accomplishments and themes invited criticism and ridicule and undermined their efforts to remain relevant and link nature protection to building West Germany's new democratic state. In general, the reconstituted organizations avoided direct confrontation with their complicity in the Nazi regime, preferring instead to simply reemphasize their traditional objectives and their nature protection work during the Nazi era.

By the late 1950s, the social movements that would later become allied with West German environmentalism were emerging. Massive demonstrations drew large numbers of supporters to new organizations, such as the World League for the Protection of Life and the Fighting Alliance against Atomic Dangers. The World League, founded by Austrian conservationist and author, Günther Schwab, combined its anti-nuclear activism with other environmental issues and a sometimes racist-tinged emphasis on population control in developing countries to mobilize a mass membership in thirty-one countries. The German branch, led for many years by the former Nazi and Hess associate Werner Haverbeck, also attracted a large membership and was one of the most visible organizations in the German anti-nuclear movement of the 1960s. Its connections to the right, however, remained little known, and it won widespread respect for its tireless campaigning.

West German Environmental Organizations in the Postwar Era

In addition to the obstacles posed by the Allied occupation, the nature protection organizations were plagued by lack of funds, wartime destruction of their offices and records, and difficulties communicating with their former members.[13] Those that had most willingly acceded to Nazi control and ideology also faced the burden of association with a failed ideology and regime. Yet despite the obstacles, the nature protection organizations were successful in reestablishing themselves in West Germany, albeit generally with much reduced membership and without those leaders who had been most compliant with the Third Reich.

[13] General discussions of the reemergence of nature protection organizations after World War II are available in the sources from note 1 and the following: Schenkluhn (1990); Bramwell (1994); Oswald von Nell Breuning Institut (1996); Riordan (1997); Brand, Eder, and Poferl (1997); Bammerlin (1998); Olsen (1999); Geden (1999); Rohkrämer (2002); and Engels (2003).

Except for a spurt of growth at the end of the 1950s and early 1960s, the nature protection associations grew slowly during the postwar years, and many years passed before they regained their former membership levels. They resumed their old strategies of public education and lobbying for relatively minor concessions from government and business and focused on their traditional goals: protecting forests and endangered species, establishing new nature reserves and parks, protecting existing reserves and parks from destructive activities such as military exercises, and protecting scenic areas—along with a newer emphasis on combating land development. They continued to seek out connections with and financial support from government, but overall their influence was even more modest than before the war. This was the consequence not only of their diminished memberships, but also the reduced representation of government officials among their leaders, which limited their ability to rely on informal contacts to influence government policy and decisions. Local chapters often had more the character of social clubs engaging in practical nature protection projects than of activist organizations.

The West German nature protection organizations played no role in emergent social movements focused on issues such as nuclear war and the population explosion, for these dealt with issues far removed from the nature protection organizations' familiar terrain and used confrontational strategies with which they had little experience. For the most part, the nature protection organizations also remained on the margins of local battles against emerging environmental issues, such as pesticides and water or air pollution. Rather than attempting to modify their objectives or strategies to attract members from these new social movements, they chose to continue to pursue their traditional goals and modes of operation.

The Founding of the German Nature Protection Ring

The nature protection groups' preference for working within the systems was symbolized by the 1950 founding of the German Nature Protection Ring (Deutscher Naturschutzring: DNR).[14] The impetus for its creation came, in large measure, from the national nature protection office and from volunteer nature protection officials at the local and regional levels. They believed that having a single umbrella organization for the nature protection organizations would reduce competition among the organizations, simplify relations between them and the government, and provide more unified and visible public support for nature protection to complement government efforts. Creating an umbrella organization for nature protection was also a logical step in a society where industry and labor were organized into powerful national associations, and there was precedent for such an organization in regional associations of nature protection groups, such as the one in Berlin and Brandenburg, that had existed before the war.

---

[14]In addition to sources in notes 1 and 11, see the following: Deutscher Naturschutzring (1976, 2000); Wey (1982); Leonhard (1986); Rucht (1991a, 1993a, 1993b); Hey and Brendle (1994); Wolf (1996); Oswald von Nell Breuning Institut (1996); Bergstedt (2002); Engels (2003); and Röscheisen (2006).

The DNR began with nineteen member organizations, but by the late 1960s it had grown to over 100 member groups with a total of about two million members. During its first two decades, it placed a great deal of emphasis on promoting instruction about nature protection in the schools and informing the public about nature protection issues, especially through its widely circulated newsletter, which was published until the end of the 1960s. It also worked to expand the narrow emphasis on protection of specific species or sites to a broader focus on protecting the landscape. It sometimes opposed specific projects that it saw as particularly destructive, but it rarely challenged prevailing pro-growth ideologies head-on, and its objectives and program were coordinated with the national nature protection authorities, with which it maintained close contact. It also worked to increase the expertise of volunteer nature protection officials and engaged in lobbying, mostly behind the scenes. Finally, it served as a service and support organization for its member groups.

In comparison to its powerful counterparts in business and labor, the DNR was a politically weak organization, and it gained effectiveness only very slowly. It was always underfunded, did not establish a permanent business office until 1964, did not move its headquarters to the capital in Bonn until 1968, and lacked a formal statement of basic principles and goals until the mid 1970s. However, even if it had been blessed with more resources, it is doubtful that the DNR could have been anything other than a relatively timid organization. To avoid repeating the Nazi pattern of state control, it had been purposefully organized as a relatively loose coalition of member groups with very general goals rather than a tightly organized, centrally directed organization with unambiguous goals. Its strategy of seeking as broad a membership as possible meant that its members were an uneasy amalgam of nature protection organizations, such as the BfV, nature-using organizations, such as the hunters, anglers, and hikers, and organizations only secondarily concerned with nature protection, such as the youth hostel association and animal protection associations. This limited its ability to address tough issues, such as reducing the impact of recreational use of nature, and laid the groundwork for later internal conflict issues like hunting and nuclear power.

The League for Homeland Protection

The challenges of postwar rebuilding were most serious for the BH.[15] Its emphasis on preserving the homeland and the German *Volk* had proved especially compatible with Nazi ideology, and its leaders had been the most willing to ally themselves with National Socialism. Consequently, the BH had to be rebuilt almost from the ground up after the war. In its postwar programs and publications, it attempted to reclaim its credibility by portraying itself as having been seduced

---

[15] The most complete information about the BH for this period comes from three internal histories, Zuhorn (1954), Klausa (1979), and Fischer (1994). See also Leonhard (1986), Wüst (1993), Lorenz (1996), and Uekötter (2003c).

and misused by the Nazis. It emphasized efforts by some of its leaders to resist the Nazis and pledged allegiance to the principle of strict nonpartisanship.

Nature protection had been sliding down on the BH's priority list even before the war, and the postwar period, with its emphasis on rebuilding and economic prosperity, did not provide much support for adoption of an aggressive nature protection or environmental agenda. Under the leadership of Adolf Flecken, a strong social and religious conservative, the BH evolved into an organization heavily dependent on government subsidies and focused on preserving the customs, architecture, and monuments of the past. Occasional efforts on the environmental front, such as its expressions of concern about deteriorating air quality in the late 1950s, remained isolated events. By 1970, it had even dropped environmental protection from the list of goals in its bylaws, and an overview of the organization's past and future published on the occasion of its 75th anniversary (Klausa, 1979) gave minimal attention to nature protection.

The League for Bird Protection

Although somewhat less tainted by associations with Nazi ideology, the BfV also faced difficult times in postwar Germany.[16] By the war's end, it had lost about half its members, closed most of its offices, and been forced to drastically cut back its activities. Nevertheless, its prewar leaders began rebuilding immediately after the war. It received quick approval from the Allied occupation authorities to reestablish itself throughout the western zone; however, its membership remained concentrated in the regions closest to its Stuttgart headquarters. In an effort to reconnect to its pre-Nazi past, Hermann Hähnle, the son of the founding president, assumed the presidency. With few exceptions, such as its campaign against the overuse of pesticides in the early 1960s, the BfV retained its moderate and rather apolitical approach and resumed its traditional activities—public education, acquiring and maintaining nature reserves, and nonpartisan, behind-the-scenes lobbying. Dues remained low, and most of the organization's support came from small government subsidies and donations from business, private benefactors, and the Hähnle family. Membership growth was slow and erratic, and the BfV did not regain its previous membership peak until the mid 1960s. With an aging membership and diminished public support, it was poorly positioned to reach the level of influence it had previously enjoyed.

The BfV was further weakened in 1965 by Hermann Hähnle's death and a reorganization that strengthened its regional units at the expense of the national organization. It was almost 1970 before increases in membership dues made possible the beginnings of professionalization, including employing a paid executive secretary, establishing a publishing subsidiary, and establishing a member magazine.

---

[16]Considerable information about the BfV in the postwar period is available in the organization's own histories, Hanemann and Simon (1987), Naturschutz Heute (1999a), and May (1999). See also the sources cited in note 1, especially Dominick (1992), as well as Wöbse (2003a, 2003b).

Like the BH, the BfV was slow to come to terms with its role during the Nazi era. An official history, written by two of its leaders and published in 1987, describes the BfV's absorption into the Nazi bureaucracy, the annexation of other bird protection organizations, and the organization's expansion into territories annexed by the Nazis briefly and almost without editorial comment. It notes only that resistance would probably have resulted in dissolution of the organization and that the wartime leadership tried to avoid political involvements and focus on nature protection (Hanemann and Simon, 1987). A 1999 history is much more open about the organization's role in the Nazi era (May, 1999).

The Bavarian League for Nature Protection

The BN also survived the war; however, its membership also declined by almost half and its paid staff to one, and many local chapters dissolved.[17] With the approval of the US occupying authorities it quickly reconstituted itself after the war, with bylaws similar to its prewar ones. To satisfy the occupying authorities and refurbish its tarnished image, leaders with the closest ties to National Socialism were shuffled off of the board of directors onto its larger advisory board, but like the other organizations, the BN generally avoided discussion of how ideas about nature protection had been misused by the Nazis. Indeed, by 1958 one of its leaders, who had served as the nature protection advisor for Nazi-era autobahn construction was back in office as chairman of its board.

The BN resumed its traditional tasks, developing nature reserves, working to reduce the destructive effects of tourism, and protecting open space from settlement. It also fought several battles against the development of hydroelectric power in Bavaria and was involved in a long effort to block construction of a new airport in an ecologically sensitive area near Munich. Nevertheless, it continued to work mainly behind the scenes in cooperation with government agencies. It did not seek a mass membership and did not regain its prewar membership level until the mid 1960s. The process of reestablishing local groups that had dissolved during the war also proceeded very slowly, especially in northern Bavaria. Its administrative office was operated by a single business manager with limited clerical assistance. Like the BfV, it again began receiving a small government subsidy. By reverting to its historical focus and mission, the BN left itself ill-prepared to deal with emerging problems like air and water pollution, nuclear power, and traffic.

The Friends of Nature

The Friends of Nature had been disbanded by the Nazi regime.[18] Nevertheless, after the war many of its local chapters requested permission to reconstitute

[17] In addition to the sources listed in note 1 (see especially Dominick, 1992), additional information about the postwar BN is available in Hoplitschek (1984), Weiger (1987b), Wolf (1996), and Geden (1999), as well as in histories prepared by the organization: Blätter für Naturschutz und Naturpflege (1931); and Lense (1973).

[18] General sources, such as those in note 1, often ignore the Friends of Nature, but a good bit has been written about the organization in the postwar period, including the following: Dulk (1984);

themselves, and—unlike most of the other prewar workers' organizations—it was able to reestablish itself as a significant national organization. It was headed by one of its former leaders, who had ridden out the war in a leadership position for a politically acceptable hiking organization. It resumed publication of its magazine and was soon readmitted into the international Friends of Nature; however, a long and difficult legal struggle was required to reclaim its houses and other property.

Despite its successful reestablishment, the Friends of Nature struggled to find a unified voice in the postwar period. The 1959 decision of its old ally, the Social Democratic Party, to transform itself from a workers' party with a socialist agenda to a party of the moderate left was echoed by the Friends of Nature's internally contested decision to affiliate with regional sports federations and cooperate with other hiking clubs rather than insisting on separate workers' groups.

In subsequent time periods and in different regions, the Friends of Nature functioned variously as a politically uninvolved federation of hikers and hobbyists, an ally of the now moderate left Social Democratic Party and the labor unions, as an ally of far-left organizations, and as a part of the anti-nuclear and countercultural movements. Nature protection and environmental issues remained among its objectives, and it continued to issue statements, hold conferences, and distribute educational materials about environmental problems. Consistent with its prewar emphases, the organization took a special interest in protecting the Alps and in reducing the damage caused by commercialized mass tourism. It also engaged in occasional protest actions against environmental threats, such as a successful effort in the 1950s to halt construction of a cement plant in the Swabian Alps and a crusade against proposed strip mining in Hesse. It sometimes lobbied the government for measures like the preservation of roadside trees and coastal areas, and even before the environmental movements of the 1970s it sometimes reproached industry for environmental damage.

Nevertheless, environmental goals continued to be only one aspect of the Friends of Nature's multifaceted agenda, and its environmental activities in the postwar period were never visible enough to give it a major role in environmental politics or earn it recognition as a major environmental organization, especially in comparison to organizations that had environmental goals as their sole objective.

The Founding of the Worldwide Fund for Nature

Despite the quiescence of the nature protection movement in the immediate postwar era, the period did see the establishment of several new nature protection organizations. The German Forest Protection Association (Schutzgemeinschaft Deutscher Wald), for example, was founded in 1947 in response to growing con-

---

Zimmer (1984); Dulk and Zimmer (1984); Schmitz (1984); Erdmann (1991); Englehardt (2000); and Uekötter (2003a). There is also a useful history published by the organization itself (George, 1955).

cerns about the German forests.[19] The forests, already damaged by Nazi policies, were very destructively exploited in the immediate postwar period for rebuilding needs, fuel, and revenue for reparations payments. The Association's membership contained many foresters and government employees, and it received modest government subsidies. Its activities were oriented toward a vigorous public education effort, lobbying the occupation administrations and German government, and fending off threats to forests.

More significant in terms of this book's focus on large environmental organizations that address multiple environmental issues was the founding of the German branch of the Worldwide Fund for Nature (WWF).[20] The WWF, known originally and still in the US as the World Wildlife Fund, was originally established in Britain in 1961 by scientists, prominent conservationists, wealthy donors, and political elites who wanted to develop a vehicle to raise money for projects to protect wildlife throughout the world. The donors provided the working capital to fund administrative expenses so that further donations could be devoted to its projects. In its early years, it devoted itself mainly to establishing wildlife reserves and protecting impressive species in less developed countries.

The German branch was founded in 1963 by ten leaders from business, government, and science. Like the other chapters, it was set up as an independent foundation with a self-perpetuating board of directors, but it cooperated closely with the international organization and helped to support it financially. During its early years, WWF-Germany—under the leadership of a relatively inactive board filled with wealthy donors, prominent politicians, and the like—functioned mainly as a fundraising organization for its international partner, although it also made grants of funds to other organizations in Germany for nature protection projects. Its work was almost entirely in the hands of volunteers, it did not publicize its efforts widely, and its fundraising success lagged far behind comparable national chapters elsewhere. It differed from other nature protection organizations in having not members, but simply donors without voting rights. Its nonconfrontational approach and noncontroversial goals dovetailed nicely with the conservative tenor of the postwar period.

Summary and Conclusions

As the ill-fated Weimar Republic entered its final years, the environmental impacts of industrialization and urbanization were being addressed at the national level by only a handful of organizations. All of these focused their attention on

---

[19] See Leonhard (1986), Dominick (1992), Bergstedt (2002), and Engels (2003).

[20] Brief treatments of WWF's early years in Germany include Blühdorn (1995) and Bergstedt (2002). For additional information see Haag (1986), Cornelsen (1991), Rogall (2003), and the following organizational histories: WWF-Journal (2002f); Wünschmann (2003); and Wünschmann et al. (2003). Information about WWF-International during this period is available in Dalton (1994), Wapner (1996), and WWF-Journal (1999d).

nature protection—the protection of specific species and of scenic or ecologically sensitive areas—as opponents of water and air pollution had not organized nationally. The nature protection organizations themselves were somewhat specialized, with separate organizations focusing on bird protection, protection of scenic rural landscapes, and creation of nature parks. All but the Friends of Nature drew support mainly from the well-educated middle class. Their supporters included some citizens who simply wanted to protect nature; however, others were influenced to varying extents by an anti-modern ideology that glorified rural life and the beauties of nature and saw urbanization and industrialization as undermining Germany's well-being and identity.

The organizations had achieved a significant membership, but their efforts to attract mass public support were largely unsuccessful. All were inclined toward working within the system. They emphasized public education and worked closely with government agencies, both supporting their efforts and lobbying them behind the scenes. Most also received small government subsidies. They enjoyed occasional modest successes but were unable to secure strong national legislation to protect nature. In short, they functioned as institutionalized but relatively uninfluential interest groups.

The exception to these generalizations, the Friends of Nature, was tightly linked to movements of the left by its anti-capitalist ideology and working-class support base, but the left itself was divided about the importance of environmental issues. Lacking access to government or strong support from movements of the left, the Friends of Nature was unable to boast of a long record of victories for the environment, and it recruited the majority of its members mainly through the other rewards it offered them.

For all of the organizations except the Friends of Nature, the rise of National Socialism posed the same troubling dilemma. Nazi ideology contained some strains that resonated with the worldviews of many of the nature protection organizations' more conservative supporters, demonstrating once again that environmental concerns can be easily linked to ideologies of the right. The organizations could reasonably hope that a National Socialist government would have the will and the power to pass and implement legislation for which they had fought so long with so little success, and there was the prospect of increased state funding as well. Moreover, the mass constituency the Nazis had attracted with their framing of Germany's problems and their skillful propaganda offered the chance to attract many new supporters to the nature protection cause. As a result, allying themselves with National Socialism was a seductive opportunity for the organizations.

On the other hand, supporting the Nazis could easily have backfired. First, there was the obvious risk of forming an alliance with a movement that was unpopular in some quarters and might not succeed. This could have cost the organizations public support and access to government. Some of the organizations' leaders and members also, no doubt, viewed the Nazis with distaste. Moreover, it was clear from the beginning that—lacking the clout to significantly influence the National Socialist movement—a nature protection organization that allied itself with the Nazis was destined to be a bit player in a much larger movement that it could not control.

Almost no information about the internal deliberations of the nature protection organizations or the private views of their leaders in the early 1930s has been preserved, but they must have worried about how to respond to the rising Nazi movement. In the end, drawing on their traditions of nonpartisanship, they opted to keep their options open. None of them supported the Nazis publicly before 1933, although some individual leaders of the BH did embrace the Party. On the other hand, none publicly criticized the Nazis. For the Friends of Nature, the options were grimmer. Overt support of National Socialism would have represented a denial of their core values and cost them most of their support base.

After Hitler's accession to the chancellorship, all of the organizations found themselves confronted with a starker and more immediate choice. The Nazis quickly and radically altered the political opportunity structure, making conventional interest group strategies, social movement mobilization, and criticism of the state all but impossible. Nor could the organizations freely exercise the educational or practical nature protection functions of civil society organizations without the approval of the state—a state that now demanded that they adapt their goals and operations to its needs as the price of their continued existence. The new realities were illustrated by the fate of the Friends of Nature. Despite efforts to save it by disavowing its ties to its international parent and deemphasizing its connections to the Social Democrats, the Nazis moved swiftly to abolish the organization and seize its assets. Being a part of the wrong social movement cost it its very existence.

The dilemma the organizations faced was clear. The sacrifices required of them were significant: loss of organizational autonomy and the potential of facing additional government demands in the future. Yet the only options to *gleichschaltung* were to face abolition or simply walk away. In fact, none of the organizations chose to cease operations, and relatively few leaders or members exercised the second option. Evidently their commitment to the cause, the rewards they received from their participation, and their hopes for the future outweighed their fears. They elected instead to make the best of the situation. They accepted *gleichschaltung,* celebrated the regime's nature protection achievements, continued their traditional work, and enjoyed the benefits of state subsidies or state-granted rights to be the sole national organization in their fields. Their decision was probably made easier by the fact that their assimilation into the state represented an extension of already institutionalized patterns of close cooperation with government nature protection offices, not a total break with the past. Moreover, the loss of independence was not total. All the organizations worked with some success to maintain as much independence as possible.

Nevertheless, the obvious drawbacks meant that all, even the relatively enthusiastic BH, also displayed some ambivalence, and subsequent events showed that there were good reasons to harbor doubts. The Nazi state could and did ram through ambitious nature protection legislation. Yet it ignored most other environmental problems, and when it suited its purposes, it could also ignore nature protection with impunity. The early record of nature protection achievements soon gave way to the realities of war preparations and war, and the gains proved more illusory than real.

The challenges the national environmental organizations faced in postwar West Germany were almost as great as those involved in adjusting to National Socialism. Indeed, it is hard to imagine a less hospitable climate for maintaining nature protection organizations—not to mention social movement mobilization for the environment. The country was shattered by war and social dislocation, administered by occupying powers intent on denazification, and focused on rebuilding the economy—and later on enjoying its newfound prosperity. Environmental concerns were barely on the radar screen.

It would have been easy to give up, but the leaders and members of the nature protection organizations did not. They overcame all of these obstacles—not to mention their association with a failed ideology and the destruction of their offices, facilities, and records—to reconstruct their organizations and resume their work. Their efforts testify not only to their devotion to their cause, but also to their commitment to the organizations in which they had labored for so many years. This commitment is obvious in postwar publications that describe with evident pride the almost heroic effort that went into reconstructing organizations that had been decimated or abolished.

All of the nature protection organizations adopted essentially the same strategy as they struggled to reestablish themselves: they returned to tried and true goals and strategies, and reoccupied the niches they had inhabited before the war. Indeed, it is likely that their leaders never actively considered other options. By conforming to institutionalized expectations about their social roles that were valued by their supporters and known and accepted by other social actors, they could reclaim a degree of legitimacy and support. Reverting to prewar patterns thus let them both accomplish something for the environment and provide rewards that had motivated leaders and supporters in the past.

This strategy allowed the reconstructed nature protection organizations to take root in the inhospitable soil of postwar West Germany, but it could hardly be said that they flourished. For most of the organizations, this was so partly because the war experience had undermined the agrarian romanticism and nationalism that had provided a motivational framing for some of their strongest supporters, leaving them ideologically adrift and deprived of old allies. The problems faced by the Friends of Nature at war's end were different in content, but similar in form. When the other workers' organizations failed to survive the war and the Social Democrats elected to break their ties with the Marxist past, they too found themselves struggling to define a new and persuasive rationale for their existence and the role of nature protection in their mission.

Whether for Friends of Nature or the other organizations, the simple desire to protect nature remained an important motive for many supporters. Moreover, supporters could enjoy the social rewards of being part of a group working together on practical nature protection projects. Still, nature protection was perceived by many outsiders as the harmless, rather quaint preoccupation of people out of touch with postwar realities, and it received little media attention. Consequently, many years passed before the organizations reached their earlier membership levels. The nature protection officials who had once been among their core members did not return in large numbers, and the organizations did not receive significant subsidies from national or *Länder* governments with other priorities.

These realities, along with the economic stresses of the early postwar period, made it impossible to fund ambitious programs or large, professional staffs through small state subsidies or member dues, and the technology of mass fundraising did not yet exist. Therefore, the organizations remained underfunded and understaffed. Though they were not completely bereft of public visibility or influence, their impact was even weaker than in the prewar period. Neither was mass mobilization of supporters for protest a viable option for organizations operating without a persuasive ideology and in such a difficult climate.

Mass mobilizations did occur eventually, but they involved issues, such as struggles against the threats of war and nuclear weapons, that had little appeal to the relatively conservative and apolitical nature protection advocates. Efforts to promote planning of the landscape as a new frame for environmental action also fell short, in part because the organization feared that they would shift the focus away from their traditional concerns. Their members were not much motivated by these issues, there was more than enough work to do, and confrontation was not part of their action repertoire. They remained lodged instead in their familiar niches as marginally effective interest groups fighting the good fight with limited public support. In the competition with powerful interest groups with seats at the table in West Germany's emerging neocorporatist system, they lost more battles than they won, and efforts to increase their power by allying themselves with nature using organizations under the umbrella of the DNR did not succeed in strengthening their hand.

By clinging to their narrow focus on nature protection and working apolitically within the system, the nature protection organizations failed to recognize or take advantage of the latent social movement mobilization potential manifested by public concerns about the growing pollution of air, water, and soil, population growth, unchecked consumerism, and natural resource exhaustion. It would require the upheavals of the 1970s and 1980s to dislodge them from their old patterns and convince them to take these new concerns seriously.

Chapter 5

CONFRONTATION AND COUNTERCULTURE

Ecology from the Left in a Turbulent Era

The postwar period of relative quiet on the environmental front in West Germany came to an end in the 1970s, when a confluence of events brought environmental issues to the fore and called forth a large and, at times, radical environmental movement centered around opposition to nuclear power.[1] The long-established nature protection organizations played almost no role in bringing about this movement, but the period of polarization and confrontation that ensued confronted them with both enormous challenges and opportunities to which they had to respond. It also led to the establishment of new environmental organizations born out of the movement.

The Rise of Environmentalism

The German "economic miracle" of the 1950s and 1960s produced not only soaring prosperity, but also growing environmental problems.[2] Clouds of air pol-

[1]The following sources provide excellent overviews of the period of confrontation and the environmental organizations during this period: Rucht (1991a); Dominick (1992); Hey and Brendle (1994); Oswald von Nell Breuning Institut (1996), and Bammerlin (1998). Several sources cover the general environmental history of the period well, but without much reference to the large environmental organizations: Dominick (1988); Joppke (1993); Rucht (1994); Koopmans (1995); Chaney (1996); Brand, Büsser, and Rucht (1986); Rat von Sachverständigen für Umweltfragen (1996); and Brand, Eder, and Poferl (1997).

[2]For discussions of Germany's escalating environmental problems in the 1960s see, in addition to sources cited in note 1, Wey (1982), Müller (1986), Hucke (1990), Jänicke and Weidner (1996),

lution hung over highly industrialized areas like the Ruhr Valley, causing health problems and damage to forests. Uncontrolled waste dumping and pollution of rivers became increasingly obvious. Despite cleanup efforts, the Rhine became so polluted by oil spills, industrial discharges, detergents, and pesticide residues that it was episodically covered with polluted foam and swimming was prohibited. There were major fish kills and warnings that the river might soon be unable to support life. Beyond Germany's boundaries, dramatic events like a major oil spill from the tanker *Torrey Canyon* in the English Channel attracted press attention. The loss of open space to new roads, housing developments, factories, and parking lots—already a focus of concern in the 1950s—continued apace. Moreover, growing prosperity and the desire to escape the cities accelerated the trend toward mass tourism, which in turn threatened the Alps and other scenic aeas. Channelization of streams, increasing reliance on monocultures in farming and forestry, and draining of wetlands and removal of hedges to expand the area under cultivation and allow for mechanized agriculture also accelerated the loss of plant and animal habitats. Finally, increased use of fertilizers, herbicides, and pesticides, and the conversion of animal husbandry to concentrated animal feeding operations, introduced additional pollutants into the environment. The effects were dramatized by the 1971 publication of the first "red list" of endangered birds, which showed a frightening number of threatened species.

These and other environmental problems were dramatized in the 1960s by a series of books and reports pointing up the dangers of rapidly increasing consumption, water and air pollution, worldwide population growth, nuclear weapons, and nuclear power.[3] These included Paul Ehrlich's *Population Bomb,* the 1964 German translation of Rachel Carson's *Silent Spring, The Plundering of a Planet,* a best seller by CDU Bundestag member and environmentalist Herbert Gruhl, and the earlier best seller by Günther Schwab, *Dance with the Devil.* The latter portrayed the devil gleefully recounting the environmental damage he was bringing about by deceiving humanity about the effects of chemical emissions, pesticides, atomic weapons, nuclear power, and the population explosion. Schwab was also the founder of the World League for the Protection of Life, whose German chapter was the most visible opponent of nuclear weapons and nuclear power during the 1960s.

The themes in these books were later echoed by the widely circulated 1972 Club of Rome report, *The Limits of Growth.* Its critique of high-consumption lifestyles and predictions of eminent exhaustion of natural resources received a great deal of press and public attention. The European Nature Protection Year and the

---

and Fritzler (1997). The problems were most severe in Germany's most industrialized and densely populated *Land,* North Rhine-Westphalia. See Oberkrome (2003a) and Maxim and Degenhardt (2003) for details.

[3] In addition to sources cited in note 1 see the following for discussions of the role of literature, social criticism, and scientific reports in publicizing environmental problems: Kneitz and Kley (1986); Küppers, Lundgreen, and Weigart (1978); Hermand (1991); Bramwell (1994); Fritzler (1997); Geden (1999); and Rohkrämer (2002).

first US Earth Day, both in 1970, the wave of environmental legislation in the US that began at the end of the 1960s, and a number of conferences, culminating in the 1972 United Nations environmental conference in Stockholm, also attracted attention to environmental problems.

Germany's environmental problems had long been under discussion among interested Bundestag delegates and in relevant government bureaus;[4] however, no political party had made environmental problems a key concern, and the postwar governments led by the Christian Democratic Union did not develop a comprehensive environmental program. Legislative successes were scattered and partial, and enforcement fragmented and weak. In 1969, however, a new coalition government comprised of the Social Democrats and the small Free Democratic (liberal) Party took office. The left wing of the Free Democrats, including Interior Minister Hans-Dietrich Genscher decided to make environmental issues a signature issue, and Willy Brandt, the new chancellor, set out to emphasize quality-of-life issues, including the environment. The political climate favored new initiatives, and a prospering economy provided the resources needed.

Acting primarily on its own initiative, the new government moved quickly. It introduced an "immediate action program," expediting plans that were already in progress or under consideration, to begin cleaning up the most visible pollution. This was followed the next year by a more comprehensive program. The new coalition also pushed through a constitutional amendment that greatly expanded the federal government's authority to legislate in the areas of air pollution, radiation, and solid waste disposal; however, efforts to acquire similar authority in other domains, including water resources and nature protection, could not overcome resistance from the *Länder.* In these areas, the federal government remained limited to passing "framework legislation," which established only broad guidelines for more detailed *Länder* legislation. Using its new authority, the government passed a series of laws in areas such as solid waste disposal and recycling, regulation of DDT and other pesticides, and controlling air pollution from factories and leaded gasoline. Expenditures for new sewage treatment plants and improving the quality of treatment also increased.

To better coordinate its efforts, the Brandt government consolidated numerous offices relevant to environmental protection, including those concerned with air and water quality and waste disposal, in Genscher's interior ministry and greatly

---

[4]There is a very large literature about German environmental politics during the late 1960s and the 1970s. In addition to sources cited in note 1, Erz (1987), Brand (1999a, in press), and Schreuers (2002) offer brief overviews. The most comprehensive general treatment, focusing heavily on details of legislation and implementation, is Müller (1986). Other very useful general treatments include the following: Küppers, Lundgreen, and Weigart (1978); Ewringmann and Zimmermann (1978); Paterson (1989); Weßels (1989); Hucke (1990); Weidner (1991a); Wilhelm (1994); Malunat (1994); Loos (1995); Jänicke and Weidner (1996); Pehle (1997, 1998); Jänicke, Kunig, and Stitzel (1999); Müller (2001), and Hünemörder (2005). Voss (1990) focuses on the media and politics, Schlipköter and Winneke (2000) on health issues, Wey (1982) on local politics, and Schenkluhn (1990) and Englehardt (2000) on environmental organizations and politics.

expanded its staff. Nature protection and protection of the landscape, however, remained in the agriculture ministry, where it had to compete with powerful agricultural interests, and some responsibilities for water, soil, and air pollution remained in the health ministry. Regional planning was even transferred from the interior to the construction ministry. A new Federal Environmental Office was established in the interior ministry to fund environmental research, prepare technical reports, and provide expert advice, and a panel of independent environmental experts was set up to monitor and report on environmental problems and progress.

The government consulted regularly with the opposition CDU, business, and labor, and it established a Working Group for Environmental Issues that included representatives from all of these groups, as well as environmental organizations, scientific associations, and the *Länder* governments. The opposition Christian Democrats quickly adopted an environmental program of their own, and by the mid1970s, all the major parties had well-developed environmental platforms. Business and agricultural interests—and the government ministries linked most tightly to them—did contest some of the provisions of the government's program, such as penalties for pollution, that involved clear and immediate costs to them, and they succeeded in weakening some of them. Nevertheless, it was clear to business that they could not completely disregard the pressure for change. The economy was prospering, business received assurances that it would not be saddled with unmanageable costs, the full costs of environmental cleanup were not yet evident, and business viewed unified national regulations as a considerable advantage and hoped that progressive environmental laws would make the economy more competitive. Hence, it did not contest environmental reforms in principle. Labor unions expressed significant concerns about job losses, but they too acknowledged the need for environmental protection. The new environmental legislation thus passed virtually without dissent, and environment was not a hotly debated issue in the 1972 or 1976 elections.

The government's program focused on the environmental problems of an industrial, urban society rather than traditional nature protection. It shifted the focus of fighting these problems to the national level, and it represented the most comprehensive attack on these problems in Germany's history. The new laws succeeded in curbing some of the most pressing problems, including threats to public health from air pollution in the most impacted areas and high lead levels in children's blood. Yet, the government's program was, in many respects, a scattershot response to crisis. It emphasized problems that posed clear threats to public health and neglected others, relied mainly on technological fixes—especially end-of-pipe filtering—to mitigate the effects of specific problems, and did not include efforts to guide or limit economic growth. Except for the creation of Germany's first national park, nature protection received relatively short shrift, and government-sponsored efforts to promote mechanized agriculture continued to destroy wildlife habitats. Still, the greatly increased level of government activity in itself drew press and public attention to environmental problems.

Another foundation stone for the development of Germany's environmental movement was the wave of protests—with concerns similar to those in the US and other European nations—that swept Germany during the late 1960s.[5] From 1966 to 1969, the Federal Republic was governed by a resolutely centrist "grand coalition" of its two major political parties, the Social Democrats and Christian Democrats. With only the small Free Democratic Party in opposition, dissenters from an emerging countercultural left channeled their efforts to effect change into protest outside the system. Organizing as the so-called "extra-parliamentary opposition," they staged a series of protests focused on the Vietnam War, university reform, and critiques of Germany's consumerist bourgeois culture and the power of political and economic elites. In comparison to its US counterpart, the German movement was slow to incorporate environmental themes, but the protest movement's critique of consumerism, which ironically echoed conservative criticism from the beginning of the century, was later assimilated into environmentalism, as were the extra-parliamentary opposition's unconventional protest tactics.

In the early 1970s these countercultural themes and growing environmental concerns began to coalesce into a more specifically environmentalist critique of German society.[6] In 1972, the "ecology group," a loose coalition of environmental thinkers and leaders from across the political spectrum, drew up an "Ecological Manifesto." Echoing themes from the Club of Rome report, it criticized Germany's pro-growth ideology and advocated a redirection of effort from protecting specific species or limited areas toward an attack on more deeply seated environmental issues, such as population growth, resource exhaustion, consumerism, and overemphasis on economic growth.

Arguments like these found broad resonance among a new generation of well-educated and materially secure young Germans, who had not experienced wartime and postwar deprivation and manifested what Inglehart called "post-materialist values" (Inglehart, 1990). Environmentalist themes were also prominent among younger cohorts of "Spontis," who adopted lifestyles similar to US hippies and squatted in unoccupied apartment blocks, as well as among young people who founded rural communes in search of an alternative to urban life and the mainstream economic system.

Responding to these political and cultural developments, newspaper and television reporting about environmental problems and politics increased markedly

---

[5]In addition to sources cited in note 1, the development of countercultural protest and the extra-parliamentary opposition has been chronicled by Offe (1981), Mayer-Tasch (1985), and Brand (1987). Sieferle (1984) and Olsen (1999) examine its connections to conservative ideologies from earlier in German history, and Renn (1985) connects it to the later flowering of countercultural ideology.

[6]In addition to sources from note 1, the following contain useful discussions of the early intellectual and ideological foundations of the environmental movement's new framings of environmental problems: Kaase (1986); Paterson (1989); Cornelsen (1991); Hermand (1991); Bramwell (1994); Wilhelm (1994); Fritzler (1997), and Müller (2001). Frederichs (1980) and Renn (1985) examine links to the later development of environmental thought, and Wüst (1993) and Bergstedt (2002) note commonalities with right-wing environmental thought from earlier and later periods.

during the 1970s.[7] Public opinion polls[8] showed dramatic increases in the percentage of citizens in all segments of the population who rated environmental protection as a very important problem, and the issue ranked near the top in responses to questions about the most important political issues. A 1972 survey, for example, showed that the public rated air and water pollution as the most serious problems facing Germany and that about three-quarters of respondents believed that little or very little had been done about environmental protection and wanted immediate action, even if Germany's economic effectiveness was affected (Bunz, 1973).

None of the issues that took center stage in the 1970s—air and water pollution, toxic waste, resource depletion, population, and economic growth and consumerism—were new. They had been addressed, to varying extents, by various writers, government bureaus, local protests, and, occasionally, even by the traditional nature protection organizations. What was new was their increased prominence and their unification with nature protection under the rubric of "environmental protection," a framing that emphasized not just saving specific species or scenic areas, but the impact of human activity on ecosystems and the entire planet, and the resultant effects on human health and quality of life (Brand, in press).

The Growth of Citizens' Initiatives

The new environmental concerns manifested themselves organizationally, not so much via the traditional nature protection organizations as in the form of a growing number of local "citizens' initiatives" (*Bürgerinitiativen:* BIs) focused on environmental issues.[9] Citizens' groups working for the betterment of local conditions or against threats to local communities had a long tradition in Germany, but they had traditionally been led by elites. The wave of BIs that originated in the 1960s, by contrast, was comprised of grassroots groups. They concerned themselves not only with environmental issues, but with a wide variety of other topics, including mass transit fares, housing and tenants' rights, improvement of schools and playgrounds, better treatment of minorities, and improving social services. Many BI members also demanded a more active voice in decisions that affected their well-being, and the impetus for their formation was often the unresponsive-

---

[7] For overview discussions of media coverge during this period see Müller (1986), Voss (1990, 1995), and Hermand (1991).

[8] See Noelle-Neumann (1976, 1977, 1983), Kaase (1986), Müller (1986), Dierkes and Fietkau (1988), and Dalton (1994) for overviews and specific examples of polling data from the period.

[9] A great deal has been written about the rise of the citizens' initiatives. The most comprehensive discussion is in Mayer-Tasch (1985). Other significant contributions include the following: Borsdorf-Ruhl (1973); Lange (1973); von Kodolitsch (1975); Andritzky (1978); Andritzky and Wahl-Terlinden (1978); Ebert (1980); Rieder (1980); Rüdig (1980); Wey (1982); Thaysen (1984); Ellwein (1984); Schiller (1984); Jäger (1984); Müller (1986); Mez (1987); Klingemann (1991); Rucht (1993a); and Wilhelm (1994).

ness of political parties and government bureaucracies closely tied to powerful, well-organized interest groups from business and labor. Nevertheless, political parties, foundations, and other existing associations did assist some BIs in getting underway. Most of the early BIs were nonpartisan, not particularly ideological, and inclined to work nonviolently within the limits of the law.

The limited available data about the characteristics of BI members suggest an overrepresentation of men, young adults, the well-educated, holders of middle-class and upper middle-class occupations, and the politically active, but the membership base was more diverse than the student protests of the later 1960s. The core members were often citizens most directly affected by an unmet need or threatening project; however, the environmental groups sometimes attracted members with more general environmental interests. Many BIs were small, loosely organized, and short-lived, terminating after achieving or failing to achieve their immediate objectives. A significant percentage, however, organized themselves more formally, sometimes as associations officially registered with the government, and pursued a variety of goals over a period of years. Most were sustained by a core of committed supporters, supplemented by a wider circle of sympathizers who could be called on to sign petitions or appear at an occasional meeting or protest. Relatively few had paid staff.

The Brandt government, which had called on Germans to "dare (more) democracy," was generally receptive to the BIs and sought to increase their input into decisions; however, the political opposition was more skeptical, and the playing field remained tilted in favor of established interest groups. Local government typically welcomed BIs that aimed to improve local facilities and services through self-help programs or volunteer work and sometimes supported them financially. On the other hand, initiatives that attempted to move inert bureaucracies into action or demanded increased input in policy making often encountered foot-dragging, exclusionary tactics, or sharp criticism from officials unaccustomed to working with such groups or to being challenged, so conflicts with the bureaucracy were common. The BIs, in turn, sometimes turned to lawsuits or demonstrations.

No truly adequate data about the number, goals, and activities of the BIs are available, but there is consensus that the number of BIs grew rapidly during the 1970s, that the BIs achieved an impressive total membership, and that initiatives concerned with environment were among the most common. Two 1973 surveys, using somewhat different question wordings, found that 3 and 12 percent of German adults, respectively, had participated in activities sponsored by a citizen's initiative, while much higher percentages expressed willingness to join or sympathy for BIs and confidence in their work. A 1976 survey showed that over 1 percent of respondents claimed to be full-fledged members of a BI (Andritzky and Wahl-Terlinden, 1978; Rüdig, 1980), and the number of environmental BIs listed with the Bavarian government increased fivefold between 1972 and 1979 (Hoplitschek, 1984). The environmental BIs covered a wide range of problems, including new environmental concerns: air and water pollution, highway and airport construction projects, traffic problems, airport noise, energy issues, and proposed nuclear power plants.

The Founding and Early Years of the Bundesverband Bürgerinitiativen Umweltschutz

BIs opposed to aircraft noise had organized a national alliance as early as 1965, and numerous regional and national networks of environmental BIs, some focused on specific problems, sprang up in the early 1970s.[10] The most important, however, was the Federal Alliance of Citizens' Initiatives for Environmental Protection (Bundesverband Bürgerinitiativen Umweltschutz: BBU). Its lack of formalization makes it a borderline case in terms of the criteria used to select organizations for attention in this book; however, its role as a key social movement organization during the 1970s and early 1980s made it a very important influence on other environmental organizations. The BBU was founded in 1972 by a group of fifteen BIs centered in the Rhine Valley to coordinate their efforts and share information, educate the public, and represent their interests in the media and national politics.

Although the BBU later played down the connection, perhaps in an effort to distance itself from the government, there is good evidence that the interior ministry encouraged its founding—probably in hopes of creating an environmentalist counterweight to powerful lobby groups from business and agriculture. Several ministry officials attended the founding conference, an FDP-linked foundation helped to fund it, and the ministry paid travel costs for some delegates. Some of the founding members also had close ties to the Evangelical (Lutheran) church. The right-oriented World League for the Protection of Life, a national organization that had played a leading role in peace demonstrations in the 1960s, was also among the founding groups, but it soon withdrew.

The BBU's founding charter clearly reflected the influence of countercultural environmentalism. It cited the danger that untrammeled economic expansion would lead to ecological catastrophe, emphasized the merits of a simpler, environmentally responsible lifestyle, and called for restructuring and democratizing economic and political institutions to reduce inequality and injustice. It also contained strong critiques of atomic and chemical weapons and the dangers of industrial chemicals and nuclear power, and it called for the development of renewable energy sources, alternatives to automobile transportation, and stronger nature protection. In line with their democratic ideals, the founders set out to establish a grassroots democratic organization with a minimum of bureaucracy.

The members of the BBU were local BIs, regional coalitions of BIs—later including some that characterized themselves as *Länder* affiliates of the BBU—and

[10] Many of the references cited in note 9, especially the exceptionally thorough history of the early years of the BBU by Rieder (1980), also contain information about the BBU. Comprehensive histories of the BBU are also available in Kazcor (1986) and Markham (2005b). Other important sources on the BBU include Kempf (1984), Rauprich (1985), Kazcor (1989, 1990), Leonhard (1986), Schmid (1987), Jahn and Wehling (1990), and Pehle (1998). Useful articles by BBU leaders include Wüstenhagen (1975) and Sternstein (1980, 1981).

other environmental organizations. Individuals could join but not vote. A board of directors elected at an annual meeting of representatives from the member groups supervised day-to-day operations. By 1975, the BBU was publishing a member magazine and a newsletter for its member groups. Beginning in 1976, it set up working groups in areas like nuclear energy, traffic, and water pollution; however, some of these proved to be more or less "one-person shows" or were inactive and short-lived.

The BBU eventually became best known for its role in coordinating demonstrations, but it engaged in a wide variety of other activities. It organized conferences about environmental problems, issued press releases, staged news conferences, lobbied the government, testified at various hearings, and participated with representatives from business and government in the Working Group for Environmental Issues. Still, its frequently confrontational stance and relatively low level of technical expertise meant that these efforts were not always well received. It also provided advice, referrals to appropriate experts, and other services to its constituent groups.

The BBU grew very rapidly in its early years, accepting over a hundred new organizations by the mid 1970s; however, it is difficult to ascertain exactly how many BIs it represented, as the board declined to release the membership list. Counting member groups was, in fact, complex. BIs formed and dissolved rapidly, and some BBU members were themselves networks of BIs. Nevertheless, its leaders' claims to represent many hundreds of BIs by the mid 1970s were clearly exaggerated, and probably knowingly so. The organization had member groups throughout Germany, but membership was concentrated in the southwest.

From the beginning, the BBU was characterized by a loose and relatively informal operating style that mirrored the pattern of its constituent groups. In this respect, it represented an alternative to the national nature protection organizations and to political parties. Its small budget was financed largely by interior ministry subsidies for conferences and educational materials; however, these funds could not be used for basic administration, and member dues were very low and often went unpaid. Consequently the organization experienced chronic budget deficits.

The BBU's effectiveness was also limited by a number of other problems. Member groups turned over rapidly, changed their names and addresses, insisted on a high degree of autonomy, and frequently failed to participate actively—often neglecting even to send delegates to the annual meeting. Underfunding resulted in understaffing and an overworked volunteer board of directors, leading, in turn, to rapid rotation of board membership. In the early years, business was conducted from the board chair's home.

The BBU also developed, somewhat ironically, into an exemplar of Michels' iron law of oligarchy. Despite the organization's formal emphasis on grassroots democracy, member groups could influence policy only at the annual meeting, so agenda setting fell mainly to the board. Most board members lacked the time or geographic access to participate actively or chose to devote their efforts mainly to their own BI, but this was not true of the BBU's second president, H.-H. Wüstenhagen, a dynamic, hardworking leader who gave up his job to work full-time for the BBU. His commitment, knowledge, and range of contacts were so great that much influence devolved to him. Although his hard work was a major factor in keeping the organization afloat, the result was complaints about centralization of power and some board resignations. The criticism arose, in part, because Wüstenhagen and his allies defined their task not only as providing information and service to the member groups, but also as representing the BBU to the media and in national politics, and the BBU was, in fact, able to exert some political influence through public information campaigns, protests and lobbying.

Polarization and Confrontation

As the 1970s progressed, several factors combined to produce a powerful social movement centered on nuclear power and other environmental issues.[11] It engaged in a series of confrontations with the government and established interest groups.

By the mid 1970s, oil shortages and energy price shocks, along with attendant unemployment and economic recession, were threatening to undermine the "economic miracle" that had brought social stability and a rising standard of living to postwar Germany. Concerned about the costs of environmental cleanup, business and labor unions—supported by the economics and transportation ministries—mounted a vigorous attack on environmental regulations, which they claimed hindered investment, cost the economy jobs, and stood in the way of much-needed increases in energy production.[12] In 1975, the government, headed after Brandt's resignation by Helmut Schmidt, responded by calling a conference, at which it decided to scale back environmental efforts.

New environmental legislation did continue to be passed under Schmidt, including a law regulating detergents and a 1976 revision of the Nature Protection Law, but these measures represented the completion of legislative processes already underway rather than new initiatives, and some of their provisions were weakened before passage. The Nature Protection Law, for example, increased the emphasis on general land use planning, provided for a review of the soundness of public works projects that might damage nature, and replaced the problematic sys-

---

[11] In addition to sources cited in note 1, brief overviews of the period of intense confrontation that bridged the 1970s and 1980s are available in Schreuers (2002), Dryzek et al. (2003), and Brand (in press). For more complete overviews see Ebert (1980), Rucht (1980, 1990), Sternstein (1980), Schiller (1984), Müller (1986), Paterson (1989), Brand (1993, 1999a), Rucht and Roose (1999), and Dannenbaum (2005). Voss (1990) focuses on the role of media during the period.

[12] In addition to sources cited in notes 1 and 11, discussions of the contribution of government policy changes and refusal to compromise to escalating conflict with the environmental movement include Offe (1981), Meroth and von Moltke (1987), Weidner (1991a), Wilhelm (1994), Malunat (1994), Pehle (1997), Jänicke, Kunig, and Stitzel (1999), and Müller (2001).

tem of volunteer nature protection committees with professionalized government offices; however, efforts to include provisions to end destructive agricultural practices and give environmental organizations the right to appeal government decisions that might damage nature were defeated. The Schmidt government also extended deadlines for compliance with some environmental regulations and "simplified" approval processes. Environmental bureaus were denied staffing increases needed to cope with increased responsibilities, and enforcement of environmental legislation remained fragmented. New money was generally available for environmental cleanup only where construction of new facilities might create jobs.

Concerned about rising oil prices and working from unrealistic projections of energy needs, the government—in close collaboration with the energy industry and with support of opposition parties—also initiated a push to speed up Germany's hitherto modest program of nuclear power plant construction. It hoped that increased nuclear power generation would increase Germany's energy independence, provide cheap power to fuel continued growth, and support the development of a leading-edge technology.

The government's new direction undermined the consensus that had previously surrounded environmental policy and stirred strong opposition. Many environmentalists and anti–nuclear power activists, including the left wing of Schmidt's own party and its affiliated youth organization, viewed these moves as environmentally untenable and as a reversal of the positive trend toward openness to citizen input.[13]

The nuclear power issue, in particular, proved to have broad, powerful resonance. It attracted local groups of relatively conservative farmers and shopkeepers, who were alarmed about the effects of nuclear power plants on their health and economic well-being. It attracted BIs with other environmental concerns, which had also become concerned about nuclear power and government high-handedness. It appealed to ideological environmentalists and counterculture advocates, who saw nuclear power as symbolizing the problems of a technology-heavy, increasingly centralized and oppressive consumer society. And it attracted far-left groups, who saw in the issue an opportunity to break the power of industry and government. Although other environmental issues did not disappear, the broad appeal of the nuclear power issue transformed it into the key issue, not only for anti–nuclear power activists, but for the environmental movement as well.

The Schmidt government and its allies in business and labor were taken aback by the burgeoning environmental movement's use of administrative appeals, court cases, public information campaigns, and protests to slow Germany's march toward nuclear power, as well as its opposition to major projects in road construc-

[13]The genesis and course of the wave of confrontational protests that ensued have been the subject of much reseach and commentary. In addition to sources in notes 1 and 11, the following are among the more important contributions: Andritzky, Walter, and Wahl-Terlinden (1978); Fredefichs (1980); Sternstein (1981); Trautmann (1984); Welz (1984); Mayer-Tasch (1985); Offe (1985); Rauprich (1985); Leonhard (1986); Kazcor (1986, 1990); Kretschmer and Rucht (1991); Bramwell (1994); Fuchs and Rucht (1994); Kriesi and Giugni (1996); and Vandamme (2000).

tion and other areas. Postwar German governments had operated in a climate where economic growth had received top priority, established interest groups from business and labor enjoyed privileged access to policy making, and politically weak nature protection groups offered minimal opposition. Being forced to slow down, or even cancel, projects already planned or underway was a new and costly experience for both business and government.

The determined, strident opposition from environmentalists with their countercultural, anti-consumerist ideology, and from the far left, did more than just threaten existing power arrangements. It challenged underlying ideological assumptions and evoked memories of the paralysis wrought by extremist parties in the Weimar Republic. The Weimar legacy had led postwar Germany to adopt a relatively repressive orientation toward dissent from both extremes of the political spectrum—exemplified by the repeated forced dissolution of far-right parties and the 1972 ban on employment of members of far-left groups in government positions.

The government's response was consistent with this tradition. After a mild initial response to the anti–nuclear power/environmental movement—including a public relations campaign that featured hiring a research institute to find out more about the anti-nuclear groups and engage them in dialogue—failed, the government became steadily more heavy-handed. It offered local authorities financial incentives to cooperate, bent administrative rules to move projects forward, and excluded BIs and other anti–nuclear power groups from access to information and participation in decisions. The push for nuclear power was supported not only by the energy industry and other business interests, but also by conservative elements in the press and the majority of labor unions. Some unions staged counterdemonstrations against anti-nuclear protests, and the German Federation of Labor actively resisted an effort by some of its affiliates to join an anti–nuclear power alliance with the BBU.

Frustrated by the government's tactics and finding itself excluded from decision-making circles, the environmental movement undertook to mobilize public opinion against nuclear power through a series of massive protests and occupations of nuclear power plant construction sites. These continued from the mid 1970s through the early 1980s. Some included construction of temporary villages on the sites, which were occupied over a period of weeks or months. In these protests, local BIs opposed to nuclear facilities were frequently joined by out-of-town supporters from other BIs and students. At times they received support from other groups, including doctors' organizations, local labor unions, and church groups. Radical left groups had initially showed little interest in the environmental movement; however, once they realized the movement's mobilization potential, they became heavily involved, often seeking to infiltrate and gain control of the BIs or instigate violent confrontations. Nevertheless, despite some isolated successes, they were never able to control the movement.

The protests were typically coordinated by networks of BIs, including the BBU, or by ad hoc committees. Differences in the membership composition and goals of the participating groups led to recurrent debates over goals and strategy and to swings between peaceful protest and more confrontational strategies. At times there were even simultaneous demonstrations by different groups at the same site. In some instances, demonstrations and plant-sites occupations led to violence and repressive police action, as the government—alarmed by violence and countercultural challenges to dominant ideas and simultaneously struggling with murders and kidnappings of business and government leaders by left-wing terrorists—responded forcefully to protests it viewed as instigated by the far left. The repressive measures, in turn, radicalized some of the more moderate participants and produced charges that nuclear power was associated with police-state tactics.

Protests against nuclear energy dominated the headlines, but other environmental BIs continued their activities. They pursued a wide range of other issues, occasionally in alliance with more traditional nature protection groups.[14] The most comprehensive information about their activities comes from a 1977 mail survey of groups that belonged to the BBU and another umbrella organization (Andritzky, Walter, and Wahl-Terlinden, 1978; Rüdig, 1980); however, the survey suffered from a low response rate and a bias toward longer-lived organizations, so the results must be interpreted with caution. They indicated that environmental initiatives focused most frequently on energy issues, including opposition to nuclear plants, traffic problems, nature protection, city planning issues, and industrial emissions. Most of them claimed to have goals that extended beyond one specific controversy, and about a third had changed their area of interest, often broadening their goals as they matured. They pursued their goals chiefly through demonstrations—including not only participation in massive demonstrations over nuclear power, but many local ones as well—leafleting, press conferences, press releases, petitions, meetings with local administrators, preparation of reports, and participation in hearings. Groups working on topics other than nuclear power also sometimes became involved in confrontations, most notably around airport expansions and the building of a canal linking the Rhine and the Danube. Not all BIs were part of networks; some pursued their work without much contact with other groups.

Despite government opposition, the movement did enjoy some successes. The movement was able to block or stall some nuclear plants and nuclear waste disposal facilities—sometimes simply by buying enough time for the courts to act or politicians to reconsider. And it successfully enlisted experts to prepare research reports challenging the government's claims.

The dramatic confrontations between environmentalists and police received extensive press coverage and polarized the German public, and opinion surveys showed divided and ambivalent opinions about both nuclear power and the pro-

[14] For overview information about BIs with goals other than combating nuclear power see Andritzky (1978), Andritzky, Walter, and Wahl-Terlinden (1978), Rüdig (1980), Hoplitschek (1984), Kempf (1984), and Rucht (1984).

tests and growing concern about the safety of nuclear power plants. Support for the environmental/anti-nuclear movement was stronger among those who shared the concerns of the more moderate protesters, saw the demonstrations as justified, and were repelled by the government's response. Others, however, saw the confrontations as a symbol of the breakdown of order and the protesters as allies of left-wing terrorists (Noelle-Neumann, 1976, 1977, 1983; Fuchs and Rucht, 1994).

The confrontations over nuclear power and other environmental issues occurred within the context of a period of broader social ferment and polarization during the late 1970s and early 1980s. Indeed, the environmental movement was only one of a broader set of "new social movements," including the peace and feminist movements, that flourished during this period and strongly shaped the environmental movement's development.[15] Embedded in an interlocking network of environmental, anti-nuclear, peace, and feminist groups centered in urban neighborhoods such as Berlin's Kreuzberg and in university towns, and supported by a lively countercultural press, many environmentalists gravitated toward the so-called second culture. Its worldview was characterized by a strong critique of consumer capitalism and an economy dominated by giant industrial firms, concerns about the dangers of runaway technology and militarism, an apocalyptic view of environmental problems, and a penchant for self-realization, alternative lifestyles, and craft or communal production systems. Also prominent were commitment to grassroots democracy, skepticism of science and modernism, a jaundiced view of the political system and political parties—which were seen as unresponsive and fixated on meeting the needs of well-established interest groups—and an unwillingness to compromise.

Participants in the environmental movement varied in the extent to which they embraced these countercultural views and participated in other new social movements, but the alternative milieu provided a natural home for activists who had been radicalized by their experiences in the movement and probably contributed to the exit of some more conservative supporters of the environmental movement. The anti-establishment views that came to characterize many members of this new generation of environmentalists made the movement even more of a threat to the German establishment by challenging its ideological underpinnings and linking it to other movements. This countercultural, anti-establishment view influenced the views of a whole generation of environmental activists, some of whom remain active in today's environmental organizations.

The German Nature Protection Ring

Germany's traditional nature protection organizations, with their emphasis on nature protection and relatively conservative memberships, did not provide an

[15] In addition to those cited in notes 1 and 11, the following sources treat the involvement of the environmental/anti-nuclear movement, the counterculture, and other new social movements of the time: Offe (1985); Ellwein (1984); Renn (1985); Brand (1987, 1995, 1999b); Rucht (1991b); Bramwell (1994); Kriesi and Giugni (1996); Christmann (1996); and Hoffman (2002).

inviting home for environmental activists interested in new environmental issues, oriented to confrontational protest, or attuned to countercultural lifestyles.[16] This was, if anything, even more true of their umbrella organization, the DNR. Its inability to assume a leadership role in the environmental movement was an important contextual factor for both the new environmental organizations founded during this period and older nature protection organizations, which were struggling to adapt to the era of confrontation.

The growth of an environmental movement centered around independent BIs and the BBU represented a major challenge for the DNR because it had long thought of itself as *the* voice for nature in German society and the chief negotiating partner for the government. As the environmental movement began to mobilize, the DNR thus responded with efforts to broaden its mission. It added the subtitle "Federal Organization for Environmental Protection" to its name in 1974, and its public education efforts began to link nature protection to environmentalism, pointing out that pollution threatened not only plants and animals, but also the ecosystems that provide the basis of human health and well-being. The DNR responded to government initiatives by participating in working groups and submitting position papers on new legislation concerning air pollution and other environmental problems, and in 1972 it opened a "hotline" citizens could use to report environmental problems. It even made an overture to the BBU, inviting it to join the DNR.

Still, the DNR remained hesitant to stake out strong environmental positions, and its political role was more reactive than proactive. Even as the confrontations of the 1970s and early 1980s swirled around it, it was focusing on such activities as coordinating German participation in the 1970 European Nature Protection Year, pushing for revision of the nature protection law, organizing an environmentally conscious auto club, and carrying out inoffensive public education efforts.

There were numerous reasons for the DNR's timidity. As an umbrella organization without a mass circulation magazine, it could communicate with individual environmentalists only via its member organizations and the press, so it was poorly positioned to mobilize mass support. Second, because its dues were low and sometimes went uncollected, it was chronically underfunded. It did not employ its first full-time executive secretary until 1980. Third, it had a long history of close work with government bureaus, and government officials frequently attended its meetings. Moreover, it was financially dependent on government for financial subsidies and for the kind of legitimacy it received from being appointed as Germany's representative to European nature protection bodies. Fourth, its longtime leader, though not closed to innovation, had close ties to conservative politicians and a very cautious approach.

---

[16] Important discussion of the DNR during this period include, in addition to sources in note 1, Wey (1982), Leonhard (1986), Müller (1986), Schenkluhn (1990), Rucht (1993a), Deutscher Naturschutzring (2000, 2002), Bergstedt (2002), and Röscheisen (2006).

Internal divisions also limited the DNR's ability to function as an actor in the environmental movement. Some of its leaders and member organizations worried that devoting too much attention to new environmental issues would divert energy from its traditional focus on nature protection. Moreover, its strategy of seeking a very broad membership base had made it an uneasy amalgam of nature protection organizations—including many small local or narrowly focused organizations, scientific societies, organizations only secondarily concerned with nature protection, and nature-using organizations, including hunters, anglers, hikers, and the Alpine Association. The latter category was especially problematic, because—until bylaw changes in the early 1980—their large memberships and dues payments gave them a substantial voting majority and let them obtain seats on key DNR committees. The influence of these organizations, with their recreational goals and politically conservative memberships, kept the DNR from taking strong stands on the environmental effects of industrialized agriculture, hunting, recreational use of nature, and nuclear power.

The period of polarization exacerbated these long-simmering conflicts, as new environmental organizations joined the DNR and some of its longtime member nature protection groups adopted a more activist stance. The result was increasing internal conflict with the dominant block, the nature-using organizations. These included not only arguments over hunting and recreational use of nature, but also conflicts over nuclear power and over whether environmental organizations should be allowed to sue the federal government for failure to follow nature protection regulations. Even the appointment of a full-time, paid executive secretary was strongly opposed by the more conservative groups. The conflicts led several groups more oriented to confrontational environmentalism to resign or threaten resignation. In 1976, these groups used threats of resignation to force replacement of a board member who was particularly strongly opposed to the DNR involving itself with environmental issues or challenging the recreational use of nature. At times, the DNR responded to these conflicts by attempting to steer a middle course. For example, it expressed concern about nuclear power and pointed to the availability of better alternatives, but avoided unqualified opposition. Such equivocation seldom satisfied either side.

The balance of power shifted after representation rules were changed in the early 1980s, giving the smaller nature protection organizations and new environmental organizations the majority of the votes. The result was complaints from the nature-using groups that their views were being brushed aside and demands that their dissenting views be included in DNR position statements they opposed. These disagreements came to a head in a series of bitter conflicts in 1986. The BUND, the most influential of the more activist environmental organizations, left the DNR only to return in 1988. More consequential in the long run was the DNR's vote to eject the hunters' association, which set off a wave of resignations by other nature-using organizations, including other hunting and fishing groups. These resignations were partly offset by the accession of new organizations; however, many of the organizations that resigned were among the largest.

The changes in membership composition further shifted the balance of power toward the nature and environmental protection groups, and the DNR began to move toward a broader environmental agenda and a more confrontational approach. These were reflected in a new 1987 goals statement. Nevertheless, groups more oriented to the enjoyment and use of nature than its protection, including canoeists, hikers, and divers, remained, and they were sometimes able to block DNR initiatives that threatened the recreational interests of their members. This occurred, for example, when the Alpine Association used the threat of resignation to force a weakening of DNR proposals to protect the Alps from recreational use.

The BBU as an Environmental Movement Organization

Partly as a result of the unsuitability of the traditional nature protection organizations and the DNR for leadership roles in the environmental movement, the BBU emerged during the mid 1970s as the movement's most important social movement organization and became very visible in the media.[17] The extent to which the BBU embraced the agenda of the environmental and other new social movements is illustrated by its 1976 goals statement and a 1977 orientation document. In these documents, the BBU strongly criticized Germany's commitment to economic expansion and "progress" and the accompanying waste of resources and pollution. It advocated an end to nuclear power, a major reorientation of transportation policy to mass transit, organic farming, simplified lifestyles, reductions of the scale and concentration of industrial production, reducing centralization of political power to reduce citizen alienation, and grassroots democracy. As a highly visible organization with a relatively radical agenda and a grassroots model of social movement mobilization, the BBU was simultaneously threat and exemplar for the nature protection organizations as they struggled to adapt to changed conditions.

Yet despite its key role in the movement, the BBU continued to be plagued by difficulties resulting from its own structure and conflicts within the movement itself. Some of these conflicts involved key strategy issues. The BBU's top leaders were generally willing to negotiate with the authorities and opposed to violence, and they sought to exclude BIs they believed to be communist front groups. Yet a sizeable minority of BBU activists favored working outside the system and were more willing to cooperate with communist-dominated groups. Related arguments broke out over whether the BBU should focus on influencing national politics or supporting local BIs, whether it should accept public funds, and whether it should cooperate with groups that were willing to use violence. A particularly virulent controversy erupted over moving the organization's headquarters to Bonn so that it could be more active in lobbying.

---

[17] Many of the sources cited in note 10 also cover the peak period of confrontation during the late 1970s and 1980s. For this period see especially Rieder (1980), Kazcor (1986), and Markham (2005b), as well as Kempf (1984), Leonhard (1986), Müller (1986), Kazcor (1990, 1992), Rucht (1993a), Wilhelm (1994), and the following BBU documents and articles by BBU leaders: Bundesverband Bürgerinitiativen Umweltschutz (1980) and Sternstein (1980, 1981).

Other internal disagreements more closely approximated garden-variety power struggles. One was the conflict between the BBU's board and its working group on transportation, which was coordinated and dominated by a very active citizens' initiative in Berlin and included many groups. Conflicts over the working group's demands for more autonomy and the right to approve press releases concerning transportation were factors in its decision to leave the BBU. Another struggle broke out between the board and the editorial staff of the BBU's magazine. The magazine frequently criticized the BBU from the left, and the majority of its staff wanted to be an independent voice that addressed the general public rather than a BBU house organ. The two parties reached a compromise after long negotiations by granting the magazine quasi-autonomy.

With so many controversies and the emphasis on grassroots democracy, annual BBU meetings sometimes became chaotic, with heated debates and various groups presenting proposals for discussion. The chaos was exacerbated by chronic underfunding and understaffing. Due to lack of funds, a formal business office was not established until the end of the 1970s, and even then it had only a handful of paid employees overseen by the treasurer, who received only a small salary.

The organization's very visible and dominating president, Wüstenhagen, who had worked for more involvement in politics and against cooperation with the far left, resigned precipitously in 1977. He cited the board's refusal to provide additional office assistance, but many believed that conflicts over the organization's direction were the underlying cause. After his resignation, there was strong pressure to avoid a recurrence of concentration of power in the hands of a single leader. A new set of bylaws was passed, which reemphasized grassroots democracy and the autonomy of the BIs, replaced the board chair position with a three-member committee, reserved more decisions for the entire board, and established a non-hierarchically organized business office. These changes were followed by a move to require candidates for the board to have the endorsement of their home citizens' initiative.

In the end, however, the effort to prevent centralization of power was not successful. The pressure to make daily decisions, issue press releases, and respond to requests for interviews made it impracticable for the entire board to exercise day-to-day control, and some board members were heavily committed to work within their own BIs. The three-member executive committee thus acquired great influence, and one of its members, Jo Leinen, who served from 1978 until 1984, emerged as the informal leader and the BBU's principal spokesperson. This development was reinforced by pressure from the media, which preferred a single spokesperson.

The Rise of the Greens

Another important contextual factor for the nature protection organizations in the late 1970s and early 1980s was the rise of the Green Party, which provided citizens who wanted to influence environmental policy with an alternative to both the traditional nature protection organizations and protest outside the system.[18] The party grew out of "green" or "rainbow" slates of candidates for local or regional offices, which appeared in various areas at the end of the 1970s. The slates ran on platforms that, depending on the locale, combined environmental issues and opposition to nuclear power with other new social movement issues, including feminism, grassroots democracy, human rights, and peace. The early organizers and candidates of the "green lists" were ideologically diverse. They included persons from the counterculture, radical leftists, more moderate environmentalists, and a contingent of politically conservative ecologists. The latter included the well-known author and Christian Democrat Herbert Gruhl and members of the conservative Action Group of Independent Germans. The various slates sometimes competed with one another, but in other instances they formed loose coalitions—sometimes only to then splinter again later.

By the end of the decade, green and rainbow slates were attracting enough voters to win seats in some *Länder* parliaments and have an impact on local politics, and in 1980 they coalesced into the national Green Party. Some members of the new party, especially the conservatives, argued that ecology transcended old distinctions of left and right and should be the party's only concern, but they did not prevail. The Greens adopted a left-leaning platform that emphasized environmentalism and other new social movement issues and values, but endorsed neither the traditional concerns of the union-oriented Social Democrats nor the hard-left agenda of uncompromising conflict with the state. Its program included both environmental protection and nature protection, but it placed more emphasis on the former.

Characterizing themselves as the "anti-party party," the Greens initially adopted a series of rules designed to ensure grassroots democracy in party affairs, including limiting the number of offices a single individual could hold, insisting on equal representation of women in their delegation, and setting term limits on key party positions. Yet despite symbolic gestures, such as eschewing conventional dress for legislators and carting plants into the Bundestag chamber, they participated in the Parliament in fairly conventional ways.

Some of the Greens' early supporters were drawn from the BIs and the BBU, and the Greens provided a good mechanism for the movement to introduce its concerns into the normal channels of political debate. Yet because of its tradition of nonpartisanship, its concerns about losing its leadership role in the movement, worries that becoming too closely identified with the Greens might drive away supporters from other parties, skepticism about the Greens' potential effectiveness, and the desire to avoid being coopted into the established system, the BBU held the Green Party at arm's length. The traditional nature protection organiza-

---

[18]The origins and early development of the Green Party have attracted an enormous amount of attention from researchers. In addition to sources from note 1, brief overviews are available in Hermand (1991), Bergstedt (2002), Schreuers (2002), and Dryzek et al. (2003). The following provide more detailed information: Trautmann (1984); Guggenberger (1984); Mayer-Tasch (1985); Mez (1987); Zeuner (1991); Müller-Rommel (1993); Frankland (1995); and Hoffman (2002).

tions, and even some environmental activists—a good many of whom already had connections to existing parties—adopted the same stance.

Many of the more conservative members of the Greens, dissatisfied with the party's direction, withdrew in 1981 to organize a new party, the Ecological Democratic Party (ÖDP), under the leadership of Herbert Gruhl;[19] however, the new party never enlisted many members or attracted enough voters to win seats in a *Land* parliament or the Bundestag. It did, however, become an important part of a network of right-oriented groups that had taken up the ecology theme. These included the World League for the Protection of Life and intellectuals of the so-called "New Right." The latter group sought to develop an environmentalism of the right without explicit connections to National Socialism. They linked environmentalism to combating overpopulation, limiting immigration, adapting social structure to the so-called laws of nature, and preserving the distinct identities of German culture and those of immigrant groups in Germany.

Nature Protection Organizations and the Environmental Movement

The rise of the environmental movement posed a serious dilemma for Germany's traditional nature protection organizations, which had entered the 1970s as the only well- established, nationwide citizens' organizations concerned with environmental themes. Like their counterparts in the US and elsewhere in Europe (Dalton, 1994; Dowie, 1995; Kline, 2000), they stood to benefit greatly from the movement if they could find a way to adapt to it and harness its energy. Yet, their membership composition, history of close work with the authorities, historical lack of concern with the new environmental issues, and worldview made them ill-suited for participation in the movement. Moreover, the movement, with its new themes and organizational forms, challenged their role as *the* representatives of environmental issues in the society and created a confrontational atmosphere that made it hard for them to pursue their traditional strategies. The movement also radicalized some of their members and created new and more confrontational environmental organizations, which competed with them for press attention and citizen support.

The organizations reacted in various ways. Some, such as the Nature Protection Parks Association, the German Forest Protection Association, and the Isar Valley Association in Bavaria, elected not to make major changes in their agendas or strategies.[20] They continued to pursue their historical goals using famil-

[19] Discussions that focus especially on ideological conflict in the early years of the Green Party and and the resulting departure of most of the right wing include Jahn and Wehling (1990), Wüst (1993), Statham (1997), and Geden (1999).

[20] Several sources cited in note 1, particularly Dominick (1992), Oswald von Nell Breuning Institut (1996), and Bammerlin (1998), discuss the varying reactions of the nature protection organizations to the environmental movement and the decision of some to remain with their traditional goals and strategies. See also Schenkluhn (1990) and Cornelsen (1991).

iar strategies, often with stagnant or declining membership. Other organizations responded to changed conditions—including pressures from some of their own members—with changes in their missions and strategies. For some, the changes bordered on the cosmetic, but others began a long and painful transition toward a more activist stance and broader set of goals.

The League for Homeland Protection

The BH responded to the new environmental movement by rather hesitantly increasing the emphasis on protection of the natural environment in its programs and publications, portraying nature and environmental protection as an important element in caring for the homeland.[21] Nevertheless, its historical commitments and conservative membership base, the significant representation of current and retired government officials in its leadership, and its dependence on government subsidies precluded active participation in the environmental movement or the adoption of regime-critical stands. And despite verbal affirmations of the importance of environmental protection, it continued to give more attention to its other goals. It received little recognition of the steps it did take, in part because other organizations had claimed center stage in the environmental movement, and it has never since figured as a central actor on the environmental scene.

The Friends of Nature

The Friends of Nature's efforts to adapt to changing times also failed to gain it much recognition as a leading environmental organization.[22] This is somewhat ironic because some of its stands were very much in line with the movement's objectives. For example, it strongly opposed nuclear energy, endorsed energy conservation and regenerative energy sources, was one of the first organizations to speak out against global warming, and mounted a major campaign to protect the North Sea coastal regions. Yet like the BH, its efforts on the environmental front were offset by an unclear identity and the subordination of environmental protection to other goals, so it failed to achieve a clear public identity or a leadership role as an environmental organization.

Worldwide Fund for Nature

WWF-Germany passed through the period of confrontation without major changes in its mission or strategies, but it benefited from the movement nonetheless.[23] Until the end of the 1970s it continued to function mainly as a fundrais-

---

[21] See note 1 and Leonhard (1986), Fischer (1994), Wüst (1993), and Rollins (1997).

[22] See Schmitz (1984), Zimmer (1984), and Hermand (1991).

[23] Sources that cover the history of the WWF during this period include those listed in note 1, as well as the following: Haag (1986); Rucht (1989); Schenkluhn (1990); Cornelsen (1991); Kazcor (1989, 1992); van der Heijden, Koopmans, and Giugini (1992); Katz, Orrick, and Honig (1993); Blühdorn (1995); Rucht and Roose (2001c), and Bergstedt (2002). The WWF itself has also pub-

ing organization for WWF-International, although it also sponsored projects—almost entirely those of other organizations—and purchased small nature protection areas in Germany. As late as 1978, it still had only 4,500 donors.

The organization was reinvigorated in the early 1980s under the leadership of a former Volkswagen executive. He professionalized its operations, added additional full-time staff, and increased funding for WWF-initiated and conducted projects in Germany, focusing especially on protection of wetlands, tidal flats, and rivers. The WWF also initiated some modest efforts to influence policy through public education working for protection of North Sea tidal mud flats, and, somewhat later, climate change. Although it played only a marginal role in the environmental movement, the WWF did have a mission clearly related to environmental concerns, considerable public visibility, and a good reputation with the general public. It was therefore in a good position to profit from growing public interest in environmental issues through rapid growth in membership and donations. It was particularly attractive to environmentally oriented citizens who were especially interested in nature protection and uncomfortable with confrontation. The result was very rapid growth in the donor base, which reached 100,000 in 1988.

WWF-International also underwent changes in mission during this period. After concluding that its original strategy of promoting and funding wildlife reserves was not adequate, it began to move toward protecting entire ecosystems and providing economically viable options for local citizens who might otherwise destroy wildlife. It also began efforts to combat global threats to ecosystems and wildlife, such as lobbying on issues like damage to the ozone layer. At the same time, however, it continued its highly visible work for the protection of large impressive species, such as panda bears and tigers, which sustained public visibility and support.

None of these changes altered WWF-Germany's reputation as a conservative, nonconfrontational organization with close ties to business. Throughout the era of confrontation, the WWF had many business leaders on its board, remained willing to cooperate with and accept donations from business, and avoided confrontational tactics. Indeed, its business ties frequently earned it criticism from more confrontational elements in the movement. They cited a brewery executive who served briefly as its president even though he had opposed Germany's returnable bottles system, an executive secretary who had worked in the nuclear power industry, and several big game hunters on the WWF board. The organization's failure to oppose nuclear power also isolated it and led to criticism. Finally, activists oriented to grassroots democracy noted that, as a foundation, the WWF has no true members, and its donors have no influence over policy. Such criticism also played a role in keeping it at the margins of the environmental movement.

---

lished several brief histories that include this period: WWF-Journal (1999d, 2001, 2002e); von Treuenfels (2003), and Wünschmann et al. (2003). A good history of WWF-International during this era is Wapner (1996).

The League for Bird Protection

The WWF, with its primarily international focus, and the BH and Friends of Nature, for which nature protection was only one assignment among many, could afford to respond to the environmental movement with only minor changes to their programs.[24] The BfV, by contrast, faced the danger of being marginalized by the environmental movement and new social movement organizations, such as the BBU and Greenpeace. It thus responded to the environmental movement with a more dramatic, albeit somewhat belated, transformation of goals and strategies.

The dilemma posed by the environmental movement was a painful one for the BfV. It had always portrayed itself as having broader interests than simple bird protection, and it had crusaded against agricultural chemicals in the 1960s. Nevertheless, it was filled with conservative bird watchers who felt little kinship with counterculturally tinged environmentalism. For most of its history, the BfV's leadership had had many overlaps with government and business, and it had generally pursued a nonconfrontational strategy. Consequently, change came haltingly and only after considerable pulling and hauling.

Nevertheless, change did occur. In a series of incrementally broader mission statements adopted during the second half of the 1970s and early 1980s, the BfV expanded its formal mission to include protection of all types of species, protection of the landscape, and general environmental protection. By accommodating itself step by step to changing times it was able to achieve significant growth, yet its rather nonconfrontational approach meant that it remained at some distance from the environmental movement throughout the 1970s.

New leadership was required to effect a more complete reorientation, and the change was accompanied by major internal conflict. The transformation began in the early 1980s, when Jochen Flasbarth and a circle of leaders of the organization's newly founded youth group began pushing the BfV to refocus its agenda on new environmental issues and adopt a more politically activist stance—including opposition to rearmament and to nuclear power.

Steps like breaking relations with the Hunters' Association over the hunting of endangered bird species had already produced controversy within the BfV in the early 1970s, but the new conflicts were even more heated. Elections of board members and officers and new goal statements became hotly contested issues. One president was forced out of office, and the split became so deep that Flasbarth was threatened with expulsion and the youth group with dissolution. Relocation of the organization's headquarters to the capital was strongly criticized as unduly politicizing, and the BfV suffered 10,000 resignations when the board finally voted in 1986 to oppose nuclear power.

---

[24] Two histories produced by the BfV itself, Hanemann and Simon (1987) and May (1999), provide the most detailed information about the organization during this period. In addition to sources cited in note 1, the following also describe the BfV during this period: Rauprich (1985); Schenkluhn (1990); Cornelsen (1991); Kazcor (1992); Engelhardt (2000); Rucht and Roose (2001c); and Bergstedt (2002).

A number of leaders of the organization I interviewed, as well as other persons familiar with it, spoke with feeling about these confrontations and the importance of the resulting reorientation for the organization. Despite the conflicts, the BfV attracted far more new members than it lost. In the end, the infusion of new members with perceptions and goals shaped by the environmental movement and the departure of older members by resignation or aging out allowed a substantial redirection of the BfV's goals and strategies.

The League for Environment and Nature Protection in Germany

The changes that occurred at the BN were even more dramatic than the ones at the BfV, for they involved not only a change of mission and strategy, but also an expansion from the *Land* to the national level.[25] As in the case of the BfV, the reorientation was associated with new leadership; however, the BN's transformation occurred considerably earlier.

Under the leadership of a conservative board of directors and a chairman drawn from the Bavarian bureaucracy, the BN continued to pursue its historical goals and strategies throughout the 1960s. It focused on protecting small areas with unusual beauty or endangered species, worked closely with the authorities, and avoided confrontation. By the end of the 1960s, some members were beginning to question this approach, and their concerns came to a head when the chair of the organization's board accepted the directorship of the Bavarian Office for Nature Protection. An oppositional group calling for a more modern, activist approach gained support within the organization, and in 1969 the old board chair resigned under pressure. He was replaced by a new, youthful leader, Hubert Weinzierl, the first chair without a background as a state official. An independently wealthy environmentalist and essayist who had trained as a forester, Weinzierl is today president of the DNR and one of Germany's best-known environmentalists.

During the early 1970s, Weinzierl and his supporters undertook a successful effort to refocus the BN on combating the environmental problems of an urban industrial society, including air and water pollution, loss of green space and forests, and waste disposal. Adopting ideas from the environmental movement, the BN began to question unbounded consumerism, resource consumption, and population growth and to argue for an economy that respected nature's limits. In 1971 it symbolized its reorientation by changing the title and content of its member magazine to give nature and environmental protection equal play.

---

[25] By far the most complete accounts of the BN and its evolution into a national organization are in two doctoral dissertations, Hoplitschek (1984) and Wolf (1996). Numerous other sources also deal with the topic, including most of those listed in note 1 and Rauprich (1985), Leonhard (1986), Weiger (1987b), von Alemann (1989), Cornelsen (1991), Kazcor (1989, 1992), Rucht (1993a, 1993b), Blühdorn (1995), and Rucht and Roose (1999, 2001c). Wolf (1996) and Maxim and Degenhardt (2003) focus on events at the *Land* level, while Gruhl (1987), Geden (1999), and Oberkrome (2003a) emphasize ideological cleavages between left- and right-wing factions. There are also two short histories produced by the organization itself: BUND (1995) and BUNDMagazin (2000b, 2005d, 2005e).

Weinzierl and his associates also adopted a more politically activist strategy. The BN began to press for a new nature protection law, comprehensive land planning, and the right to file lawsuits against the government. It also undertook to quiz political candidates about their environmental views and publish the results. More importantly, during the decade of the 1970s, it moved by stages from mild support of nuclear power to strong opposition. By the early 1980s it was participating in anti-nuclear power protests in Bavaria, although it continued to avoid the most confrontational ones.

The transformation of the BN's goals and strategies thrust it into an increasingly adversarial relationship with the government agencies to which it had long been closely tied, as the Bavarian government remained strongly committed to nuclear power. The escalating conflict was symbolized by the BN's decision to cease holding its annual meeting in the office of the Bavarian interior ministry and by the Bavarian government's decision to no longer allow its employees to participate in the BN in their official capacities. Associated with these changes was a period of organizational renewal, including a new constitution and mission statement, the establishment of new local chapters and regional offices, redesign of the magazine, employment of additional staff, and founding of working groups. Most of the new working groups focused on various environmental issues, but one group was designated to cultivate ties with the media, issue press releases, hold press conferences, and monitor the organization's presence in the press. The significant dues increases required to institute these changes precipitated some resistance, but the leadership was able to overcome it.

By altering its course, the BN was able to assume the role of a social movement organization for the environmental movement in Bavaria and capture some of the energy and members that had been flowing to the BIs. Between 1969 and 1987, its membership increased from 19,000 to 77,000 (Weiger, 1987b).

The recruitment of many new members with ties to the environmental movement and the BN's occasional cooperation with BIs and participation in anti–nuclear power protests in Bavaria were not, however, always viewed with equanimity by old-line members. There were numerous resignations over the nuclear power issue, and in the early 1970s the BN's more conservative members were able to secure passage of a controversial revision of bylaws, allocating voting rights at the annual member meeting to delegates from local groups rather than to all individual members in attendance. This measure was meant to block any effort by radicals to organize and take over one of the often poorly attended annual meetings.

The Founding of the League for Environment and Nature Protection in Germany

In 1975, the ongoing changes in the BN led it to join its smaller counterparts in Baden-Württemberg, Lower Saxony, the Saarland, and elsewhere, as well as a number of well-known individual environmentalists who had joined in the so-called Ecology Group, to establish a new national organization, the German League for Environment and Nature Protection (Bund Umwelt und Naturschutz Deutschland: BUND). The founders envisioned a more activist, politically effective organization with broader appeal than existing nature protection organizations such as the BfV or the DNR. They wanted it to address a broad range of environmental concerns, including the newer ones emphasized by the BIs; however, they were reluctant to work through the BBU, which they saw as too radical and too chaotic to be effective.

BUND's early leadership included a wide spectrum of political opinion. Many of the organization's early leaders had been strongly influenced by the Club of Rome report, with its emphasis on the population explosion and resource exhaustion. Indeed, for some, concerns about overpopulation in less developed countries—sometimes tinged with conservative ideologies—were a central theme. BUND's second president was Herbert Gruhl, the author of *The Plundering of a Planet* and a former Bundestag representative from the conservative Christian Democratic Union, who later became a founding member of the Greens. He had hoped that the organization might evolve into an ecological party, but resistance to this plan and to his conservative leanings led to his defeat after a single term. He then resigned, taking many of BUND's more conservative members with him.

Early BUND campaigns focused on issues such as rail line closures and solar power, but at the end of the 1970s and in the early 1980s it became deeply involved in anti–nuclear power demonstrations and protests against aircraft runway construction projects, and launched a series of strong verbal attacks on the government. While BUND never became as confrontational as the BBU, its more conservative members sometimes opposed participation in the protests and the collaborations BUND sometimes forged with radical BIs, and some additional members departed. In North Rhine-Westphalia, there was even resistance to emphasizing environmental themes and including the word "environment" in the organization's name, as some members feared being associated with the far left. Nevertheless, BUND continued to maintain a nonpartisan stance, and it retained a good many politically moderate supporters and a few connections to the right.

Despite its commitment to the agenda of the environmental movement, BUND continued to pursue traditional nature protection projects, buying nature reserves and participating in local nature protection efforts, and its campaign against forest death combined both nature protection and environmental concerns. The larger, well-established chapters in southern Germany, in particular, placed great emphasis on traditional nature protection and preserving their autonomy, insisting that individuals be able to join the *Land* chapters without direct membership in the national organization.

BUND also continued to emphasize grounding its positions in expert opinion and research. It drew on the skills of its well-educated membership to develop independent expertise on environmental issues and publish well-respected technical reports and brochures, and it was also regularly consulted by government agencies concerning draft legislation and regulations.

The internal conflicts made for a somewhat bumpy beginning, and BUND's growth was at first slow, especially in northern Germany. In the end, however, the organization was able to consolidate its position. By 1981, it was represented in every *Land*, and it grew very rapidly in the late 1970s and 1980s, attracting a constituency comprised mainly of the well-educated professional middle class. By working for both nature and environmental protection, BUND provided a bridge between traditional nature protection concerns and newer environmental issues, attracting members with both orientations.

Although BUND became a member of the BBU and sometimes joined it in cooperative efforts, there was also considerable competition, including efforts by BUND to convince some of the BBU's less confrontational member groups to switch their affiliation to it. These sometimes met with success, especially after BUND's adoption of a formal position against nuclear power. Especially during its early years, BUND was also able to poach some of the BfV's more activist members. In North Rhine-Westphalia the competition escalated to the point that the BfV strongly criticized BUND for usurping its role and stealing it supporters.

Greenpeace Germany

The environmental movement also gave birth to Greenpeace Germany.[26] Greenpeace originated in Canadian-based protests against US nuclear testing, but branches formed almost immediately in the US and several European countries, and Greenpeace International was organized in 1979 to coordinate their efforts. The international organization soon broadened its agenda to environmental issues, including whaling and industrial emissions. The German branch was established in 1980 by volunteer activists from BIs in several cities with assistance from Greenpeace International and Greenpeace Holland. Its founders were concerned about reducing chemical industry pollution and saving marine mammals and were impressed by Greenpeace's approach to these issues.

Like its international partner, Greenpeace Germany's trademark was staging spectacular actions to point up environmental abuses. These included dramatic protests against dumping acid into the North Sea, carried out by protesters using the rubber rafts that had become a Greenpeace trademark, and climbing the smokestacks of polluting industries to hang banners calling attention to the problem. Almost from the beginning, these actions were professionally planned and executed events designed to garner media attention and raise public consciousness.

Greenpeace's approach fit well with the confrontational tenor of the times, attracted much media attention, and resonated with the German public. As a result, the organization achieved almost exponential growth during the 1980s. By the early 1990s it had reached over 700,000 supporters and 140 employees and

---

[26]Greenpeace's high visibility and success created a wave of publications about it. This section relies on sources in note 1 and the following: Rauprich (1985); Reiss (1988); Kunz (1989); Der Spiegel (1991a); Kazcor (1992); van der Heijden, Koopmans, and Giugini (1992); Rucht (1993a, 1995a, 1995b); Blühdorn (1995); Reichert and Schmied (1995); Krüger (1996); Streich (1986); Rucht and Roose (1999, 2001c); Bergstedt (2002); and Felbinger (2003). Sources that emphasize Greenpeace International include Eyerman and Jamison (1989), Rucht (1995a, 1995b), and Wapner (1996). Histories by members of Greenpeace include Flechner (1999), Greenpeace Magazin (1996c), and Eitner (1996).

had a budget of almost €30,000,000, making it by far the largest environmental organization in Germany (Der Spiegel, 1991a; Rucht, 1993a). It was also by far the environmental organization most trusted by the public (Voss, 1990).

In one respect, however, Greenpeace was out of line with the predominant direction of the environmental movement. The BIs and BBU were purposefully set up as loosely organized, decentralized networks to maximize member input, and BUND and the BfV, although more tightly organized, were nevertheless formally democratic. Greenpeace, on the other hand, was from the beginning a highly centralized organization. Its campaigns were tightly coordinated with international efforts, and local groups of Greenpeace activists were closely controlled from the Hamburg headquarters. The German branch had the legal form of a governmentally chartered association, but there were always fewer than forty voting members (Greenpeace Deutschland, 2000), including large contingents from the paid staff and Greenpeace International. Hence the vast majority of Greenpeace supporters had no say in its governance.

Not surprisingly, it was not long before criticisms of lack of democracy and closed decision-making were raised by Greenpeace activists recruited from the environmental movement. Almost immediately after Greenpeace's founding, there was a schism within the staff over this issue, and the international organization had to step in to reassert centralized control. Two years later there was another schism, when activists dedicated to grassroots democracy and pursuit of their interest in countering damage to Germany's forests joined other activists with a background in the BIs to start their own purposefully decentralized organization, Robin Wood (an English pun never translated into German). It quickly attracted a core of committed supporters and launched its own program of hanging banners on the smokestacks of polluting industries and other spectacular demonstrations (Rauprich, 1985; Scholz, 1989; Bergstedt, 2002; Rucht and Roose, 2001c).

Partly because of these differences in operating style and its emphasis on maintaining a positive public image, Greenpeace often seemed to hold the environmental movement that spawned it at arm's length. For example, it was so concerned that nuclear protests in the early 1980s might escalate into violence and undermine its public image that it refused to take part in them. Moreover, its emphasis on quick action tended to frustrate efforts to collaborate with organizations like the BfV and BUND, whose democratic decision-making process operated more slowly. As a result, although Greenpeace benefited greatly from the environmental movement and contributed to it by raising public consciousness, it was never networked tightly enough with other elements of the movement to play the role of a central social movement organization.

Summary and Conclusions

In the decades after the war, the major environmental organizations in Germany, all of which remained focused on nature protection, had settled again into a fairly stable configuration. The appeal of ideologies of the conservative right and socialist left had diminished, both for Germans in general and for nature protection advocates in particular, and the Friends of Nature and BH no longer figured as major nature protection organizations. The BfV, BN, and some of the smaller nature protection organizations had reoccupied their historical roles, joined by the WWF, which had found a distinctive niche of its own. All functioned as organizations of civil society, educating the public, raising funds for nature protection, and working to set aside and preserve nature protection areas. Most also attempted to influence politics and policy, but they lacked the resources to be more than weak interest groups. The DNR, for its part, attempted to function as the single voice of nature protection in Germany's quasi-corporatist politics, but lack of resources and internal divisions also made it a weak organization.

During the early 1970s a confluence of events called forth a powerful new environmental movement in Germany. The movement did not originate through the efforts of Germany's existing nature protection organizations. It was, instead, a product of worsening environmental conditions, books, articles, and international events that awakened public concern about the environment, legislative initiatives that called attention to environmental problems, and a countercultural movement based in student protest and the extra-parliamentary opposition that contributed to the movement's ideological underpinnings and strategy.

The main vehicle for the movement, especially in its early stages, was a new wave of grassroots BIs and their loosely organized regional and national networks—including, most prominently, the BBU. Although the BIs and their networks were underfunded and sometimes chaotic, their emphasis on grassroots democracy and new "environmental" problems, such as air and water pollution, traffic, and massive construction projects, made them well suited to take the lead in the new movement. Local governments were often skeptical of their activities, especially when they challenged existing arrangements, but the new Brandt government that assumed office in 1969 provided a more supportive political opportunity structure at the national level.

The new and rapidly growing environmental movement experienced a decisive course change in the mid 1970s, when the government responded to oil price shocks, economic downturn, and growing resistance to environmental measures from business and unions by slowing Germany's environmental efforts and expanding its nuclear power program. These steps forged a powerful opposition movement that united environmental concerns and opposition to nuclear power. It brought together local BIs worried about nuclear facilities and other massive construction projects, environmentalists from other BIs, protesters from the youth counterculture, and activists from the radical left. When the Schmidt government undertook to push through its nuclear power program despite the opposition and exclude the BIs and their networks from participation in decisions, environmental activists—finding the prospects for success by working within existing structures much reduced—turned to social movement mobilization and protest. An extended period of polarization and confrontation ensured, extending into the mid 1980s. Demonstrations and site occupations were often countered by aggressive government responses, and some turned violent. These confrontations created a cycle of escalating movement mobilization and conflict and radicalized many movement supporters, leading them to embrace countercultural ideas and forge coalitions with other new social movements.

The confrontational movement's ideology included conceptual links between nature protection, anti-consumerism, and skepticism of technology that had been forged during the prewar period; however, the new framing replaced conservative longings for a vanished premodern social order with new elements: protecting human health and nature from industrial pollution, concerns about population growth and resource exhaustion, and countercultural ideologies. This new framing, with its links to feminism, the anti-war movement, and the like, emphasized not only political action but lifestyle change and personal identification with the movement and its ideology. Advocates of right-wing ecology, with ideologies that combined prewar conservatism and new ecological themes like overpopulation and overconsumption, found themselves struggling to obtain a hearing.

When the environmental movement arose, no large national citizens' organizations representing the full range of its concerns were available to serve as social movement organizations. The goals of nature protection organizations coincided with the movement's objectives only in part, and national public membership organizations to fight air and water pollution, work against highway and runway expansions, or combat growing traffic problems had not evolved. During the movement's early years, the DNR, acting as the established voice of organizations advocating nature protection and care of the landscape, was able to exert some influence on the government's legislative program. However, its disunited and diverse membership, cautious approach, reliance on government subsidies, and history of participation in Germany's corporatist system made it unsuited for leadership of a movement that incorporated many countercultural ideas and was becoming steadily more confrontational. The DNR's efforts to chart a compromise course often satisfied no one, and it fell into internal strife.

The BBU thus emerged almost by default as a key social movement organization. Like the founding of the DNR in the 1950s, the BBU's founding illustrates the key roles of interest group and social movement entrepreneurs, along with outside supporters and funding sources, in getting new organizations off the ground. The BBU's embrace of countercultural ideologies and protest tactics probably gave its supporters in government frequent heartburn, but for almost a decade it was sustained by modest financial support from government.

The BBU played a key role as an environmental movement organization for a decade, but it was never fully satisfactory. Its frequent internal conflicts over issues like involvement in national politics and alliances with the far-left groups, its factional power struggles, its struggles to contain tendencies toward oligarchical leadership, and its chronic underfunding illustrate the disadvantages of informal networks as a mode of coordinating social movements. Many of its more moderate sympathizers viewed its combative pronouncements and disorganization with concern and wished for a better alternative. Yet its activist, countercultural, and grassroots emphasis was in tune with the times, and its members' commitments to these social movement ideologies made it almost impossible for it to change course.

At the end of the 1970s the environmental movement acquired an additional node of organization in the form of the Green Party. Although environmental issues were a central part of its program, the Greens also embraced other new social movement issues and grassroots democracy, and neither right-wing ecologists nor the far left were able to seize control. As an alternative mechanism for channeling environmental concerns into the political arena, the Greens functioned as both a complement and a competitor for the nature protection and environmental organizations. While they welcomed the party's help, most of the organizations also held it at arm's length because they feared that it would fail or that they would lose their supporters to it. In the long run, the Greens' success gave the party a key role in the environmental movement, but it was not suited to take over the environmental organizations' functions as civil society organizations or environmental interest groups.

The rise of the environmental movement was simultaneously a great opportunity and a major challenge for the nature protection organizations, since its goals overlapped with theirs even more than had the goals of the homeland protection, socialist, and Nazi movements. The environmental movement emphasized the protection of humankind more than the protection of animals or plants, but it insisted that humans were parts of interconnected ecosystems that had to be protected for the sake of human welfare. These were arguments that nature protection advocates and the DNR had been advancing for a decade.

Paralleling their previous experience with the Nazis, the relationship of the nature protection organizations to the environmental movement had all the elements of an approach-avoidance conflict. In a social context so strongly polarized by social movement mobilization and confrontation, it was clear that the organizations' conventional interest-group strategies were unlikely to be effective. Allying themselves with the new movement or functioning as social movement organizations for the movement could bring them more members, public support, donations, influence, and effectiveness. Failure to affiliate with it might leave them isolated, viewed as irrelevant, and unable to compete for support. Still, the movement had a considerably broader agenda and set different accents than the traditional nature protection organizations, and its loose organizational structure, countercultural ideology, and confrontational strategy were unfamiliar and threatening. Moreover, the gap only widened as the environmental and antinuclear movements moved closer together, attracting supporters from social circles quite different from the nature protection organizations. Certainly the nature protection organizations could not hope to control the environmental movement, and active participation might divert them from their traditional concerns, drive away supporters committed to traditional goals and strategies, and strip them of their valued legitimacy and influence with government and business—which had quite different ideas about what they should be doing. In short, a strategic choice to ally themselves with the movement meant wagering the loss of the support, funds, legitimacy, and influence they gained from their existing approach against possible gains from adopting new goals and strategies.

In contrast to 1933, the social context of the early 1970s left the nature protection organizations free to choose their own strategies for adapting to the movement. Some of the smaller ones resolved the dilemma by electing to pursue their traditional goals via their usual strategies. This approach did not prove very successful; most of the organizations that chose it stagnated even as the movement grew. Other organizations, including the Friends of Nature, BH, and WWF made only minor adjustments in goals and strategies. The increased emphasis on environmental protection by the Friends of Nature and BH proved too small to earn them a leadership role in the movement, especially in light of their continued pursuit of other goals unrelated to environmental protection. Only their members' continuing support for their numerous other objectives prevented them from losing ground. The WWF, with its conservative, business-oriented support base, also made only minor changes in its mission; however, its clear identity as a nature protection organization allowed it to benefit from growing general public interest in environmental issues and nature protection by attracting additional supporters and donations, especially among those not inclined toward movement ideologies and strategies.

Greater benefits accrued to two nature protection organizations that modified their missions and strategy in more radical ways. The earlier and more dramatic change occurred at the BN. During the 1970s, it transformed itself from a small organization with narrowly defined goals and a cautious approach, a conservative membership, and an ingrown relationship with the Bavarian nature protection authorities into an independent, dynamic, and much larger organization that pursued both environmental protection and nature protection and played a significant role in the environmental movement.

The BN also became the key founder and largest regional affiliate of BUND, a new national organization that mirrored its orientation and grew by the end of the decade to be one of Germany's most important environmental organizations. BUND prospered because it filled the need for an organization that shared the goals and action repertoire of the environmental movement but was also better organized, more stable, and more moderate than the movement's extreme elements. As such, it provided a home for more moderate middle-class supporters who wanted an organization that would embrace new environmental issues without giving up its nature protection goals and that could function not only as a participant in nonviolent demonstrations, but also as an interest group in the policy-making process and a civil society organization protecting nature and educating the public.

The transformation of the BN and BUND's ascent to its position as one of Germany's leading environmental organizations required less than a decade, but they were not accomplished without resistance and conflict. The BN's transformation cost it its comfortable relationship with the Bavarian government and the influence that went with it, its legitimacy in some circles, and the support of some of its more conservative members. Soon after its founding, BUND was riven by conflict between its conservative cofounders and those with more sympathy for other new social movement issues and confrontational tactics. These events illustrate well the difficulty of transforming an institutionalized organization with an established social role and members committed to its existing goals and strategies.

The parallel transformation of the BfV occurred somewhat later and was marked by even more rancorous conflict. Changes in mission were hotly disputed, advocates of change were threatened with expulsion, and the decision to oppose nuclear power led to mass resignations. Nevertheless, by the late 1980s the former bird protection organization had become, if not a social movement organization, then at least a very active and relatively confrontational interest group committed to both nature protection and the core issues of the environmental movement.

By making major changes in their mission and strategies, the BN/BUND and the BfV entered the field of national organizations devoted to the goals of the new environmental movement that had been previously occupied only by the BBU. By broadening their goals to include environmental protection, adopting strategies more attuned to the movement, and maintaining governance structures that were democratic in principle but less chaotic than those of the BBU, they occupied previously unfilled niches within this field and competed with the BBU for members. By holding on to many of their traditional supporters and attracting new ones—including moderate environmentalists who had eschewed the BBU as too radical and chaotic and some former BI supporters—they achieved major increases in membership and financial support. BUND, in particular, found itself trying to steer a course that was radical enough to attract some BI members, or even entire BIs, while holding on to members who considered the BBU to be too radical or too chaotic.

In 1980, the organizational field became more crowded when it was joined by Greenpeace Germany, which occupied a somewhat different niche. Like the BfV and BUND, Greenpeace combined nature protection and environmental issues. However, it also combined the confrontational approach of the BIs and BBU with the most centralized and professionalized organization structure of any of the organizations. By cultivating its image of a David unafraid to confront industrial and government Goliaths, and by staging spectacular actions, Greenpeace garnered a great deal of press and public attention. It was thus able to attract a small core of committed activists to participate in its demonstrations and hundreds of thousands of citizens concerned about environmental movement issues to fill its coffers. Nevertheless, it remained at some distance from other environmental organizations and the protests at nuclear power plants, and it did not play a key organizational role for the movement.

By the early 1980s, four large formally organized national organizations—the WWF, BUND, the BfV, and Greenpeace—had joined the more loosely organized BBU as key players on the environmental scene. Although they occupied different niches in terms of goals, strategies, and constituencies, all were profiting from the environmental movement in terms of rapid growth in support. On the other hand, signs of the end of the cycle of social movement mobilization were already appearing. Yet even as the environmental organizations began to consider strategies for adapting to the movement's decline, they found themselves faced with another challenge: the sudden and unexpected collapse of the GDR and German reunification.

Chapter 6

NATURE AND ENVIRONMENTAL PROTECTION EASTERN-STYLE

Environmental Organizations in the German Democratic Republic

Events in the Soviet occupation zone and German Democratic Republic followed a different trajectory from those in the West, posing different challenges for nature protection advocates.[1] East Germany emerged from the war a shattered, resource-poor land under strong pressure from the Soviets to build a socialist state and orient itself to the Eastern Bloc. Nature protection groups in the East, which had just emerged from the Nazi *gleichschaltung* experience, soon found themselves once again confronted by an authoritarian state, which forced them to sever ties with their peers in the West and subordinate themselves to state-controlled organizations. Yet environmental interest did not die. Officially sanctioned and formally organized nature protection groups persevered and grew throughout East Germany's history. And in the 1980s, new environmental groups, operating without official sanction, contributed to the growth of envi-

[1]A brief overview of GDR environmentalism and environmental groups is available in Rat von Sachverständigen für Umweltfrage (1996). The following sources offer more comprehensive coverage: Rösler, Schwab, and Lambrecht (1990); Rink (1991, 2001, 2002); Behrens (2003a, 2003b, 2003c); Chaney (2005), and Rink and Gerber (no date). Despite a focus on the GNU and on events in Berlin and Brandenburg respectively, Behrens et al. (1993) and Nölting (2002) also provide comprehensive treatments.

ronmental consciousness and played an important role in the social movement that led to German reunification, and ironically often to their own demise.

East Germany in the Immediate Postwar Period

Under the oversight of the postwar Soviet occupation forces, East Germany was transformed step by step after the war into a de facto one-party, socialist state.[2] The Social Democratic Party was forced to merge with the communists, large property holdings were confiscated and turned over to small farmers or to state farms, moves to nationalize industry began, and the secret police was established. In 1949, the German Democratic Republic was established as part of the Eastern Bloc, and in 1961 it walled itself off from the West with the Berlin Wall.

The new nation faced many challenges. Germany's natural resources and industry were concentrated in West Germany, and even as the West benefited from the Marshall Plan, the Russians dismantled many plants in the East and shipped them home as reparation payments. The war devastated farming areas and left agricultural equipment and fertilizer in short supply. The nationalization of industry, the breakup of large agricultural estates, and—somewhat later—the forced collectivization of agriculture added the burdens of a massive restructuring of the economy to the problems of rebuilding. In the immediate postwar period, the Soviet zone was flooded with refugees from former German territories ceded to Poland and Czechoslovakia, and until construction of the wall, a continual stream of the GDR's most skilled workers flowed to the prospering West. Consequently, the GDR's economic situation remained precarious well into the 1960s.

Like their counterparts in West Germany, the remnants of the prewar nature protection organizations in the Soviet zone moved soon after the war to reconstitute themselves and resume their work. As in West Germany, nature protection seemed to many an unnecessary luxury for a country beset by so many problems, and the Soviet occupation authorities and the new socialist government viewed the nature and homeland protection organizations with suspicion because of their predominantly middle-class membership and their wartime acceptance of the fascist regime. The nature protection groups—along with literary, arts, and hobby groups—were thus required to affiliate with a new state-sponsored umbrella organization, the Cultural Federation for the Democratic Renewal of Germany. Even the left-leaning Friends of Nature was denied the right to operate independently.

---

[2] For useful general histories of events in the immediate postwar years see Turner (1991) and Fulbrook (1995). In addition to sources cited in note 1, the following sources provide especially good coverage: Würth (1985); Riesenberger (1991); Behrens and Benkert (1992); Bauer (2001), and Dix (2003). I also drew on two important documents from the GDR itself: Aus der Arbeit der Natur- und Heimatfreunde (1954) and Zentrale Kommission der Natur- und Heimatschutzfreunde (1956).

The unit of the Cultural Federation charged with nature protection, the Friends of Nature and Homeland, included not only nature protection groups but also groups of professional and amateur scientists, folklore and local history groups, photographers, and even hobby groups for aquariums and terrariums. It was structured hierarchically in three levels, matching the organization of the East German state.

Many nature and homeland protection groups resisted assimilation into the Friends of Nature and Homeland because they feared loss of their autonomy and the imposition of changes in their orientation and goals. Certainly being forced to cede their wildlife reserves, museums, and other facilities to the state did nothing to improve their opinion of the new order. Some resisted passively by joining the Cultural Federation but attempting, in all other respects, to continue business as usual. The leaders of the Cultural Federation, for their part, sought to reinterpret nature and homeland protection within a socialist framework and used pressure and reeducation efforts to bring the nature protection groups into compliance with the government's agenda.

By the late 1950s these conflicts had subsided, and the Friends of Nature and Homeland included 600 local nature protection groups with over 45,000 members, mainly from the better-educated segments of the population. Prominent among the members were scientists and volunteer local officials charged by the government with monitoring and reporting nature protection problems. This close cooperation between scientists and environmentalists became a prominent, lasting characteristic of GDR environmentalism.

Like the nature protection groups in the West, the GDR's nature protection groups worked to educate the public about nature protection, establish nature protection areas, and protect endangered species and rare micro-ecosystems such as ponds or forest groves. They held conferences and seminars on nature and nature protection, prepared educational materials for schools, and conducted research, including inventorying exceptional natural features and measuring pollution levels. At the national level, they advised the government about the formulation of new nature protection legislation. Until 1962, the Friends of Nature and Homeland also published a magazine with wide national circulation, and the national organization and many local groups published member magazines and newsletters. Reading a sample of the latter, I was impressed both by how similar their activities were to those of their prewar predecessors and their West German contemporaries, as well as by the extent to which they had to verbally hew the government line. In the early years, the nature protection groups were able to maintain some contacts with their counterparts in the West, but most of these ended after the wall went up.

Despite their state-sanctioned status, the nature protection groups' objections to environmentally questionable projects, such as beach development and garbage dumps in sensitive areas, were often brushed aside by the authorities, who subordinated nature protection to reconstruction and economic growth. Faced with the challenges of rebuilding under difficult circumstances, the GDR's leaders and many citizens accorded environmental issues a priority even lower than in West Germany.

Environmental Degradation and Environmental Policy in the GDR

As in West Germany, the 1935 Nature Protection Law remained in place after the war.[3] The new government established offices for nature and landscape protection in 1949, but they were understaffed and weak. Following prewar patterns, responsibility for nature protection at the regional and local levels fell to officials with other responsibilities, assisted once again by appointed volunteer experts and committees. The 1935 law was superseded in 1954 by a new law designed to adapt to the new realities of state ownership of land; however, its narrow definition of nature protection was poorly suited to controlling damage from industrialized agriculture, forestry, and mining, and it continued preexisting administrative practices. Other environmental legislation was passed on a piecemeal basis, generally in response to crises such as land erosion or accumulation of waste from strip mining. Enforcement was understaffed and lax, and complaints from nature protection groups were sometimes contested by scientists and planners whose commitment to making socialism economically successful was greater than their devotion to environmental principles. Nevertheless, the system did succeed in designating a significant number of nature protection areas.

By the early 1960s, the GDR's emphasis on economic growth had led to both a somewhat higher standard of living and to rapidly increasing environmental degradation. The latter resulted from massive strip mining and burning of highly polluting soft coal for home heating and electricity generation, an aging chemical industry, and industrialized agriculture, with attendant problems of overuse of pesticides and fertilizer, erosion, and loss of habitat for plants and animals. The growing environmental problems, much more serious than those in West Germany, laid bare the inadequacies of the narrow focus on the protection of nature reserves, scenic areas, and specific species inherited from the prewar era. Noting these trends, leaders of the nature protection groups began to argue for allotting more attention to combating air and water pollution, maintenance of healthy ecosystems, and a more comprehensive approach to planning the use of land, water, and natural resources; however, neither their concerns nor the numerous citizen complaints eventuated in public protests or a mass movement. With the exception of the 1953 workers' uprising, protest was rare during the GDR's first

---

[3] Especially since 1989, a significant amount of information has become available about the GDR's environmental problems and policies. In addition to sources cited in note 1, see Fulbrook (2005) for an overview, and Gruhn (1972), Würth (1985), Wensierski (1985), Großer (1990/1991), Rüddenklau (1992), Kloth (1995), Jänicke, Kunig, and Stitzel (1999), Behrens (2000, 2001), Dix (2003), and Gensichen (2005) for more detail. Articles focused on the state's environmental policy and its interpretation of continued environmental problems include Timm (1985), Lausch (1986), and Wegener (2001). On state agricultural policy of the time, see Schnurrbusch (1979).

two decades, and most of the dissent that did exist originated with writers, artists, and intellectuals and focused on lack of political freedom.

By the 1970s, however, some intellectuals were expressing concerns about the environmental effects of the nation's expanding economy. In the early 1970s, the government responded to these concerns, the increasingly obvious environmental problems, and its interest in gaining respectability on the international stage by placing ambitious new environmental legislation on the books. Particularly impressive was the replacement of the 1954 nature protection law and its later amendments with a much broader land use planning law, passed in 1970. It aimed to develop the landscape in a manner that would simultaneously increase agricultural and industrial production, beautify it, provide recreational opportunities, and ensure nature protection. The law incorporated such progressive steps as the identification and preservation of large nature reserves, comprehensive land use planning in rural areas, forest protection, controls on soil, air, and water pollution, improved solid waste disposal, noise abatement, use of financial incentives to reward industries for reducing pollution, and consultations between local industries and local governments to agree on pollution reduction goals. The government also established an environmental ministry in 1972, fourteen years in advance of West Germany, made environmental protection part of its five-year plan, and improved funding for environmental protection, staffing of nature protection offices, and training of volunteers. Financial pressures and resource shortages also led to continued emphases on mass transit use, avoiding urban sprawl, shipping freight by rail, and recycling a wide variety of materials. East Germany's record in all of these areas substantially exceeded West Germany's.

Unfortunately, the initial high expectations of the new environmental legislation were only partially realized. The law treated nature protection not as an absolute priority, but as part of an ensemble of land-use planning goals, and in practice nature protection was often subordinated to agriculture and industry as the government attempted to ensure its legitimacy by providing its citizens with a standard of living more comparable to their cousins in the Federal Republic. Implementing regulations were passed for only part of the legislation, the new financial resources made available were not really adequate to the task, and understaffing of the national offices remained a problem. Emission control standards were set at low levels, and penalties for violation were small. Local nature protection officials had only minimal paid staff, and the government continued to rely heavily on volunteer experts and voluntary assistants to monitor nature protection problems and provide advice. Citizens or environmental groups could file statements of concern about environmental problems, but the disposition of these was up to the authorities, and there was no legal recourse.

A second aspect of the GDR's effort to mitigate environmental damage in the late 1960s and early 1970s was its increased purchases of petroleum to reduce its dependence on soft coal. Extraction of lignite, the East's only native fuel source, from enormous strip mines had proven terribly destructive. Strip mining led to high levels of particulate pollution and created enormous amounts of waste material, and mining residues poisoned adjacent land and water. The vast mining operations also required relocation of entire villages and disruption of natural hydrology to lower the water table. Efforts to place land devastated by mining back under cultivation proved expensive, only partly successful, and unable to keep up with the rate of destruction. Finally, because of soft coal's heavy sulfur content, burning it increased air pollution and acid rain.

Importing additional oil reduced these problems, but the improvement was short-lived. The mid-1970s oil price shocks, which led the West German government to pull back from environmental initiatives, had exactly the same effect in the East. Some energy savings were achieved via energy-saving measures like lower speed limits, but this was not enough, and high oil prices began to threaten economic growth and thereby the government's legitimacy. The GDR leadership turned back to soft coal as its prime energy source and undertook to supplement coal-fired plants by constructing nuclear reactors, built using Soviet technology to safety standards well below the Western standard.

The effort to maximize production for the consumer market also exacerbated the country's other environmental problems. Efforts to increase production diverted resources away from upgrading aging, environmentally destructive plants, and an outmoded and poorly maintained chemical industry continued to discharge toxics into air and water, threatening the health of workers and nearby communities. Pollution, smog, and damage to forests were especially severe in the southern GDR, where the mining and chemical industries were concentrated. Waste disposal problems were also serious. Most of the country's household waste was deposited in unregulated dumps and landfills. Sewage treatment was inadequate or nonexistent, and basic precautions for toxic waste disposal were often omitted. In search of hard currency, the GDR even contracted with the Federal Republic to accept some of its toxic waste. Budget problems stemming from the oil price shocks also led to reductions in government spending for environmental cleanup.

The 1970s also brought a renewed push to increase agricultural production, partly to improve the balance of payments. Among the measures adopted were the development of some of the world's largest concentrated animal feeding operations, the draining of wetlands, increased mechanization of agriculture, increasing field size through removal of woodlots, hedges, and hiking trails, and heavy use of fertilizer, pesticides, and herbicides. The results included increased erosion, damage to soils, pollution of waterways, and loss of bird and other animal habitats.

These realities stood in sharp contrast to the government's official ideology about environmental issues. It held that East Germany's environmental problems were being eliminated under socialism, which in the long run would allow both increasing prosperity and environmental protection. Remaining environmental problems were attributed to the need to compete with capitalist economies and defend against military threats from the West, inefficient production processes inherited from prewar capitalism, the inadequacies of existing ecological knowledge, the high cost of controlling pollution with existing technology, and unresolved problems of building appropriate ecological incentives into planning and production processes. Claims, like those in the 1972 Club of Rome report, that exhaustion of resources or population growth would ultimately result in an insurmountable crisis did not square well with this ideology and were greeted with reactions ranging from complete rejection to skeptical criticism.

By the end of the 1970s, environmental conditions in the GDR had entered a downward spiral that persisted until the end of communist rule. Discussion of environmental issues among scientists and some government officials continued, and there were thousands of formally registered citizen complaints. However, the government, state-controlled industry, and state-controlled agriculture, which had been consolidated into increasingly large and influential enterprises, were generally unresponsive to environmental concerns. Except in cases like recycling, where environmental and economic objectives coincided, environmental laws were weakly enforced, and continuing budget shortfalls resulting from economic stagnation ruled out investments to modernize outmoded polluting factories. Environmental sections disappeared from five-year plans in the mid 1970s.

The regime had never been forthcoming with information about the extent of environmental problems, but a 1982 law, strengthened in 1984, banned the release of all data about environmental problems on the pretext that it was being used by propagandists from the West. This action not only cut off the flow of information, it deprived those who wanted to complain of a factual basis for their complaints and intimidated persons who might otherwise have spoken out about environmental problems. Most of the population remained relatively uninformed about environmental problems, and many citizens were, in fact, more interested in enjoying the modest level of prosperity that had finally come their way than in delving into environmental issues.

The Rise of the Environmental Movement in the GDR

Despite the mounting problems, no meaningful environmental movement arose in the GDR until the 1980s. A nascent oppositional movement did germinate among younger cohorts during the brief period of relative prosperity and political liberalization in the early 1970s. Influenced by developments in the West, it focused on peace and environmental issues, grassroots democracy, and countercultural lifestyles; however, it failed to find resonance among the general public and was choked off in the later 1970s by the government's return to repression and Cold War rhetoric. When environmental activism finally did become established a decade later, it faced a repressive state that suppressed information about environmental problems, viewed environmental activism as a threat to its dominance, and provided few opportunities for expression of dissent. Consequently, it appeared only in a few protected niches within the society.

The Society for Nature and Environment

One such niche was state-sanctioned nature protection groups that had originally been part of the Friends of Nature and Homeland.[4] In 1980, the government incorporated them into a newly created organization under the umbrella of the Cultural Federation, the Society for Nature and Environment (Gesellschaft für Natur und Umwelt: GNU), and assigned homeland protection groups to a separate organization. Creation of the GNU was intended to defuse growing public concern about environmental problems, provide a state-controlled counterpart to the West German environmental organizations, broaden the agenda of state-sponsored nature protection to include problems of cities and industrial pollution, and provide a basis for building international contacts and legitimacy. Like its predecessor, the GNU was charged with developing environmental consciousness through lectures, museum exhibits, and excursions, advocating for environmental protection, assisting with the identification of nature protection areas, and conducting practical nature protection work in cooperation with the authorities and state-owned farms and industries.

Beginning with a membership of under 40,000, the GNU approached 60,000 members in over 1,500 groups by the end of the decade. Its members included many scientists and government functionaries concerned with environmental problems, but there were also many for whom nature protection was solely an avocation. The single largest number of members belonged to nature protection groups, and the second largest to hiking groups. Significant numbers were also organized in scientific groups, such as botanists and ornithologists. Research institutes and production facilities with strong environmental impact, such as power plants or collective farms, could also hold membership.

A structure of local and regional governing boards and a national board dominated by government officials attempted to steer the activities of the local groups from above. At the national level, the GNU had the recognized prerogative to advise the authorities about nature protection matters. It had no formal say in decisions, but it was sometimes able to use its informal contacts and influence to head off or modify environmentally damaging proposals. It had about twenty-five full time employees, but its budget was far smaller than its status as the largest organization in the Cultural Federation might have suggested.

The change of name to GNU had relatively little effect on the day-to-day activities of the local groups. They consulted with and lobbied local authorities about environmental problems, such as the effects of pesticides on bird life, runoff of pesticides and fertilizers into streams, illegal garbage dumps, and problematic waste discharges from factories. They also inventoried, studied, and advocated for

---

[4]The most complete treatment of the GNU is Behrens et al. (1993). Additional useful sources include the following: Würth (1985); Wensiersky (1985); Gilsenbach (1989); Behrens and Benkert (1992); Jones (1993); Kloth (1995); Neubert (1997); Bammerlin (1998); Illig, Donath, and Donat (2001); Wegener (2001); and Rutschke (2001).

preserving endangered areas and species. They also attempted to increase the public's environmental consciousness through newspaper articles, periodic lectures, excursions and nature study trips, exhibitions, conferences, and youth activities. They also staged "nature and landscape protection days," day-long programs of presentations and events intended to inform the public about environmental problems and how citizens could reduce them, and to give citizens a chance to discuss local environmental issues with scientists and nature protection officials. Local GNU groups also engaged in practical nature protection activities, including preserving sensitive areas, tree planting, renaturalizing wetlands, laying out and maintaining nature paths and hiking routes, and carrying out city beautification projects.

Although the GNU could and did work for environmental protection, it was in no position to function as a full-fledged social movement organization. It was deeply embedded in the network of government-sanctioned organizations, and the volunteer experts appointed by local or regional government to monitor environmental problems and assist with nature protection were among its most active members. Its national governing body contained numerous government officials and party representatives. As one former activist pointed out in an interview, ideological conformity and fear of taking risks were, therefore, especially high among its leaders. In its official statements, the GNU hewed the party line, and its approach was generally cooperative rather than confrontational; that is, it preferred filing complaints under the established system and lobbying behind the scenes to making waves in public. Its newsletter was aimed at GNU leaders, not ordinary members or the general public. The authorities, for their part, were usually willing to hear the GNU's concerns, and occasionally even to respond to them constructively—so long as these conversations took place through established channels and behind closed doors.

Although the GNU was not a social movement organization, some of its members were dissatisfied with their government's failure to adequately address East Germany's environmental problems, and local GNU groups did at times vigorously contest environmentally questionable practices or decisions. On occasion, they achieved noteworthy successes, especially in districts where local officials and party leaders were favorably oriented to nature protection; however, groups that became too confrontational sometimes found themselves under surveillance by the secret police or saw their leaders replaced by new leaders more willing to cooperate with the authorities.

Beginning in the mid 1980s a more activist and oppositional approach crystallized within the GNU, as new "city ecology" (*Stadtökologie*) groups were founded in numerous cities. They attracted both environmentally concerned citizens and GNU members who were most dissatisfied with the organization's centralized structure and establishment-oriented approach. By 1987, there were almost 400 informally networked city ecology groups, constituting over 10 percent of GNU membership.

In sharp contrast to the GNU's historical nature protection focus, the new groups focused their attention on urban problems and pollution by conducting research, publicizing the problems through lectures and films, and advocating for lifestyles with lower environmental impact. Like the BIs in the West, which were in part their role models, the city ecology groups adopted a more confrontational approach than had been usual in the past, calling for both immediate action to remedy environmental problems and more public input and democracy. They often criticized the GNU for its orientation toward working within the system. Nevertheless, only a minority of the groups were truly radical.

Although they shared many objectives with other GNU groups, the city ecology groups' more confrontational approach made their relationships with the GNU leadership tense, and their connection to the GNU was always tenuous. The GNU's leaders believed that working within the system was more productive than scolding from outside, and, like the authorities, they often provided the city ecology groups with only grudging support or cooperation. The city ecology groups, in turn, saw the GNU leadership as ineffectual and timid, and challenged the GDR's focus on growth at any price and lack of openness. When the opportunity presented itself at the end of the 1980s, they were ready to act as environmental movement organizations.

Church-Sponsored Environmental Groups

Critique of the GDR's environmental policies came not only from the city ecology groups and other activist elements within the GNU, but also from less formally organized environmental groups that emerged in the 1980s under the wing of the Evangelical (Lutheran) church.[5] The church, which enrolled almost half the population of the GDR—and 85 percent of its Christians—as members, had earlier negotiated an uneasy truce with the authorities. It agreed to sever its ties with the West German church and lend its general support to the government in return for a degree of independence, freedom from overt harassment, and the opportunity to speak publicly and have some input into policy making in areas of interest to it. This made the church an attractive sponsor for groups of citizens concerned about environmental problems.

During the liberalization of the early 1970s, when the government was tolerating more open discussion and showing more concern for environmental issues than it did before or afterward, church theologians and leaders initiated a discussion of environmental problems, at first within the church and later with the authorities. In 1973, the church assigned pursuit of this theme its research center in Wittenberg. Under the leadership of Hans-Peter Gensichen, this center developed into a hub for research and discussion of environmental issues. By 1980 it was publishing reports, pamphlets, and the only newsletter on ecological topics

[5] A great deal was written about church-related environmental groups after the fall of the wall, by both scholars and some of the groups' former members. In addition to sources cited in note 1, see the following: Würth (1985); Knabe (1985, 1988); Becker (1990); Kühnel and Sallmon-Metzner (1991); Gensichen (1991, 1994, 1998, 2005); Rüddenklau (1992); Jones (1993); Halbrock (1995); Fulbrook (1995); Kloth (1995); Naumann (1996); Neubert (1997); Bammerlin (1998); Hoffman (1998); and Weinbach (1998).

in the GDR. The newsletter included articles on theology and the environment, scientific discussions of environmental problems, reports about workshops and events, information about government environmental decisions and policy, and book reviews. The center also organized conferences and traveling exhibitions about environmental problems. Later it functioned as a networking center for local church-sponsored environmental groups and hosted annual meetings of their representatives.

The first of these local groups appeared around 1980 in the northern GDR, but they soon became more numerous in the southern GDR, where industrial pollution was most serious. The groups' memberships often overlapped heavily with church-sponsored groups working for human rights and world peace. During the early 1980s, groups often combined several of these themes, but after Chernobyl, more groups with a solely environmental agenda appeared, including some anti–nuclear power groups. The church provided the groups with meeting rooms, use of duplicating machines and office equipment, and the opportunity to publish newsletters and reports in its name.

Discourse about the causes and cures for environmental problems in church circles emphasized environmental ethics and criticized consumerism and the GDR's heavy reliance on alienating and polluting technologies. The church-affiliated groups emphasized personal lifestyle change, projects that demonstrated ecologically oriented lifestyles, and reforms of the existing socialist system as the preferred means to meliorate environmental problems. They collected information about environmental problems, held lectures and seminars, organized meetings with similar groups from other regions, and sponsored symbolic events, such as auto-free weekends in cities, bicycle rides through nature protection areas threatened by new highways, mass tree plantings, stream clearings, rural holidays for children from heavily polluted regions, and demonstrations of solar power. To inform one another and the public about environmental issues, they published newsletters, reports about environmental problems, proposals for more environmentally sound policies, and brochures offering practical advice on environmentally friendly lifestyles, all carefully labeled "only for internal church use" to protect them from censorship. The publications had limited circulation, primarily within church circles, but they were symbolically important and often the only source of information about environmental problems.

The church groups were also able to mobilize mild public pressure for reforms through letters, petitions, and meetings with the authorities, and they were sometimes able to derail environmentally unsound industrial, public works, or agricultural projects or obtain modifications of them. On occasion, they even staged mild protests. However, until the last months of the GDR, they did not function as a confrontational social movement. As one activist who had been a member of a church group explained to me, members of the church groups were cast in an oppositional role more by the government's view of them than their own inclinations.

In the second half of the 1980s, a number of local church groups set up environmental libraries, which archived relevant reports and documents from both

East and West Germany. The most important of these was the Berlin Environmental Library, located in the basement of the parish house of a church with a sympathetic minister.[6] Although the library's collection was small, its very existence was a symbolically important statement. Moreover, by providing a meeting place and a mail drop for environmental and peace activists from Berlin and beyond, it emerged as a new contact point for church-sponsored ecology groups and a link to similar groups in other Eastern Bloc nations. The Library's newsletter became widely circulated and influential in environmental group circles, and it held periodic programs on environmental and related topics. The library was operated by a loosely organized but tightly networked group of activists subdivided into several working groups. With help from Western allies, it was able to acquire computer and printing equipment that was otherwise unavailable to local environmental groups.

In general, the groups centered around the Berlin Library adopted a more confrontational stance than those oriented to the Wittenberg research center, and in 1988 the Berlin Library spawned the more tightly organized Green Network–Ark, which served as a central coordinating point for the most confrontational elements of the emerging environmental movement. The Green Network–Ark had working groups focused on topics such as air pollution, water pollution, and waste disposal, and a network of affiliated groups throughout the GDR. Its journal was very critical of the government and emphasized making scientific data about the extent of the East's environmental problems publicly available. Some of its articles were subsequently reprinted or summarized in the West German media. With assistance from supporters in West Germany, the network even succeeded in making a film about damage wrought by the East German chemical industry and smuggling it out for broadcast by the West German media. Finally, the network sought out contacts beyond East Germany. It was an associate member of Friends of the Earth and a member of the Greenway network of Eastern Bloc environmental groups.

The tighter organization and more confrontational approach of the Green Network–Ark eventually led to a contentious split with the Berlin Environmental Library, whose leaders were wary of possible trends toward centralization or formalization or formation of a political party. In fact, however, resistance and infiltration by the authorities and its loose organization kept the network from achieving a high level of integration or central control.

Despite their disagreements, the Green Network–Ark and the Berlin Library were both part of a general trend toward increasing confrontation with the authorities. The church-sponsored environmental groups had always included members whose motivations were not primarily religious. As East Germany's environmental problems mounted and the government lost credibility, they attracted even

---

[6] In addition to the sources cited in notes 1 and 5, information about the Berlin Environmental Library and related organizations can be found in Jordan (1993, 1995), Cooper (1995), Kloth (1995), Neumann (1995), and Moritz (1997).

more such members, who saw participating in them as a chance to express their environmental concerns, have an area where they could act independently, and protest against the regime. The result was the development of a more activist approach less grounded in the church's discourse. Actions or protests initiated by the church groups sometimes took on a life of their own, attracting many citizens who were not church members and extending to other issues. While this new approach rarely led to immediate results, it had significant symbolic and motivational value.

The relationship between the church and the environmental groups became a topic of lively discussion in church circles during the mid 1980s, but overall the church hierarchy consistently maintained its interest in environmental issues and its supportive orientation. Frictions between some individual churches and their environmental groups were more pronounced. Local groups sometimes pushed local churches to be more active than they wished to be in environmental matters, and some groups criticized the church's lack of action and unwillingness to criticize the state or Party. Some pastors and church members lacked sympathy for environmental concerns, disapproved of some group members' unconventional lifestyles or irreligiosity, or feared retaliation from the state. A few parishes dissolved environmental groups, and others exerted pressure to rein them in. For their part, the groups typically resisted church control but sought to maintain good relationships in order to preserve the safe harbor the church offered.

Estimates of the number of church groups and persons involved in them vary considerably, but according to the most commonly mentioned figures, there were over forty church-affiliated environmental groups by the mid 1980s, and the number had increased to more than sixty by the end of the decade. The groups ranged in size from a handful of members to about a hundred; however, the number of sympathizers was much higher, as many church members who were interested or active in the environmental arena were not members of a church ecology group. Unlike the GNU, most of the church groups lacked a formal organizational structure, emphasizing instead grassroots efforts and relying on informal coordination among small numbers of committed members. Some groups were continuously active, while others centered their attention only on specific activities. Differences in goals and structures made building networks with the GNU difficult, but some collaborations did occur, mainly with city ecology groups.

Like the members of the city ecology groups, the core members of the church groups tended to be young. They included many students, countercultural dropouts, artists, intellectuals, and well-educated young professionals who were concerned about social issues and were not party members. The latter group, sometimes called the "new middle class," was never as large in East Germany as in the West, but the growth of the GDR's service sector, the improved standard of living, and reductions in work hours all increased the pool of potential activists. On the other hand, environmental action could be a dangerous course for government employees, scientists, and teachers, and the lack of countercultural neighborhoods and newspapers, the straitened economic conditions in which most citizens lived, the constant state monitoring of citizen activities, and the pressure to be in the labor force all hindered the development of countercultural milieus.

Government Reaction to the Environmental Movement

The East German government found coping with environmental activism difficult.[7] Until the last months of the GDR, the groups lacked the ability to publicize their cause widely. Their goals were never as self-evidently in conflict with those of the regime as those of the human rights advocates, and environmentalists developed considerable finesse in presenting their case as simply an effort to help the state live up to its own stated goals. Moreover, the groups included far more reformers than real revolutionaries. Only a few city ecology groups and the Green Network–Ark were overtly political, and these did not appear until the late 1980s. Still, from the beginning, the environmental groups focused attention on embarrassing shortcomings of the GDR, and only the GNU operated under the kind of tight control that the government preferred. Consequently, the government viewed environmental activism as a potential threat.

As a result, the government's approach to environmental activism was characterized by considerable ambivalence. Conscious of its image abroad and fearing the creation of martyrs, the authorities usually shied away from overt repression. Local governments sometimes agreed to meet with environmental group representatives, to attend their forums, or to cosponsor tree plantings or stream cleanings—especially when these activities were conducted by the state-sponsored GNU or were seen as a chance to steer the groups toward a moderate course. And while the authorities were generally not very responsive to complaints about environmental degradation, this was not invariably true.

On the other hand, the government closely monitored and often forbade contacts with the West even for GNU groups, and it sometimes blocked relatively harmless events like tree plantings or bicycle rides to protest highway projects. The authorities sometimes refused to charter new city ecology groups, and they pressured the Cultural Federation to rein in or replace leaders of GNU groups that became too aggressive or cooperated with church groups. One former activist I interviewed recounted an incident in which an effort to publish a catalog of bikeways compiled jointly by a church and a GNU group was temporarily blocked due to fears that associating the GNU with the church group in print would legitimate the latter. The regime also exerted informal pressure on the church to keep the focus of its environmental work on theological and moral

---

[7] Since 1990, the government's efforts to monitor and infiltrate the environmental groups have come under scrutiny by researchers and former members of the groups. In addition to note 1, see the following sources: Knabe (1985); Becker (1990); Kühnel and Sallmon-Metzner (1991); Gensichen (1991, 2005); Rüddenklau (1992); Jones (1993); Fulbrook (1995); Jordan (1993, 1995); Halbrock (1995); Naumann (1996); and Neubert (1997).

issues and to self-censor environmental publications; however, it usually stopped short of direct censorship. Influencing the content of church-sponsored publications was an effective way to control information because it was almost impossible for other nongovernment organizations to acquire printing equipment.

The regime was more likely to openly oppose groups that were not under state supervision or that engaged in confrontational protest, and the authorities were unwilling to become involved in a dialog with more critical groups. The secret police compiled lists of such groups and their members and closely monitored the activities of local church and city ecology groups and the church's research center. Secret police agents sometimes overtly observed public meetings or protests in order to intimidate the participants, but they also secretly infiltrated the Berlin Library and other groups to report on their activities or sow dissensus. The most active participants in environmental groups sometimes found themselves under surveillance and were denied promotions and admission to universities. Still, there were few arrests or imprisonments. One former activist explained that so long as one was not seeking career advancement at work or a political career and avoided overtly political activities such as leafleting, an active member of a church environmental group had little to fear beyond being denied a visa to travel abroad.

There were, nevertheless, also a few instances of direct repression. The authorities initiated criminal action against the leaders of a demonstration against a chemical factory, arrested Potsdam activists who tried to display an acid rain–damaged tree from the mountains, and beat up demonstrators protesting construction of a potentially polluting factory in Dresden. The most publicized incident occurred in 1987, when the police raided the Berlin Environmental Library, charging that it had allowed the printing of a subversive publication by another group. The confiscation of the library's printing equipment and arrest of its leaders led to official protests from the church, street demonstrations by hundreds of citizens, and the eventual return of the presses and release of the prisoners.

The government's strategies produced uneven results. Fulbrook (1995) provides considerable evidence that the government was able to steer the activities of the Wittenberg research center onto a more moderate path during the mid 1980s, and the authorities were clearly able to limit the nascent movement's size, reduce its freedom of action for a time, and intimidate some of its members. As a result, openly seeking to mobilize public opinion for mass protest—always a possibility in the BRD—did not become a realistic option in the East until the last months of the GDR. On the other hand, the government was unable to halt the movement's growth, and the more repressive measures probably contributed to its radicalization.

The Environmental Groups and Reunification

In the late 1980s, economic stagnation, the Honecker regime's refusal to contemplate reforms similar to Gorbachev's, and the government's use of repression against reformers led to a powerful social movement aimed at overhauling the ossified system. Concerns about environmental problems were a significant element of its agenda, and some demonstrations focused specifically on environmental abuses; however, they were never its primary driving force.[8] As one former activist explained, he viewed the main goals of the movement as civil liberties and political and economic reform; however, he and other environmental activists assumed that implementing these changes would make it possible to accomplish environmental reforms. The movement succeeded dramatically with the fall of the wall and the replacement of the old regime, but the street protesters were soon calling for new goals: reunification with West Germany and a market economy.

Even as this movement was beginning to gather strength, many environmental activists, especially those in church and city ecology groups, reached the conclusion that environmental problems could not be addressed without a thorough overhaul and democratization of the system. They became part of opposition groups, such as the New Forum and the Citizen's Movement for Democracy Now. Like other activists in these groups, environmentalists more often envisioned reform within the GDR than radical change or unification with the Federal Republic. After Honecker's resignation and the fall of the wall, the government entered into dialog with these groups, and representatives of the environmental groups were invited to participate in "round tables," the committees of opposition and government representatives that shaped policy before the first parliamentary elections.

In sharp contrast to the Honecker government, the round tables and the freely elected East German government that followed made information about the country's environmental problems public and took the first steps toward reform. Some members of the environmental groups played key roles in the new elected government, which shut down nuclear power plants, closed polluting facilities, and designated extensive new nature reserves. The transition period was also marked by a flowering of citizen interest in environmental issues and the founding of many new local environmental groups. Nevertheless, as events progressed,

---

[8] The complex sequence of events that ensued during the last months of the GDR and the first years after reunification is not easily summarized. Brief overviews can be found in Oswald von Nell Breuning Institut (1996), Bammerlin (1998), Bergstedt (2002), and Brand (in press). Among the sources cited in note 1, the most comprehensive and systematic is Rink and Gerber (no date). Sources that provide more detail about specific aspects of the transition include the following: Behrens (1993); Weinbach (1998); Gensichen (1991, 2005); Jordan (1993, 1995); Kühnel and Sallmon-Metzner (1991); Turner (1991); Neubert (1997); Hirche (1998); Schmitt-Beck and Weins (1997); Rink (1999); and Lebrecht (2003). A number of articles focus on the activities of various organizations in the closing years of the GDR and thereafter. For Greenpeace, see Reiss (1988), Naumann (1996), Eitner (1996), and Greenpeace Magazin (1996c). For NABU, the former BfV, see NABU (2000b). BUND's activities are described in Enders (1995) and Bodenstein et al. (1998), and WWF's in Knapp (2003). For information about the fate of the GNU see Behrens and Benkert (1992), and for information about the founding of a new East German network, the Green League, see Grüne Liga (1999, 2003). Especially useful articles about the Green Party in the East include Hampele (1991), Schmitt-Beck and Weins (1997), and Poguntke (1998).

environmental concerns lost ground in comparison to other issues, especially the calls for reunification.

Reunification of the two Germanys followed in short order, with the GDR voting to annex itself to West Germany and adopt the Federal Republic's political and economic institutions wholesale. After reunification, the East German movement organizations supported by pre-1989 activists, such as the New Forum, which had envisioned reform of the GDR, not unification, failed to hold their own in competition with well-funded political parties from the West. The Western parties quickly emerged as the major vote getters, with significant competition only from the Party of Democratic Socialism, which represented the core supporters of the former socialist party. Members of the church groups centered around the Green Network–Ark and the city ecology groups were instrumental in the founding of a Green Party in the East. The new party united almost immediately with the New Forum and other citizens' groups to form Bündis 90, which in turn later merged with the Western Greens. In the first round of all-German elections, this coalition managed to elect Eastern representatives to the Bundestag, but its support faded rapidly in subsequent elections.

Following reunification, decisions about the East were made by a national government dominated by politicians from the former Federal Republic, which had over three-quarters of the nation's population. Many East German industrial plants and coal mines, burdened by outdated equipment and faced with pressures to quickly bring wages up to Western levels, were judged unable to compete in the world market and to pose environmentally unacceptable risks. These were shut down. Much of the remaining industry was sold to firms from West Germany and Europe. The plant closings and upgrades of old facilities greatly reduced visible pollution in the East—although new problems such as increased auto use and sprawling development partially offset the gains. The federal government directed massive infusions of cash into infrastructure, housing, and cleanup of environmental damage in hopes of stimulating a wave of new investment. Unfortunately, in the context of Germany's stagnating economy, this did not occur. The former East German territories continue to suffer from an unemployment rate about twice that of the rest of Germany, and from steady out-migration to the West.

As in many other spheres of life, reunification resulted in the shouldering aside of most elements of GDR environmentalism, with most of its remnants assimilated into the institutions and organizations of the Federal Republic. West German environmental law, with all its strengths and limitations, was adopted wholesale, and the government's program for speedy reconstruction of the East's infrastructure contained provisions that limited the public's right to complain about the environmental consequences. The flowering of citizen interest in environmental issues during the transition period proved relatively short-lived, and neither the GNU and its city ecology groups nor the church groups managed to survive the dislocations that accompanied the fall of communism and reunification intact.

The GNU, with its older, establishment-oriented membership, its accomodationist traditions, and its connections to the government-controlled Culture Federation, proved to be very poorly adapted to the increasing radicalization of the environmental movement during the waning months of the GDR. Many of its leaders and more conservative members believed that the confrontational strategies of the city ecology and church groups generated more heat than light. They sought continuity with the past and reform and democratization of the GNU and the GDR. Their refusal to take an active role in the environmental movement was viewed with skepticism by both the church groups and the GNU's own city ecology groups. Open conflict broke out between the city ecology groups and their allies in the GNU, on the one hand, and the organization's conservative leadership on the other. In the end, the city ecology groups were forced out, but this proved to be a Pyrrhic victory for the leadership. Surrounded by the tumult accompanying the transition to a democratic government and reunification, the GNU began to unravel. Many of its scientific groups merged with their Western counterparts, and the hiking groups affiliated with the Friends of Nature. Most of the nature protection groups affiliated with West German organizations, became independent, or dissolved.

Many of the church groups and some city ecology groups also dissolved. The church's research center held its last meeting of church environmental groups in 1991, and the church gradually withdrew from environmental involvement. Some of their activists, especially those affiliated with the Green Network–Ark, affiliated with the new Green Party, but many others resisted working within the established political system. In 1990, a number of groups that had been unenthused about party politics or affiliating with West German environmental organizations, primarily city ecology groups, formed the Green League (Grüne Liga), a loose and purposefully decentralized network of left-leaning local environmental groups. The Green League undertook projects such as restoring the polluted Elbe, protesting toxic waste dumps and expansions of strip mines, and advocating for turning the no-man's land that had bordered the Berlin wall into a park. Headquartered in Berlin, the Green League established regional offices in all five states of the former GDR and began with over 100 local groups. Other local environmental groups in the East continued to operate independently; they were joined by new groups and coalitions protesting environmentally questionable facilities like the huge airport expansion planned for Berlin. Some of these profited from temporary job creation measures in the early 1990s, which provided them free staff.

The demise of most of the East German environmental groups was, in part, the consequence of efforts by West German environmental organizations to establish themselves in the East. Greenpeace had begun to concern itself with the environmental problems of the GDR almost a decade before reunification. Working from its West Berlin branch office, it advocated an end to nuclear testing in the Soviet Union, the elimination of nuclear power in East Germany, and a cleanup of the Elbe. It was even able to stage occasional demonstrations in the

East. These included flying a hot air balloon into East Berlin to protest nuclear weapons, dumping 100 kilograms of salt extracted from river waters downstream from a potash mine on the steps of the GDR's environmental ministry, hanging a banner protesting pollution of the Elbe on a bridge in Dresden, and cosponsoring a letter writing campaign against air pollution and damage to forests with the Green Network–Ark. Before reunification Greenpeace actively sought partners among East German church and city ecology groups. For example, it provided the test equipment that church-based ecology groups in Dresden used to expose water pollution. Greenpeace established an East German office in Berlin in 1990, which was merged with the West German office shortly after reunification. However, after reunification it was able to establish only a handful of local groups and attracted few donors.

The BfV also worked actively to establish itself in the East. Shortly after reunification, some of the more scientifically oriented GNU groups, under the leadership of Michael Succow, a well-known East German environmentalist who had been a GNU leader and served as deputy environmental minister in the transitional government, effected a merger with the BfV, which offered them substantial financial backing in return. According to one very well-placed interviewee, these GNU groups were also actively courted by WWF and BUND; however, their desire to preserve the tradition of local groups and their distaste for confrontational stands made a merger with the BfV the logical choice. The resulting organization took the name German Nature Protection League (NABU). This "coup," as one environmental journalist I interviewed described it, provided NABU with a base of very committed members in the East. The newly merged organization was also able to strengthen its scientific expertise by adopting the GNU's model of appointing working groups of experts on specific topics and bringing many such experts from the East into the organization.

Other GNU nature protection groups formed a new umbrella organization, the Federation for Nature and Environment. It explored a merger with the West German BUND, but in the end only its Brandenburg chapter and some local groups affiliated with BUND. BUND itself had failed in an earlier attempt to absorb the Green League, so it suffered from a late start in forming its own local groups or recruiting existing local groups. The Federation for Nature and Environment's remaining elements were unable to maintain a viable organization, and it soon faded away. At the end of the process, the majority of GNU members had not affiliated with any environmental group.

WWF moved quickly after the fall of the wall to support East German activists who were working to secure former Party hunting reserves, military exercise areas, and other lightly populated areas as national parks. It also established an office in Potsdam to coordinate its efforts in the former GDR and other former satellite states.

Despite their best efforts, the West German environmental organizations have found it difficult to attract supporters in the East. As a former East German environmental leader pointed out to me, the former East German *Länder* have a smaller percentage of well-educated, middle-class citizens than West Germany. Moreover, the region's weak economy, high unemployment, and significant out-migration to the West contributed to feelings of insecurity, diverted people's attention away from environmental problems to more immediately pressing issues, and heightened concerns that environmental measures might contribute to unemployment. Adapting to the rapid changes associated with reunification proved to be a stressful process, and many former environmental activists, including two I interviewed, were disillusioned by the results of reunification. One former East German activist I interviewed, for example, spoke with considerable passion about the disorientation, materialism, and lack of spiritual values he saw as characteristic of present-day German life and the calamitous results for the environment. The environmental organizations' efforts to recruit members in the GDR were also hindered by the absence of a tradition of charitable or environmental groups supported by voluntary donations, and by widespread public suspicion of large centralized organizations. Finally, there were differences in the kinds of environmental problems facing the East and the West, and former activists from the East knew little about the environmental regulations and organizational forms of the West.

As a result, the number of supporters the national environmental organizations attracted in the East proved very disappointing. Greenpeace, for example, had only 8,500 supporters in the East five years after reunification—compared to a half-million in the West (Naumann, 1996; Rink, 2001). At the end of the decade, NABU had only about 19,000 and BUND about 7,500 members in the East—gains from 5,000 and 2,500 members respectively in 1993—and both organizations were growing only slowly in the East (Behrens, 1993; Rink, 2001). One activist I interviewed told me that the relatively weak BUND chapter in his city attracted mainly immigrants from the West. And in a recent environmental ministry survey, 10 percent of respondents from the West versus 5 percent in the East said they were members of some type of environmental organization. The corresponding percentages for having donated to an environmental organization in the last year were 17 and 9 percent (Bundesministerium für Umwelt, 2004). However, as several interviewees pointed out to me, a far higher percentage of members in the East than in the West are active. One interviewee, a former leading East German activist, went on to explain that East Germany had had no tradition of passive membership in environmental organizations, so people assume that membership means active membership.

Summary and Conclusions

East German environmentalism offers not only a unique case for analysis, but many instructive comparisons to developments in the postwar Federal Republic. Moreover, differences between the histories of environmental struggles in the old and new *Länder* continue to shape the goals and programs of German environmental organizations today.

Like their West German counterparts, members of nature protection groups in the East set about rebuilding their organizations almost immediately after the war. They faced obstacles to attracting supporters, financial support, and legitimacy that were, if anything, even more daunting than the ones facing their Western counterparts. A weak prewar industrial base, reparation payments to the Soviet Union, and the effort to build a socialist economy from the ground up engendered an emphasis on economic development at almost any price even more pronounced than that in the West. As in the West, the legitimacy of the nature and homeland protection organizations had been undermined by their collaboration with the Nazi regime, and the de facto one-party socialist state that emerged at the end of the 1940s was unwilling to tolerate open challenges to its policies and insisted that nature and homeland protection groups operate under its close oversight in state-sponsored organizations.

It would have been easy for the nature protection advocates to give up, but they did not. Renewed state control and socialist ideology were bitter pills for some, but in the end the groups chose the only viable way to continue their work. Like many of their predecessors who experienced Nazi rule, most accommodated to political realities, electing to work within the new order in return for the legitimacy and access to members and financial support they needed. Like the history of postwar West Germany and environmentalism's previous encounters with economic disaster, war, and dictatorship, the history of East Germany demonstrates the capacity of environmental themes to retain their advocates' commitment in the face of adversity and accommodate to diverse political and social conditions. Environmental concerns, which had already found a place in the writings of pioneering socialists and proved compatible with Social Democracy, were now selectively assimilated by state socialism—just as they had been by National Socialism.

The reemergence of nature protection groups pursuing their traditional goals and modes of operation also illustrates again the staying power of institutionalized patterns of organizational activity and the attachment of environmental group members to these patterns and the rewards they provided. Even as they set about rebuilding after the war, East German environmental groups adopted most of the goals and patterns of action that had characterized their predecessors. These goals and strategies no doubt seemed "right and normal," and following them afforded environmental groups a measure of legitimacy. Moreover, these patterns possessed demonstrated potential for meeting the needs of the groups' supporters—individuals who cared both about nature and the survival of the groups in which they worked to protect it. Following them attracted supporters not only on the basis of their commitment to the cause, but also via the social rewards, hobby-like participation in practical nature protection projects, and chances for self-development, prestige, and leadership that participation in nature protection groups afforded.

By the end of the 1950s, environmental protection had settled into stable patterns. The state-sponsored Friends of Nature and Homeland (later the GNU) had emerged as the GDR's only large, formalized environmental organization. Working under the oversight of a state for which environmental issues were at best a secondary concern, it labored to save as much of nature as possible and to curb some of the worst excesses of government policy. If its accomplishments were modest, it must be remembered that nature protection organizations in West Germany during the immediate postwar years were scarcely more effective.

There were, however, differences between the GNU and environmental organizations in the Federal Republic. Operating outside state control and with little or no state funding, the West German organizations were free to use the full range of interest group strategies and, if they chose, assume the role of social movement organizations. They could organize protests to underline their concerns, seek out media coverage, openly criticize public decisions, lobby formally, form alliances with other interest groups or movements, solicit donations from a broad public, and hire as many professional staff as they could afford. The GNU had fewer options, being almost completely dependent on an authoritarian state for funding, legitimacy, and influence; however, it could and did function in some respects as an interest group. Like its Western counterparts, it brought together people who shared an interest in environmental protection, worked to educate the public about environmental problems, and lobbied, albeit behind the scenes, at both the local and national levels. Also like the Western organizations, it engaged in some activities, such as caring for nature reserves and practical efforts to protect endangered species, that lacked direct political relevance but contributed to the public welfare in almost the sense that contemporary scholars attribute to civil society organizations.

Events in the two German nations diverged markedly at the end of the 1960s. The economically thriving West, with its growing number of well-educated, prosperous citizens and greater political freedoms, experienced a modest greening of national policy, the emergence of new counterculturally tinged framings of environmental issues, and a flowering of BIs focused on environmental problems. The GDR's slower economic growth and repressive state produced a different result. The brief political liberalization and greening of environmental policy in the early 1970s there proved short-lived. Until the 1980s, only the GNU represented environmental concerns in the East, and it continued down well-trodden paths.

In the second half of the 1970s, the GDR responded to oil price shocks, which threatened the fragile public acceptance it had earned by providing consumer goods and full employment, with policy changes that greatly exacerbated environmental problems. Consequently, by 1980 the situation there had come to resemble the one in the Federal Republic a dozen years before. An industrial economy with minimal environmental precautions produced apparent environmental threats, while the government's response was a combination of half-measures and policy innovations—nuclear power in the West and increased reliance on lignite, nuclear power, and industrialized agriculture in the East—that exacerbated environmental problems and risks.

In a closed society where economic conditions were only slowly improving, this situation initially provoked no wave of mass protest. The GNU was deeply embedded in the system, and its leaders were convinced that their strategy was the only viable way to proceed. It thus offered little basis for social movement activity. When a nascent environmental movement finally did appear in the 1980s, it was centered not in the mainstream GNU, but in the city ecology and church groups. Both of these settings provided marginal, but somewhat protected, niches where mild dissent could be voiced by the small but growing number of citizens who were willing to take the risks of participation in new groups with less conformist framings of environmental problems and a broader set of goals and strategies.

Worsening environmental problems were only one factor in the rise of these new groups. By contributing meeting space, use of duplication equipment, and a degree of shelter from criticism, the church was instrumental in getting some of the new groups off the ground. And although many GNU leaders harbored reservations about city ecology groups, they benefited nonetheless from their GNU affiliation. The emergence of new environmental groups was also pushed ahead by new framings of environmental problems that called into question not only the efficacy of the GNU and state environmental programs, but also the GDR's reliance on production and consumption levels as primary indicators of societal well-being. The new ideologies promoted ecologically sound lifestyles as a moral imperative and grassroots action to raise public consciousness. Some of their ideas were borrowed from the West, but others were developed at the church's research center or by dissident activists themselves. The characteristics of participants in the new groups—young, well-educated, and imbued with postmaterialist values—also paralleled earlier developments in the West.

The new groups' propensity to organize themselves as loose networks and forge alliances with groups working for other new social movement goals was analogous to similar tendencies present among the West German BIs of two decades before, and their skepticism of the GNU was reminiscent of the BBU's difficult relationship with the established West German nature protection organizations. But regardless of the organizational form environmentalism assumed, the limited pool of potential members, the obstacles to obtaining press coverage, and the surveillance and hostility of a still potent authoritarian state stood in the way of large-scale, public mobilization centered around environmental issues. So too did the new groups' reluctance to abandon reform strategies for fear of forfeiting the support of the church and the grudging tolerance of the authorities. Consequently, large-scale mobilization did not occur until the final months of the GDR. Until then, the church and city ecology groups recruited so few persons and received so little public attention that they can be characterized as a social movement only in the most expansive definitions of the term.

In the end, a mass movement did occur, but the city ecology and church groups played only a minor role in its genesis, chiefly by contributing some of their ideas and activists to new groups. Instead, it was the breakdown of Soviet hegemony, the example of resistance movements elsewhere in the Eastern Bloc, and the government's declining legitimacy that precipitated a wave of public discontent and mobilization and the demonstrations that brought down the Honecker regime and led the way to East Germany's voluntary self-annexation to the Federal Republic. This movement was, in the first instance, a movement for political freedom and a higher standard of living, not an environmental movement. Environmental ills were included in its catalog of grievances against the old regime, and the transitional government took key steps to reduce them, but environmental problems were never more than one concern among many.

Like earlier generations of environmentalists who had seen their themes incorporated into homeland protection, Social Democracy, National Socialism, and the anti-nuclear crusade, the GDR's environmentalists stood to gain support and legitimacy by casting their lot with this broader social movement. Yet they had to worry that their own cause would be submerged in the process or that the movement would fail.

This dilemma was most acute for the GNU. Like the West German BfV and BN in the 1970s and 1980s, it found itself in competition with new environmental groups full of enthusiastic activists. Moreover, like the dissident factions in these organizations, the GNU's city ecology groups aimed to introduce a new and more confrontational agenda into the GNU. As had been the case for the BfV and BN, the result was sharp conflict between the more conservative old guard and the advocates of change. Assuming an active role in the movement offered the GNU the promise of accomplishments far beyond those it had achieved by working within the system, yet active participation in the movement might have threatened the resources it needed to survive, including especially the legitimacy, freedom from repression, cooperation, and financial support it received from the authorities. Moreover, active participation would also have required precipitous changes in organizational goals, strategies, and identity—changes resisted by members who had learned to work within, and gain rewards from, the existing system.

In the end, the forces arguing for a major reorientation within the GNU did not succeed. Perhaps the pace of events was simply too rapid. The GNU had hardly begun to adjust to Honecker's resignation, the removal of the wall, and the growing influence of organizations like the New Forum and Democracy Now when it was confronted with another new situation—reunification and competition for members and financial resources from the Green Party and West German environmental organizations. Then too, in a somewhat ironic rerun of events at the beginning of the GDR, the GNU's legitimacy was undermined by its history of accommodation to the previous system. Elements of the conservative core of the GNU did survive the transition, but only by annexing themselves to existing Western organizations.

The city ecology and church groups, with their more radical framings of environmental problems, more confrontational strategies, and supporters who were less integrated into the establishment, proved better adapted to play a role in the movement and in initial efforts to remake the GDR. They furnished the ideas and personnel basis for the environmental segment of the movement, just as the West German BIs had done almost two decades earlier. Yet in the long run, the results of the movement proved no more favorable for them than the denouement of the environmental and anti-nuclear movements had been for the BIs in the West. Contrary to the wishes of most members of the environmental groups, forces pressing for reunification quickly won the day. Rapid implementation of changes that quickly reduced the most visible environmental problems in the GDR followed, as did speedy adoption of government and business practices from the Federal Republic, where environmentalism was already institutionalized. These changes favored working within the system versus mobilizing against it, and unemployment and the uncertainties surrounding the rapid economic and political transformation worked against mobilization of an environmental movement. Some of the pre-unification environmental groups live on today as independent organizations or members of the Green League, but their situation is precarious.

These same factors, together with skepticism of large formalized organizations of all types, resistance to Western domination, and lack of a tradition of contributions to public interest groups, contributed to the initial failure of NABU, BUND, the WWF, and Greenpeace to make major inroads in the East. The years since reunification have, predictably, seen a fading of some of the differences between the old and new *Länder.* Yet as described in the next two chapters, differences in the social contexts of West and East remain, and the largest environmental organizations still find the East less fertile ground for their efforts.

Chapter 7

New Challenges at Century's End

The period of polarization and confrontation described in Chapter 5 continued into the 1980s. Yet by the beginning of the decade, new trends were already emerging, trends that both signaled the end of the period of polarization and shaped the challenges facing the large national environmental organizations as they entered the twenty-first century. This chapter discusses six such trends: the institutionalization of environmentalism, reduced coverage of environmental issues in the media, the decline of polarization and protest, the rise of competing political issues, the rise of new types of environmental issues, and changing public opinion about environmental issues. Although these trends did not begin at exactly the same time or develop at a constant rate, they have tended to be mutually reinforcing. By the end of the twentieth century, they had radically transformed the context within which the environmental movement and environmental organizations operated.

The Institutionalization of Environmentalism

Probably the most significant trend of the late twentieth century was the ongoing institutionalization of environmental concerns in political, economic, and other institutions. This trend, which originated during the early 1980s, accelerated throughout the decade. Indeed, by the 1990s, environmental protection was being taken into account in almost every aspect of German life.[1] While there is

[1] Good discussions of the trend toward institutionalization, together with supporting evidence, appear in the following: Brand (1993, 1995, 1999a, 1999b, in press); Oswald von Nell Breuning

some evidence that the process of institutionalization has slowed in recent years, there are few indications of reversals of past gains.

Institutionalization of Environmentalism in Government

Growing public concern about environmental issues and the visible success of the BBU and the Greens during the late 1970s and 1980s created considerable pressure on the government to act on environmental issues.[2] The Greens steadily gained strength in local and *Länder* governments during the early 1980s and surpassed the 5 percent of the national vote required for entry into the parliament in 1983. Once in the Bundestag, the Greens provided environmentalists with more direct access to the federal government, called attention to environmental issues by raising a stream of inquiries in the Bundestag, and used the funding that political parties represented in the national legislature receive from the treasury to present environmental concerns in a scientifically well-grounded and persuasive way. A number of independent environmental research institutes were established during the 1980s to provide expert reports about environmental problems to the Greens and other parties.

The other three parties responded to the successes of the Greens and to public opinion by increasing their attention to environmental problems and publishing position papers on environmental issues. The effects were especially visible in the case of the Social Democrats, who remained the strongest party in the governing coalition until 1982. After a government commission report at the end of the 1970s indicated that nuclear power was not really necessary for the country's future, the party began a gradual evolution from almost unqualified support of nuclear power to opposition. Shortly before the Social Democrats left office, their leader, Chancellor Helmut Schmidt, set aside the hard feelings precipitated by bitter confrontations over nuclear power and reopened channels of communication with environmental groups. The government launched several new legislative initiatives during its final years, including legislation to make chemical production safer and less polluting and measures to reduce the flow of phosphates into

---

Institut (1996); Rat von Sachverständigen für Umweltfragen (1996); Eder (1996); Brand, Eder, and Poferl (1997), and Bammerlin (1998).

[2]Most every treatment of German environmental politics written over the past quarter-century discusses or provides evidence of trends toward the institutionalization of environmental concerns, although some more recent works also note instances of backsliding. The account here draws on the following: Müller (1986); Weßels (1989); Padgett (1989); Hucke (1990); Voss (1990); Weidner (1991a, 1991b); Stoß (1991); Wilhelm (1994); Hey and Brendle (1994); Malunat (1994); Loos (1995); Jänicke and Weidner (1996); Pehle (1997, 1998); Jänicke, Kunig, and Stitzel (1999); Reutter (2001); Schreuers (2002); Bergstedt (2002); Rüdig (2002); Dryzek et al. (2003); Newig, (2003), and Blühdorn (2000, 2006, 2007). In addition, numerous articles discuss the evolution of the Green Party as a component of institutionalization. Examples include Zeuner (1991), Doherty (1994), Frankland (1995), O'Neill (1997), Hoffman (2002), and Blühdorn (2002). Sources that discuss the institutionalization of environmentalism in politics in the context of broader discussions of the evolution of the environmental movement include Mayer-Tasch (1985), Kazcor (1989), Schmid (1987), Rucht (1990, 1991a), Joppke (1993), Koopmans (1995), and Rucht and Roose (1999).

waterways. It also negotiated the first voluntary pollution reduction agreements with industry in areas like phosphates in detergents, minimization of cadmium use, and phasing out asbestos. Budgets for environmental cleanups and support of nature protection were increased, and the DNR began receiving a regular subsidy rather than funding for specific projects only.

The more conservative Christian Democrats had adopted a more cautious environmental program, and environmental issues were far from the top of the initial agenda of the Christian Democrat–led coalition that assumed power in 1982. The new government shelved a comprehensive environmental report initiated by the previous government, and it never developed a master plan of its own or showed much independent interest in nature protection. Nevertheless, it was confronted by extensive media coverage of acid rain and other environmental problems, widespread public concern about environmental issues, public support for BIs, environmental organizations, and the Green Party, as well as pressure for more ambitious measures, not only from these groups but also from the Social Democrats. In combination, these circumstances made an active approach to environmental problems unavoidable, and an improving economy made new initiatives and expenditures easier to undertake.

The Kohl government responded with new legislation and regulations, some of them based on initiatives begun by the previous government. Although some of the new measures were more symbolic than effective, they included controls on air pollution from power plants that were among the most stringent in the world, an ambitious program of recycling and reductions in use of packaging material, and improvements in water quality in the Rhine and other rivers. Expenditures on pollution control by government and industry kept pace with the legislative initiatives, and the government moved to introduce environmental education into the schools.

Some of the new initiatives reflected a change in emphasis from command-and-control measures and end-of-pipe filtering to voluntary agreements with industry and ecological modernization. Moreover, government, the major political parties, and the public began to accept the legitimacy of environmental issues and the premise that ecological reforms could generate new technologies and jobs. By the end of the 1980s, Germany occupied a leadership role in the environmental politics of the European Union and in international forums concerned with air quality, saving the ozone layer, warding off climate change, and developing sustainable energy sources. And while the Kohl government continued to support nuclear power in principle, cost overruns, faltering energy demand, opposition from other parties, and increased public and expert concerns about safety after the 1986 Chernobyl accident brought the program to a near standstill.

In the aftermath of the 1986 Chernobyl disaster, the government responded to public concerns by establishing an Environmental and Nuclear Safety Ministry.[3] It consolidated into one place numerous offices responsible for water and

[3] See Pehle (1998) for an especially comprehensive treatment.

air purity, nuclear energy, solid waste, toxics, soil conservation, and nature protection, which had previously been scattered over the Interior, Agricultural, and Health Ministries. It also absorbed the Federal Environmental Office, which had been established as part of the legislative initiatives of the early 1970s to monitor environmental quality, sponsor research, and advise the government. Parallel offices for radiation protection and nature protection research were created later.

The new environmental ministry represented a significant step in the institutionalization of environmentalism, but the results were not as unambiguously positive as environmentalists had hoped. The ministry did give environmental groups more opportunities for input into deliberations about legislation and administrative rulings; however, the input often occurred under restricted conditions, and environmentalists and ministry officials sometimes fell into mutual criticism. Moreover, few positions in the new ministry were filled by staff from environmental organizations. Finally, compared to powerful ministries linked to interest groups from industry, energy, construction, transportation, labor, and agriculture, the new ministry was perceived both by outsiders and its own staff as weak. It was understaffed and poorly funded, and it lacked the legal right or the informal power to veto proposals on environmental grounds.

The institutionalization of environmentalism in politics was echoed in the institutionalization of the Green Party. The realities of participation in *Länder* legislatures and, after 1983, in the Bundestag put pressure on the Greens to operate more as a normal political party, and doing so offered many potential benefits. The result was a long-running internal dispute between the "Fundis" (fundamentalists), who wanted to retain the Greens' image as the uncompromising anti-party party, and "Realos" (realists), who aspired to participate in the rituals of government and shaping of policy and were willing to compromise, or even seek to become part of a national coalition government. At times, the heated disputes threatened to immobilize the party, but the Realos gradually gained influence throughout the 1980s and 1990s, especially after the Greens were shocked by failing to gain enough votes to be represented in the parliament in 1990. A group of the most militant Fundis departed to form an unsuccessful splinter party in the early 1990s, leaving the Realos in firm control. They distanced the party from its most extreme positions and supporters and modified or eliminated policies, such as term limits and rotation of offices, which they saw as obstacles to efficiency and acquisition of legislative influence. The controversy continued to simmer for years thereafter, but the Greens have nonetheless evolved into a rather normal political party, albeit one with distinctive concerns.

The growing institutionalization of environmentalism in politics was also reflected in other opportunities for environmental groups to influence policy-making. Representatives of environmental groups began participating, for example, in the periodic meetings of the Council of Environmental Ministers, which includes both the federal and *Länder* ministers. A similar opportunity was provided by the Working Group for Environmental Issues, an independent discussion forum founded in 1971. Here representatives of environmental organizations met in small working groups with representatives of government agencies, industry groups, labor unions, consumer protection organizations, and scientific societies to exchange ideas about specific environmental issues. This group, frequently used by government agencies to obtain feedback about policy proposals, was funded by dues from its member organizations and a subsidy from the environmental ministry (Arbeitsgemeinschaft für Umweltfragen, 2003; Röscheisen, 2006).

Like the other changes outlined above, these increased opportunities for input evidenced the ongoing institutionalization of environmentalism in politics and policy during the 1980s. The period since the early 1990s has been marked, however, by more contradictory developments, which suggest a slowing of the pace of institutionalization.

During the early 1990s, several *Länder* dissolved their environmental ministries and allocated their responsibilities elsewhere or combined them with other ministries, and there were cuts in environmental budgets at the *Land* level. Meanwhile, national government placed several environmental initiatives, such as an update of the nature protection law and plans for requiring recycling of automobiles, on the back burner. By the mid 1990s, problems resulting from reunification and Germany's sustained economic downturn were providing new impetus for the old argument that overly rigorous environmental regulations were strangling economic growth. Relying on this rationale and citing the need to redevelop the new *Länder* as quickly as possible, the government instituted "simplifications" of environmental regulations. These changes, strongly criticized by environmentalists, allowed faster planning and permitting processes and reduced opportunities for citizen input. By the beginning of the new century, Germany also appeared to be losing its leadership role in EU environmental politics. Finally, the Working Group for Environmental Issues dissolved in 2003 due to internal conflicts and reductions in its government subsidy.

Nevertheless, not every recent political development has been negative. After serious losses at the polls in the early 1990s, the Greens rebounded to achieve 7 percent of the vote in the 1998 national elections. Because the Social Democrats—by far the leading vote getter—preferred them as a coalition partner, the Greens became the junior partner in the governing coalition, a long time goal of the party's Realo faction. As the junior partner the Greens occupied a relatively weak position, but they did achieve some successes. They were able to push through an ecology tax on fuel, a phaseout of atomic power, and a long-needed revision of the outdated nature protection law. The latter included provisions to regulate agriculture, provided for additional national parks, and allowed environmental organizations to file appeals of some administrative decisions that might lead to damage to nature (Naturschutz Heute, 2002a). After Germany's BSE crisis, the historically anti-environmental agriculture ministry was transformed into the Ministry for Consumer Protection and Agriculture and turned over to a well-known Green minister (Berliner Zeitung, 2001e; Lange, 2004).

On the other hand, the Greens had to settle for a small initial fuel tax with many loopholes, and scheduled further increases never took place. They also had to accept a thirty-year phaseout of nuclear power plants (Berliner Zeitung, 2000c), which left the environmental minister in the awkward position of defending the continued operation of nuclear plants and shipments of nuclear wastes against mass protests and critiques from moderate environmentalists (Hunold, 2001a, 2001b). Later, the environmental minister found himself embarrassed again when his plans to require soot filters on diesel automobiles were vetoed by the chancellor (DNR Deutschland Rundbrief, 2003f). Outcomes like these led to criticism from the remnants of the party's Fundi wing and from strong environmentalists outside the organization. They complained that the party was too willing to compromise and unable to accomplish much (Frankfurter Allgemeine Zeitung, 1998f, 1999g; Baukloh and Roose, 2002; Hoffman, 2002; Schreuers, 2002). In the fall of 2005, the Greens and Social Democrats failed to win a majority of seats in the parliament, and the Social Democrats joined the Christian Democrats in a "Grand Coalition" government, relegating the Greens to the opposition.

By the opening years of the twenty-first century, Germany had a well-developed body of environmental law and regulations, political parties that felt obligated to lend at least verbal support to environmental protection, and an array of administrative agencies to conduct research and enforce the laws. While mutual suspicion, efforts to exclude environmentalists from decision making, and accusations of "greenwashing" have not disappeared, experts from environmental groups now consult regularly with government officials at all levels, and relations have improved in recent years. It is not clear that the changes that have occurred can actually reduce Germany's ecological footprint, but at the end of the day, it is clear there has been genuine change in the sense that environmental concerns have become well institutionalized in politics and government.

The trend toward institutionalization of ecological ideas in German politics was also evident in right-wing politics.[4] The Ecological Democratic Party (ÖDP), founded by conservatives who defected from the Green Party in 1981, suffered a split in 1990 when it tried to move closer to the mainstream and distance itself from far-right members and positions; however, the far-right splinter group proved even weaker than the ÖDP, and none of the conservative ecology parties attracted enough support to be represented in the national or *Länder* legislatures. Nevertheless, they and other right-wing groups did succeed in keeping the connection between conservative thought and ecology alive, and other right-wing groups, including the "new right," also incorporated ecological themes into their discourse, mainly by portraying population growth resulting from immigration as a threat to Germany's environment. This line of argument was prominent in the Heidelberg Manifesto, which was issued by a number of right-wing intellectuals in 1982, and the same themes found their way into the platforms of right-wing political parties like the Republicaner, which achieved several striking but short-lived electoral success at the end of the 1980s.

[4] Most treatments of German environmental politics have little to say about the far right, but a good bit has been written on the topic. See Jahn and Wehling (1990), Jaschke (1990), Wüst (1993), Geden (1999), and Bergstedt (2002).

Institutionalization of Environmentalism in Business

The institutionalization of environmental concerns in German society[5] has not been limited to government. By the end of the 1970s, German business had also taken note of public concern about environmental issues. Moreover, it had begun to realize that environmentally friendly production technologies sometimes increased efficiency and that there was money to be made in marketing environmental cleanup technology in Germany and abroad. Business thus began to back away from its more or less across-the-board opposition to environmental reforms and cooperate and consult with less confrontational environmental organizations. An increasing number of firms also initiated processes of ecological modernization, developing new technologies that require less raw materials and using efficient production techniques to reduce emissions rather than adding expensive end-of-pipe filtering measures to existing plants. Indeed, Germany has become a leader in these technologies. In 1985, a number of leading German businesses joined in the founding of the German National Working Group for Environmentally Conscious Management to promote environmentally sound practices. In 1989, the Federation of German Industry and BUND issued a joint statement acknowledging industry's responsibility for environmental protection, and even Shell, boycotted by environmentalists during the mid 1990s because of its plan to sink oil platforms in the North Sea, now invests heavily in solar power and renewable energy. Concerned about their public image, many firms also began to develop ecological plans, publish annual environmental reports, and establish communication programs to publicize their achievements and influence public opinion (Clausen, 2002). Some even began awarding prizes to environmentalists and environmental organizations (Bergstedt, 2002).

There is room for dispute about the extent to which perceptions that business has become more environmentally responsible reflect real change versus successful public relations efforts. A survey in the early 1980s showed that managers in Germany were considerably more likely than those in the US and UK to believe that business could help to solve environmental problems; however, they were just as likely as managers in the other two countries to perceive polarization between them and the environmental movement (Dierkes and Fietkau, 1988). Critics continue to characterize business' approach as primarily "greenwashing," and it can be argued that business is unlikely to give up the basic commitment to consumer capitalism that drives Germany's ever increasing ecological footprint. Still, on balance, it is clear both that some change has occurred and that business has become more adept at presenting itself as ecologically aware.

---

[5] In addition to the sources cited in note 1, especially Brand, Eder, and Poferl (1997), information about the institutionalization of environmentalism in business is available in Zillessen and Rahmel (1991), Hey and Brendle (1994), Rucht and Roose (1999), and Felbinger (2003). A number of treatments of German environmental politics, e.g. Müller (1986), Hucke (1990), Weidner (1991a), Wilhelm (1994), Jänicke, Kunig, and Stitzel (1999), Schreuers (2002), and Blühdorn (2000, 2006, 2007), also include discussions of institutionalization of environmental concerns in business.

Institutionalization of Environmentalism in the Labor Movement

The 1980s and 1990s also saw a rapprochement between environmentalists and labor unions.[6] By the beginning of the 1980s, some unions had begun to rethink their undifferentiated opposition to environmental initiatives and called for improvements in production technology to reduce pollution and make work processes safer. In 1982, the German Federation of Labor (DGB), the umbrella organization for the unions, elevated environmental protection to one of its official goals. This was followed by a formal endorsement of the environmentalists' longstanding position that only environmentally sound jobs are apt to be secure in the long run and that anti-pollution requirements result in a net gain in jobs. The chemical industry's union joined environmentalists in promoting legislation to reduce pollution and accidents, although the miners and metal workers continued to resist proposals to reduce air pollution from coal burning plants well into the 1980s.

The 1980s and 1990s also saw a spate of efforts to build cooperation between labor and environmentalists through consultations, joint conferences, and joint projects in a wide range of areas. These included transportation policy, reducing resource consumption, groundwater pollution, energy policy, jobs in the environmental sector, forest management, ecologically sound building materials, ecological tax reform, climate protection, and north-south trade relations. In 1991 the DNR and the DGB agreed on a joint statement of principles to guide future cooperation, and the two also issued a joint statement on the ecology tax. Unions and environmental groups also worked together in preparations for the 1992 Rio Conference and continued to consult after the conference, and the DGB was a cosponsor of the 1992 German Environmental Day.

These steps, however, failed to convince all union members and leaders that environmental measures were benign. Claims that environmental measures are job killers continue to surface periodically, and the increasingly difficult economic climate of the later 1990s and early twenty-first century has done nothing to reduce tensions.

Environmental Issues in the Media

The 1980s saw sustained and relatively comprehensive media attention to environmental issues, including acid rain, dying forests, repeated chemical spills in the Rhine, the Chernobyl disaster, and Germany's growing list of endangered species;[7] however, it appears that media attention to the environment began to decline in the 1990s.

---

[6]For discussions of labor and environmentalism see Teichert (1992), Sander (1992a, 1992b), S. Krüger (2000), Rogall (2003), and Röscheisen (2006). Some discussions of environmental politics—e.g., Müller (1986), Rucht (1990), and Schreuers (2002)—are also relevant.

[7]Major sources used in preparation of this section include the following: Leonhard (1986); Voss (1990, 1995); Brand, Eder and Poferl (1997); Bergstedt (2002); Börnecke (2003); and Rogall (2003).

Studies of newspaper reporting showed substantial coverage of environmental issues in both national and regional newspapers during the 1980s and early 1990s (Leonhard, 1986; Brand, Eder, and Poferl, 1997). The *Tageszeitung*, a nationally circulated daily founded in 1979, led the way by giving environmental and other new social movement issues especially heavily coverage. However, all available evidence points to a sharp decline in media coverage in recent years (Bergstedt, 2002; Börnecke, 2003; Groth, 2003; DNR Deutschland Rundbrief, 2003c; Braun, 2003).

During the environmental movement, Germany also experienced a flowering of countercultural weekly newspapers, which covered environmental issues heavily, but most have since ceased publication or moved toward coverage of lifestyle issues. General interest magazines also regularly featured environmental stories in the 1980s, and numerous new magazines focused on nature and environmental issues were founded. A 1989 inventory of nationally circulated magazines devoted exclusively to environmental issues counted thirty-seven, with a total circulation of four million, but the number is believed to have declined since then (Jänicke, Kunig, and Stitzel, 1999).

Studies of environmental reporting in the broadcast media suggest a similar pattern. During the 1980s and early 1990s television news and documentary programs dedicated to the environment proliferated; however, informed observers (Bergstedt, 1998; Rogall, 2003; Groth, 2003) believe that coverage on television has declined in recent years, and my own fairly extensive viewing of German television over the past few years indicates that—with the exception of nature shows—environmental themes are far from prominent.

Decline of Polarization and Protest

A third significant trend over the last two decades has been diminishing polarization of the population over environmental issues, accompanied by a general trend toward less reliance on protest strategies.[8] These developments can be traced to at least three major causes.

First the successes of the environmental movement, as well as its institutionalization in politics, business, and other fields, reduced environmentalists' propensity to view themselves as an isolated, embattled minority whose only real chance for garnering attention for their problems and influencing the political system was

[8]The following are among the more important of the large number of sources that discuss or document a trend toward declining of polarization and protest beginning in the 1980s: Dierkes and Fietkau (1988); Kazcor (1989); Brand (1993, 1999a, 1999b, in press); van der Heijden, Koopmans, and Giugini (1992); Joppke (1993); Hey and Brendle (1994); Rucht (1994); Koopmans (1995); Opp (1996); Christmann (1996); Rat von Sachverständigen für Umweltfragen (1996); Oswald von Nell Breuning Institut (1996); Brand, Eder, and Poferl (1997); Bammerlin (1998); Diani and Donati (1999); Rucht and Roose (1999); Rootes (1999a); Hoffman (2002); Blühdorn (2002); Bergstedt (2002); and Dryzek et al. (2003).

confrontational protest. These changes provided increased opportunity to participate in the policy formation process and consult with business and labor, in turn creating incentives for environmentalists to reenter mainstream politics and adjust their tactics accordingly.

At the same time that opportunities for input into the system were increasing, the countercultural critique of German society from the left, which had undergirded the confrontational environmental movement, was waning. As countercultural milieus, countercultural media, and networks of new social movement activist groups weakened during the 1980s and 1990s, many of the environmental movement's supporters became more mainstream. This decoupling of environmentalism from its strongest countercultural connections made it less prone to polarizing stands and strategies and less threatening to the German establishment.

Third, by the late 1980s, the close alliance between the environmental and anti-nuclear movements, which had made the two almost indistinguishable and sparked heated and violent confrontations over environmental issues, had all but dissolved. In part this occurred because more moderate environmental activists, who had learned the drawbacks of confrontational protest through hard experience, began to question the efficacy of continued confrontation. Moreover, as it became evident that Germany's nuclear power program was dead in the water, many anti–nuclear power activists began to shift their attention to the peace movement and other causes, leaving the anti–nuclear movement in the hands of a radical fringe of young activists with a strong anti-statist orientation and a propensity for violent confrontation.

The balance of the evidence suggests that these trends have resulted in a decline in the number and size of protest actions, though it is less clear that protests have become less confrontational. Statistics from two separate data series collected from newspaper reports, both of which included anti–nuclear power protests (Rucht and Roose, 2001b, 2001c; Rootes, 2002), show a noticeable decline in the number of environmental and anti-nuclear protests in the second half of the 1980s and early 1990s, a return to higher levels in 1993 and 1994, mostly due to anti–nuclear power protests, but another decline in the mid 1990s. Data for the number of participants varies more erratically but follows a similar pattern. Unfortunately, no data are available for years after 1997, and data about the level of confrontation in protests were available only until 1994. The latter do not reveal a clear pattern of declining levels of confrontation. No more recent quantitative data have been reported; however, almost all of my interviewees agreed with the president of BUND (Zahrnt, 2002) that the era when environmental organizations could regularly and easily mobilize mass protests is over, and the great majority of well-informed observers share this view.

This does not mean that environmental protest has disappeared from the German scene or is likely to disappear in the foreseeable future; however, confrontational protest has become more focused on a few issues—mainly those involving continued use or introduction of risky technologies, such as oil tankers, nuclear power, and genetically modified crops. Shipments of radioactive waste, in particular, are still regularly accompanied by major demonstrations (Kolb, 1997; Berliner Zeitung, 2001b; Hunold, 2001a, 2001b). Still, the percentage of the population mobilized around this issue is much lower than in the protest wave of the 1970s and 1980s, and the phaseout of nuclear power is likely to weaken them further. International protests against globalization at the international level also draw their share of German participants; however, their failure to frame issues in a way that makes participation seem efficacious or to present a positive alternative has made it difficult for them to attract broad public support.

The Rise of Competing Issues

A fourth major trend in the social context of the environmental organizations was the development of two new and very pressing issues that competed with environmental problems for the attention of the German public and policy makers. The first of these was the dramatic transformation of the GDR and its self-annexation to West Germany. The second is the serious and sustained economic downturn that has beset Germany since the mid 1990s.

Effects of Reunification

The reunification of East and West Germany—unquestionably the most significant single event in postwar German history—played a major role during the 1990s in shifting the focus of public and government attention away from environmental issues.[9] It required extension of West German political and economic institutions to the five new *Länder,* massive expenditures to upgrade the East's decrepit infrastructure, cleanup of the GDR's legacy of massive pollution, privatization or shuttering of East German industry, and coping with the mass unemployment that resulted.

The ecological benefits of closing polluting industries and partially cleaning up toxic waste sites were impressive, even if they were partly offset by new problems connected with skyrocketing rates of car ownership, the extension of consumerism to the East, appropriation of green space for new developments on the periphery of cities, and road construction. Unfortunately, the enormous amounts of money channeled into improving infrastructure and housing and repairing the worst environmental damage—financed in part through a tax surcharge—did not result in the hoped-for economic boom in the East. Instead, unemployment spiraled upward and remained lodged at about double West German levels. The heavy costs of redevelopment and unemployment also drained the public purse, burdened the economy, and diverted attention and funding away from less pressing environmental problems in the West.

---

[9] On the effects of reunification on German environmentalism see the following: Rink (1991, 1999); Hey and Brendle (1994); Wilhelm (1994); Brand (1995, in press); Hirche (1998); Schreuers (2002); and Nölting (2002).

Germany's Economic Difficulties

Since the mid 1990s, Germany has been beset by economic stagnation and persistent high unemployment.[10] This economic downturn led to a clear shift of government and media attention, as well as public concern, away from environmental issues to job creation and economic growth. It also gave renewed strength to the argument that excessive environmental regulation was undermining Germany's ability to compete, and discourse about environmental issues was increasingly framed in these terms. Finally, economic uncertainty tends to undermine the postmaterialist values that occupy a key role in the belief systems of many environmentalists. More recently, economic worries have become linked to rising concerns about immigration, globalization, and terrorism, further distracting Germans from environmental problems.

The Rise of New Environmental Issues

As several of my interviewees pointed out, another new challenge confronting environmentalists results from changes in the kinds of environmental problems Germany faces.[11] The gradual transition from an industrial to a service economy and implementation of pollution controls has reduced the availability of "easy David and Goliath targets" like egregious chemical spills and pollution-belching smokestacks. Improvements in visible pollution were especially dramatic in the East following the shutdown of polluting factories and power stations, but reductions in the most visible pollution have occurred nationwide. Salmon thus swim once again in the Rhine, even if their existence there remains precarious.

These changes have not gone unnoticed by the general public. Studies sponsored by the environmental ministry showed that the percentage of Germans who rated environmental conditions in West Germany as "rather good" or "very good" increased from 55 percent in 1991 to 81 percent in 2004, while the percentage rating conditions in East Germany this way increased from 2 to 48 percent. Those who rated environmental laws as sufficient increased during the same period from 23 to 41 percent (Preisendörfer, 1999; Gruneberg and Kuckartz, 2003; Bundesministerium für Umwelt, 2004). Similarly, opinion surveys conducted by the Allensbacher Institute showed that the percentage of Germans who

---

[10]Germany's economic problems are almost universally viewed as having pushed environmental issues away from center stage. See, for example, Malunat (1994), Brand, Eder, and Poferl (1997), Blühdorn (2000), Nölting (2002), and Brand (in press).

[11]Academic discussions of the changing nature of environmental issues and the resulting effects on environmentalism include the following: Beck (1993); Brand (1993, 1999a, in press); Hey and Brendle (1994); Koopmans (1995); Opp (1996); Oswald von Nell Breuning Institut (1996); Brand, Eder, and Poferl (1997); Rohkrämer (2002); Bergstedt (2002); and Dryzek et al. (2003). The phenomenon is also a subject of concern to the leaders of the organizations. See, for example, Die Zeit (1991), Zahrnt (1991), Sachs (1996), and Bode (1996a, 1996b).

viewed the environment as somewhat damaged declined from 67 to 28 percent between 1990 and 2004 (Umwelt: Kommunale Ökologische Briefe, 2004).

As the most visible problems declined in severity, they were replaced by new issues: global warming, overreliance on automobile transportation, loss of species diversity due to deforestation in faraway lands, overuse of fertilizers and pesticides, loss of open space to suburbanization, and high consumption levels in developed nations. Problems like these are often transnational; they tend to develop incrementally, frequently occurring out of the immediate view of the average citizens. Their effects can take years to materialize, they create risks that are hard to calculate and subject to competing claims, and they require complex and costly solutions, not mere end-of-pipe filtering.

Environmentalists have tried to address these problems by embracing the concept of "sustainable development." It calls for ecological modernization of industry, government initiatives to direct behavior and investment into projects that decrease resource use and pollution, and public education about needed lifestyle changes. However, setting clear goals that, if achieved, would contribute significantly to solving these new ecological problems has proven very difficult. The public finds itself instead confronted by confusing, technical disputes among environmentalists, industry, and government, all of whom claim to want sustainable development and environmental protection. As the president of one NABU *Land* chapter pointed out to me, this confuses citizens, who find it hard to distinguish genuine change from greenwashing.

Finding solutions to the new environmental problems also requires sustained involvement in the nuts and bolts of politics and policy making, along with a great deal of international cooperation. This makes these problems harder to attack than the older ones. As one of my Green Party interviewees pointed out, it is relatively easy to interest the well-informed party or environmental organization leaders in international issues, but ordinary members are mainly interested in events near home. For all these reasons, mobilizing the public to work on today's environmental issues and achieving success is more difficult than tackling the relatively "easy" issues that the environmental movement addressed in its earlier stages.

Changing Public Opinion about Environmental Issues

Public concern about environmental issues remained high during the 1980s. Pollution problems receded only gradually as measures implemented in the 1970s and 1980s began to show results,[12] and there was a continuing flow of news stories about environmental disasters. Especially important in the public mind was

---

[12] For discussions of continuing environmental problems in the 1980s and press treatments of them see the following: Meroth and von Moltke (1987); Paterson (1989); Weßels (1989); Voss (1990); Hucke (1990); Weidner (1991a); Brand (1993, 1999a, in press); Wilhelm (1994); de Haan (1995); Jänicke and Weidner (1996); Schreuers (2002); and Oberkrome (2003a).

the concern that acid rain and air pollution were resulting in *Waldsterben,* the death of the beloved German forests. There was also a rash of chemical spills, and pollution of the Rhine and other waterways continued to cause fish kills and temporary bans on drinking water from the river or from nearby wells. Pollution of the North and Baltic Seas also received considerable press attention, especially in connection with fish and seal kills. Air quality in the Ruhr continued to be a concern, and on a few occasions, limitations had to be imposed on auto travel, industrial production, and use of high-sulfur fuels. Early reports of the ozone hole also began to surface during the 1980s. Meanwhile, demonstrations at nuclear plants kept the issue of nuclear power before the public, and the 1986 Chernobyl disaster and the German government's ineffective response spawned a powerful, although short-lived, wave of protests by organizations like Mothers against Nuclear Power.

National surveys throughout the decade showed high levels of environmental concern, and sentiments that more action was needed were widely held. After Chernobyl, public opinion also swung sharply against nuclear power plants.[13] Eurobarometer polls and polls by the Electoral Research Group, for example, revealed steady increases during the 1980s in the percentage of Germans who rated environmental issues as "very important" (Dalton, 1994). Annual Ipos polls conducted for the interior ministry between 1984 and 1989 showed that, in comparison to other items in a list of political issues, environmental protection was rated as "very important" by the highest or next-highest percentage of the population in every year but one (Voss, 1990). Approval of environmental organizations also rose steadily throughout the 1980s (Noelle-Neumann and Köcher, 1993), and in a 1986 survey, 13 percent of respondents claimed to have donated money to an environmental organization and 4 percent said they had participated in a protest (Dalton, 1994). Germany's level of environmental consciousness also remained above the European average into the early 1990s (Commission of the European Communities, 1992; Dalton, 1994; European Commission, 1999; Wurzel, 2002).

Since the 1990s, the institutionalization of environmentalism, improvements in the most obvious environmental problems, the more complex nature of the new environmental problems that replaced them, the rise to prominence of other, seemingly more pressing issues, and declining press coverage have evidently combined to reduce the priority of environmental problems in the public mind.[14] In a series of Eurobarometer studies, the percentage of Germans who characterized the environment as an immediate and urgent problem declined from 89 percent

---

[13] For overviews see Müller (1986), Dierkes and Fietkau (1988), Voss (1990), Dalton (1994), Diefenbacher (1994), Brand, Eder, and Poferl (1997), and Wurzel (2002). For research reports on specific studies see, for example, Noelle-Neumann and Köcher (1993) and Billig (1994).

[14] For general discussions and additional documentation of the decline in public concern with the environment see Bundesministerium für Umwelt (1996, 1998, 2004), Wimmer and Wahl (1995), Noelle-Neumann and Köcher (1997), INRA (Europe) - ECO (1995), Preisendörfer and Franzen (1996), Fritzler (1997), Preisendörfer (1999), Hagedorn, Stawowy, and Meyer (2002), Blühdorn (2002), Gruneberg and Kuckartz (2003), and Kuckartz and Rheingans-Heintze (2006).

in 1992 to 70 percent in 1999 (Wurzel, 2002). The percentage of respondents to a series of EMNID surveys who said that they were very concerned about environmental problems declined from 61 percent in 1990 to 35 percent in 1997 (Preisendörfer, 1999), and Allensbacher Institute data for the period 1990 to 1996 showed that the percentage of respondents who saw implementing environmental protection as decidedly important declined from 81 to 61 percent in West Germany and from 82 percent to 66 percent in the East. The percent of Allensbacher respondents who said they were quite especially interested in environmental protection also declined, from 42 to 30 percent in the West and 49 to 27 percent in the East (Noelle-Neumann and Köcher, 1997). In a series of surveys sponsored by the environmental ministry, positive responses to a set of items about willingness to pay additional taxes or fees to improve the environment showed declines between 1991 and 1998, and willingness to pay higher prices or cut one's standard of living to protect the environment declined sharply in another survey (Bozonnet, 2004).

Time series data that extend into the early twenty-first century show little evidence of a reversal of this pattern. A series of EMNID and environmental ministry surveys, for example, showed that the percentage of respondents who ranked environment as a very important political issue varied erratically between 1993 and 2004 but ended the period down from just under 60 percent to 45 percent. Unemployment remained the top concern throughout the eleven years, and by 2004 ensuring the security of the pension system, stimulating the economy, ensuring social justice, health care, education, and crime had all moved from below to above environmental concerns in importance. Moreover, despite some ups and downs, a series of less direct indicators used in environmental ministry surveys between 1996 and 2004 showed no overall pattern of increasing environmental consciousness (Gruneberg and Kuckartz, 2003; Bundesministerium für Umwelt, 2004; Kuckartz and Rheingans-Heintze, 2006).

Decline in concern about the environment was even more precipitous in questions using open-ended items. In a series of EMNID and environmental ministry surveys that allowed respondents to nominate a maximum of two problems as the most important problems facing Germany, the percentage of respondents who mentioned the environment declined from 66 percent in 1989 to 28 percent in 1996. By the end of the 1990s the percentage had declined into the teens, where it has remained since. The relative importance of environmental issues did, however, remain a bit higher using this approach than in surveys with closed-ended items. In 2004, environmental issues, mentioned by 18 percent of respondents, tied for third place, ranking far below concerns about the labor markets (55 percent) and at about the same level as worries about the economic situation and concerns about social welfare and social justice (Preisendörfer, 1999; Gruneberg and Kuckartz, 2003; Bundesministerium für Umwelt, 2004; Kuckartz and Rheingans-Heintze, 2006).

There is also some evidence from recent studies that things may get worse before they get better. For many years, research consistently showed greater environmental consciousness and activism among the young than any other group; however, the generation of young environmental activists from the 1970s and early 1980s has now reached middle age, and younger cohorts have not taken up the mantle, a phenomenon some analysts describe as the "graying of the greens." Environmental ministry surveys in 1996 and 1998, for example, continued to show the higher levels of self-reported environmental consciousness among younger adults but revealed an increasing tendency toward an inverted U pattern for participation in environmental organizations, with the highest rates among middle-aged persons (Preisendörfer, 1999). By the early years of the new century, this inverted U pattern had appeared for most measures of environmental consciousness, with respondents under the age of twenty-four showing the lowest environmental consciousness of all on some items (Gruneberg and Kuckartz, 2003; Bundesministerium für Umwelt, 2004; Reusswig, 2004). By 2004, a pattern of noticeably lower environmental concern in the under-24 age group dominated the results (Bundesministerium für Umwelt, 2004; Kuckartz and Rheingans-Heintze, 2006), and other surveys of youth were showing steady declines in the salience of environmental concerns (Schuster, 2003; Kuckartz and Rheingans-Heintze, 2006).

The environmental groups are well aware of these changes. Several environmental group leaders interviewed by Brand and his colleagues during the mid 1990s said that persuading the press to pay attention to environmental issues had become ever more difficult as public interest declined (Brand, Eder, and Poferl, 1997). A column written by Jochen Flasbarth, NABU's longtime president, on the occasion of his resignation to accept a job in the federal government exemplifies the organizations' worries especially well. After noting NABU's accomplishments with pride, Flasbarth also noted the difficulties of keeping environmental concerns high on the national agenda when unemployment was high, environmentalism was being portrayed as a brake on economic growth, and proposals were being voiced to relax environmental regulation to promote growth (Naturschutz Heute, 2003a). Certainly, environmental issues have been overshadowed by economic and other issues in recent federal electoral campaigns (Blühdorn, 2002).

None of this should be interpreted as indicating that Germans have lost all interest in environmental issues. As leaders of the environmental organizations repeatedly pointed out to me in interviews, the absolute declines in public concern about the environment have not been large. Environmental concerns have simply fallen behind other issues that have ballooned in perceived importance. Crises like the discovery of BSE in German cattle herds in 2001 can also still ignite fears, attract press attention, and provide the environmental organizations with an opportunity to push for their agenda (DNR Deutschland Rundbrief, 2001a; Lange, 2004); however, public reaction to such crises tends to be short-lived. Finally, there are some signs from poll data that public interest in environmental issues might have begun to stir again as Germany entered the twenty-first century (Gruneberg and Kuckartz, 2003; Kuckartz and Rheingans-Heintze, 2006).

Effects of Changes on the Environmental Movement

The institutionalization of environmentalism and melioration of some of the most visible environmental problems has failed to still the voices of radical environmentalists. They condemn the ongoing environmental destruction, label the changes as more cosmetic than real, and bemoan the decline of environmental protest and environmental consciousness (Hermand, 1991; Bergstedt, 2002). Some more careful scholarly analyses (e.g., Blühdorn, 2000, 2006, 2007) have also questioned whether recent changes have actually set the country on a path to sustainability.

What is more important in the context of this book, however, is that the changes created a new situation for the environmental movement and the major environmental organizations. They found themselves operating in a social context where 1) every major social institution was addressing environmental problems and publicizing its claims to do so, 2) media coverage of environmental problems had declined, 3) countercultural ideologies and polarization over environmental problems had declined and mobilization for protest had become more difficult, 4) competing issues, such as reunification and economic stagnation, had displaced environmental problems as top public concerns, 5) the most threatening new environmental problems were less visible, more complex, more apt to cross national boundaries, and 6) environmental issues were perceived as less pressing by the general public.

None of these changes were favorable for sustaining environmental movement mobilization, so it is hardly surprising that the German environmental movement at the beginning of the new century was being characterized by both scholars (Hey and Brendle, 1994; Blühdorn, 1995, 2002; Brand, 1999a, 1999b, in press) and the press (e.g., Frankfurter Allgemeine Zeitung, 1997a; Die Zeit, 1999a) as becalmed and its supporters as somewhat disheartened and uncertain about how to proceed. By 2004, even Germany's Council of Environmental Experts was expressing concerns about a slowdown of progress on the environmental front (Berliner Zeitung, 2004), and the president of NABU (2005a) and the executive secretary of BUND (2005) both expressed concern over growing resistance to environmental concerns.

Effects on Citizens' Initiatives and the BBU

The changes described above represented a challenge for almost every segment of German environmentalism, but the difficulties were especially great for organizations that had grounded their approach in countercultural ideology and protest strategies. The most important of these groups were the BIs and their umbrella organizations. Most informed observers (e.g., Oswald von Nell Breuning Institut, 1996; Brand, Eder, and Poferl, 1997; Weinbach, 1998; Bergstedt, 2002; Brand, in press) agree that the number of environmental BIs has been on the decline since the 1980s; however, there has been very little research about this. Surprisingly, Rucht and Roose found an increase in the number of environmental groups in

Berlin between 1988 and 1997 (Rucht and Roose, 2001b), but there is no way to be sure whether this trend can be generalized to the rest of the country or continued beyond 1997. Nor do we know much about how the goals and strategies of present-day BIs compare with those of earlier decades.

What is beyond dispute is that national and regional networks of BIs experienced a steep decline beginning in the late 1980s. Indeed, by the beginning of the new century, the premier network, the BBU, had all but disappeared.[15] Its sharp decline is important for two reasons. First, it illustrates the problems faced by the confrontational wing of the environmental movement in the changing circumstances of the late twentieth and early twenty-first century. Second, the BBU's decline represented an important change in the social context in which the large, national environmental organizations operate.

Although it remained influential and highly visible well into the 1980s, the BBU proved poorly adapted to the trends described above. As the anti–nuclear power movement began to decline, it attempted, with some success, to build alliances with the peace movement; however, the high levels of mobilization reached by the peace movement during campaigns against the stationing of additional American nuclear weapons in Germany proved short-lived. The BBU also introduced a new emphasis on dangerous chemicals, and, somewhat later, it turned its attention to dying forests.

These new initiatives were unable to compensate for the BBU's many problems. Long-running internal disputes over toleration of violent protest, cooperation with the authorities, and the relative emphasis on national-level lobbying versus emphasizing local action and providing service to local initiatives all rekindled in the early 1980s. The conflict between the BBU board and the editorial staff of the organization's magazine, which had moved to the left and become ever more critical of the BBU, also escalated. Finally, the BBU's large and influential working group on transportation, which had been demanding more autonomy and influence, left the BBU in 1985.

The BBU also had to struggle with financial problems. In 1981 the government, responding to press criticism that the government-subsidized organization was promoting illegal demonstrations, cut its subsidies, and the election of Kohl in 1982 signaled the end of federal government financial support, which had accounted for about half the BBU's small budget. The organization never succeeded in garnering much financial support from its member groups or private donors, so the result was a downward spiral involving repeated financial crises and cuts in programs and services. These included closing down the affiliated magazine, staff reductions—ultimately to one person—and diminishing political influence. The BBU's visible and respected leader, Jo Leinen, left the organization in 1984 to become a *Land* environmental minister. BBU leaders also had to struggle against

---

[15] For a comprehensive summary see Markham (2005b). Other useful accounts of the recent history of the BBU include Kempf (1984), Kazcor (1986, 1990, 1992), Brand, Büsser, and Rucht (1986), Rucht (1991a, 1993a), Leonhard (1986), Hucke (1990), Schenkluhn (1990), Joppke (1993), Koopmans (1995), Ehlers (1995), and Bergstedt (2002).

court cases in which they were accused of instigating illegal demonstrations, and the tax-free status of some BIs was withdrawn.

A final factor in the organization's decline was competition with more formally organized and institutionalized competitors. The most active of these was the rapidly growing BUND, which attracted former and potential BBU supporters who wanted an organization that took an activist approach but also functioned more like a well-organized interest group. Some BIs reconstituted themselves as local chapters of BUND, and some individual members of the BBU transferred their membership to BUND. Additional competition for donations came from the highly visible activities of Greenpeace, while the growing success of the Green Party provided yet another alternative to the BBU.

Plagued by the combination of changing social conditions, internal conflicts, loss of funding, and competition from other organizations, the BBU began to experience significant losses of member groups during the early 1980s. New BIs did continue to join, but some of them soon ceased operating, and some long-term member BIs dissolved. The BUND and some of its *Land* chapters also joined the BBU, as did Greenpeace and Robin Wood, but rather than strengthen the organization, this merely introduced members that were not BIs. Other groups abandoned the BBU in favor of networks focused on specific topics, such as traffic control, while still others withdrew because they wanted complete autonomy.

By the end of the 1980s, the BBU had lost most of its member groups and influence, and the pattern of decline and continuing financial crisis continued throughout the 1990s. It still exists in greatly weakened form (Bundesverband Bürgerinitiativen Umweltschutz, 2005, 2007, no date), but it has only a handful of member organizations and attracts little notice.

By the mid 1990s, most smaller regional networks of environmental BIs, including those affiliated with the BBU, had also collapsed, and there was evidence that the Green League, the BBU's counterpart in the East, was experiencing difficulties.[16] At the end of the 1990s, the Green League had regional branches in all five of the Eastern *Länder* and approximately 100 local groups, but recent years have brought increasing financial difficulties and a decline in the number of member groups. The regional branch in Mecklenburg-Western Pomerania dissolved in 2002 due to insolvency. Financial difficulties caused the chapter in Saxony-Anhalt to close its business office in 2003. It was dissolved in 2004.

Similar problems have arisen in recent years for Greenpeace's more radical and less formally organized offshoot, Robin Wood. Its activities—especially its defense of the German forests against *Waldsterben*—won it high visibility during the 1980s. After a long and heated discussion of whether the step threatened its commitment to remain an activist, grassroots organization, Robin Wood initiated use of a professional fundraising agency to distribute mass-mail appeals for donations

---

[16] For information about the evolution of the Green League see Weinbach (1998), Rink (2001), Rucht and Roose (2001b), Nölting (2002), and Bergstedt (2002). Useful documents from the organizations itself include Grüne Liga (1999, 2003, 2005) and Grüne Liga Berlin (2001).

in 1988. As a result, its budget and number of donors expanded dramatically, reaching €1,000,000 and about 20,000 donors. In 1994 membership totaled about 3,000 organized into thirty regional groups, and its magazine had a circulation of 11,000 (Rauprich, 1985; Rucht, 1989; Scholz, 1989; Blühdorn, 1995; Bergstedt, 1998). However, this success proved short-lived. Membership declined during the mid 1990s (Rucht and Roose, 1999, 2001c), and as of 2003, its web site claimed 2,300 members in twenty groups (Robin Wood, 2003a), and its magazine's circulation was down to 9,000 (Robin Wood, 2003b). By 2007, only thirteen groups were listed (Robin Wood, 2007a), and the most recent budget report available showed that the organization had receipts of just under €900,000 in 2004 and had operated at a loss in 2003 and 2004 (Robin Wood, 2007b).

Consequences for Environmental Organizations

The goals, structures, and strategies of the large, national environmental organizations suited them better than the BBU, Green League, or Robin Wood to adapt to the trends described above, and their larger memberships and financial resources provided them with more leeway to do so. Nevertheless, developments over the past two decades have proved exceptionally challenging for them as well, and to succeed they had to develop an innovative and appropriate mix of goals and strategies. They proved quite adept in doing so and emerged in the new century as the most resilient and successful components of German environmentalism. Indeed, by the early 1990s the largest four—NABU, BUND, the WWF, and Greenpeace—had become key actors on the environmental scene (Brand, 1995; Koopmans, 1995; Oswald von Nell Breuning Institut, 1996; Brand, Eder, and Poferl, 1997).

After very rapid growth in the 1980s, all four of these organizations experienced a slowdown in growth during the early years of the 1990s, but only Greenpeace lost members (Hey and Brendle, 1994; Rucht and Roose, 1999, 2001c). Greenpeace's support did not begin to grow again until after the turn of the century, and BUND's membership has stagnated for several years. Nevertheless, the general pattern since the 1990s has been one of steady, if not always spectacular, growth (see Chapter 9)—a striking contrast to the experiences of the BBU, Green League, and Robin Wood. These four organizations also enroll far more members than the Greens, or any other political party, number among the more successful fundraising organizations in German society, and enjoy far more public confidence than other actors in the environmental arena. In short, their ability to cope successfully with recent developments has made them the "success story" of German environmentalism.

Summary and Conclusions

As described in Chapter 5, a number of the traditional nature protection organizations were able to prosper during the West German environmental movement by adopting framings and problem definitions, goals, and strategies from the movement. The changes required of them were not trivial, and they sometimes engendered internal conflict, but they allowed the BfV, the BN—and to a lesser extent the WWF—to grow rapidly by riding the wave of social movement mobilization, media attention, and public interest. BUND and Greenpeace, as new organizations founded during the movement, adopted the same combination of nature protection goals with new environmental ones and of traditional and confrontational strategies, albeit with more emphasis on the new. The WWF, NABU, BUND, and Greenpeace thus emerged from the 1980s as the largest and most influential environmental organizations.

The central role of these four organizations in German environmentalism has, if anything, been cemented by developments over the past two decades; however, to reach their leading position they had to master a new set of new challenges. The developments that challenged them had begun to emerge even as the wave of polarization, social movement mobilization, and confrontation crested during the 1980s. By the end of the century, these trends—the institutionalization of environmentalism in major institutions, decreasing coverage of environmental issues in the media, the decline of polarization and protest, the appearance of competing political issues, the rise of new environmental issues, and the declining relative priority of environmental issues for the public—had created a new and challenging social context for environmental groups of all kinds.

The perils of clinging to old framings, strategies, and goals under changed conditions were thrown into stark relief by the fate of the BBU and the decline of other organizations with ideologies and operating styles closely linked to a confrontational environmental movement that was reaching the end of a cycle of mobilization. Reluctance to negotiate, propensities to define government and business as the enemy, and resistance to professionalization proved poorly suited to a period when doors were opening to those who were prepared to work within the system. Confrontational ideologies and crepe hanging proved out-of-sync with an era where countercultural ideologies were fading, making a living was no longer easy, and visible environmental problems were in decline. An ideology of grassroots involvement and heavy reliance on volunteers did not fit well with widespread job insecurity, rising labor force participation by women, and declining public concern about the environment.

Nor could this style of organization easily provide the kind of continuity required to work within the system. Reliance on strategies of protest proved out of tune with a period when environmental problems were less visible and more complex, the public believed that things were improving on the environmental front and that other issues were more pressing, and a financially strapped government had few reasons to continue funding confrontational groups—even if they had somewhat ironically come to depend on it for their livelihood. Nor were strategies of protest well suited to a period in which media showed less propensity to cover environmental scandals. In short, previously successful ideologies and tactics proved poorly suited to a period when the last wave of mobilization had crested and conditions less favorable for social movement mobilization had emerged. As a result, the BBU, Green League, Robin Wood, and similar organizations found themselves short of the member support, funds, legitimacy, and influence they needed in order to remain successful.

Whether these groups' failure to successfully adapt resulted from poor strategic choices by their leaders, inability to quickly alter commitments to worldviews and patterns of action from the past, or insufficient slack resources to survive a period of adversity—or, as is more likely, some combination of these—remains to be clarified (but see Markham, 2005b). What is most noteworthy is that major environmental organizations, by contrast, did manage a successful adaptation to changing circumstances. Chapter 8 describes this new social context and the possibilities and challenges it offered in detail, and Chapters 10 to 12 are devoted to the four organizations' strategies for coping with the strategy dilemmas that the transformed social context posed for them at the beginning of the new century. How could they structure themselves to deal with a situation where addressing technically complex issues within complex systems of governance had become more rewarding, while committed volunteers were becoming harder to find? How could they procure the financial resources they needed to work within the new context of a declining economy and the need to maintain their historical commitments and image of independence? To what extent should they temper or replace ideologies, goals, and tactics from the environmental movement era with new ones? Questions like these provide the focus for the remainder of the book.

Chapter 8

THE SOCIAL CONTEXT OF ENVIRONMENTAL ORGANIZATIONS AT THE BEGINNING OF A NEW CENTURY

The changes described in the previous chapter—the institutionalization of environmentalism, decreased coverage of environmental issues in the media, diminished polarization and protest over environmental issues, the rise of competing political issues, the appearance of new environmental issues, and the declining relative priority of environmental issues in the public mind—combined to produce many new challenges and some new opportunities for Germany's major environmental organizations. They find themselves today in a complex social context in which they must adapt to demographic and geographic factors, economic and political institutions, the growing importance of the European Union, the characteristics and preferences of the German media, other environmental groups, and the views and needs of potential contributors and volunteers. Understanding the dilemmas the organizations face and the strategy choices they have made as they seek to obtain the resources they need and accomplish their goals in a new century requires a close look at this context. German readers will, predictably, find much here that is already familiar—though perhaps some new insights and information as well. For other readers, my goal is to offer a summary treatment of the external environment within which the organizations were operating as they entered the twenty-first century.

Demographic and Geographic Factors

With eighty-two million inhabitants, Germany has the largest population in Europe.[1] Like most of its European neighbors, it has a very low birth rate and an aging population, a source of considerable financial strain for its well-developed welfare state. Even after substantial immigration, slow population decline is projected for the next half-century. Consequently, any increases in the country's ecological footprint will result from new technologies or changes in consumption patterns, not population growth.

Germany is a densely settled land. At 230 persons per square kilometer, its population density is among the highest in Europe and almost eight times that of the United States. Density is greatest in the highly industrialized areas of west central Germany and lower in the former East Germany. Genuine wilderness and large animals have all but disappeared from Germany, and there are few intact ecosystems. Consequently, popular conceptions of "nature" and "nature protection" more often involve preserving beautiful rural landscapes or carefully managed forests than untouched wilderness.

The burdens of a dense population, increasing urban sprawl, and widespread industrial and agricultural activity affect most of the country directly, making loss of open space, provision of adequate park land and nature protection areas, siting of waste disposal facilities, and air and water pollution potential issues. The high impact of human activity on nature is also evident; 40 percent of all animal species in Germany are listed as threatened, ranking it among the highest in Europe for threatened wildlife (Bundesamt für Naturschutz, 2007). Unlike the US or UK, Germany, with nine other countries on its borders, is greatly affected by air and water pollution generated by its neighbors and subject to complaints from its neighbors about the pollution it generates. Under these conditions, purely national solutions to environmental problems are rather obviously of limited utility.

Economic Institutions

Germany's well-developed consumer economy is the largest in Europe. Its per capita national product is above the European average, although well below that of the US and the most prosperous European nations. A growing service sector provides almost two-thirds of jobs, but the manufacturing sector remains prominent and provides a higher proportion of jobs than in most developed economies. Mining and heavy industry have been in gradual decline for decades, but several of Germany's core industries—steel, automobiles, and chemicals—place high demands on the environment, as does its highly developed transportation network.

---

[1] For basic statistical information and comparisons to other nations see German Federal Foreign Office (2006) and Central Intelligence Agency (2007). Useful overviews of these factors and the resulting environmental problems include Jänicke and Weidner (1996), WWF-Journal (1999g), Lange (2000), Schreuers (2002), Wurzel (2002), and BUNDMagazin (2003d).

Despite high taxes on gasoline and efforts to maintain and promote mass transportation, Germany has a high and increasing rate of automobile ownership. Its per capita emissions and energy consumption compare well with those of other industrialized nations, but the sheer size of the economy necessitates a rather large ecological footprint and raises concerns about air and water pollution, waste disposal, and the nation's contribution to global warming.

Germany's economy is highly dependent on exports, making international competitiveness a prime concern and a possible lever for business to use against environmentalists; however, Germany is also a leading exporter of environmental technology, and by 2001 Germany's growing and successful environmental protection sector employed almost 4 percent of the labor force (DNR Deutschland Rundbrief, 2001d). Exports are required to finance the nation's high dependence on imported energy. Only coal occurs in significant amounts in Germany, and coal, including highly polluting soft coal, remains heavily used. In the last decade the German economy has been plagued by anemic growth, high labor costs, and unemployment. These trends have provided talking points for those seeking to characterize environmental protection as a threat to prosperity.

Business and Industry

The activities and policies of German business affect the national environmental organizations in both positive and negative ways.[2] In the first place, the activities of business firms generate many environmental problems, for despite the ongoing ecological modernization of some sectors of the German economy, the ecological impact of a highly industrialized consumer economy remains substantial. And although German business has recognized that environmental cleanup measures sometimes save money and has become more willing to negotiate with environmental organizations in recent years, it has a long history of using its influence to block, delay, or lessen the impact of environmental initiatives. Business and environmental organizations thus remain frequent opponents, as business often views environmental regulations as too costly and opposes them. Industries that contribute most heavily to Germany's pollution problems, such as the automobile and chemical industries, are major players in the German economy, exert great political influence, and have often been resistant to environmental initiatives.

German business is a formidable opponent. A very high proportion of German businesses belong to powerful national business federations, which employ about 120,000 persons to represent them in Berlin and the *Länder* capitals (Ro-

---

[2] The relationship between German business and environmental organizations has been the subject of considerable research and commentary. This section relies on the following sources: Katzenstein (1987); Weßels (1989); von Alemann (1989); Zillessen and Rahmel (1991); Dalton (1994); Hey and Brendle (1994); Rat von Sachverständigen für Umweltfragen (1996); Oswald von Nell Breuning Institut (1996); Brand, Eder, and Poferl (1997); Pehle (1997, 1998); Bammerlin (1998); Jänicke, Kunig, and Stitzel (1999); Reutter (2001); Dryzek et al. (2003); and Rogall (2003). For a history of the response of German industry to environmental problems see Wey (1982), and for commentary from the BUND's president, also a PhD economist, see Zahrnt (2002).

gall, 2003). The most important of these organizations is the Federation of German Industry, which has chapters in each *Land*, as well as crosscutting groups for each branch of the economy. By one estimate (Rogall, 2003), 40 percent of the members of the Bundestag, the more powerful house of the national legislature, belong to a business or professional association, and the more conservative political parties enroll many activists with strong ties to business. Industry is well represented in expert commissions and informal working groups that propose and shape legislation, and its influence is strongly felt by the environmental ministry. Consequently environmental organizations that choose to engage in lobbying must usually reckon with well-organized and highly professionalized opponents.

Business often tries to block or modify the environmental initiatives it opposes by arguing that they go too far and will damage the economy. For example, in 2005 the executive secretary of the Federation of German Industry mounted a strong argument that Germany's commitment to the Kyoto Protocol should be relaxed because it was injuring the country's competitive position (DNR Deutschland Rundbrief, 2005b). This is a potent argument, not only at the national level but also for hard-pressed local governments competing with one another to create jobs. Another common claim is that more study is needed to determine what steps need to be taken. This argument defines environmental problems as technical issues and often leads to extended discussions among dueling experts, thus delaying action. A third typical strategy is pushing for voluntary environmental standards as an alternative to government regulation, and business and industry groups have entered into numerous nonbinding agreements with government in recent years. This reflects, in part, the growing institutionalization of environmental concerns. Nevertheless, such agreements can also serve to avert more stringent government regulation, and they sometimes come in for criticism from environmentalists on these grounds, as occurred when the BUND criticized a voluntary agreement by industry to reduce greenhouse gases (DNR Deutschland Rundbrief, 2000i).

German business works hard to build public support for its positions. It carefully monitors environmental news and its own environmental image, drawing, for example, on detailed information about press reporting collected by the German Economic Institute (Voss, 1995; Institut der Deutschen Wirtschaft Köln, 2007). The Federation of German Industry also monitors the actions of environmental organizations and other NGOs it sees as its opponents, and develops elaborate plans to counter them (e.g., DNR Deutschland Rundbrief, 2002b). Business invests a great deal of money in well-staffed and highly professional press offices and communicates its environmental achievements through annual environmental reports and press releases (Clausen, 2002). Indeed, entire textbooks (e.g., Hopfenbeck and Roth, 1994) explain how business can present its environmental record. Critics (e.g., Zahrnt, 1991; Ehlers, 1995; Schönborn, 1995; Bergstedt, 2002) interpret these programs as evidence of German business' increasingly skillful use of "greenwashing" to defuse environmental concerns. Greenwashing strategies include press releases crafted by public relations departments, establishment of industry-sponsored groups with names that invoke images of environmental concern, insincere offers to negotiate with environmental groups on business' terms, and donations or sponsoring arrangements for selected environmental causes. Such efforts only to influence public opinion, they deprive environmental organizations of their claim to be the only advocates for the environment.

On the other hand, business can also be a partner in efforts to protect the environment.[3] About 450 businesses, including leading firms such as Airbus, Lufthansa, and Siemens, belong to the German National Working Group for Environmentally Conscious Management (Zilessen and Rahmel, 1991; DNR Deutschland Rundbrief, 2001g; B.A.U.M., 2007), which promotes green business practices; however, many of these same firms are ecologically problematic and members of environmentalists' frequent nemesis, the Federation of German Industry (Bergstedt, 2002). Two smaller organizations, the Association of Young Entrepreneurs and the Green Entrepreneurs Association, which represents mainly smaller firms, have been even more willing to cooperate with environmentalists. Specific industries have at times been willing to work with environmental organizations to develop declarations of common principles or joint programs, and individual firms are sometimes willing to cooperate or negotiate about making their operations more environmentally sound. A few firms, such as the Otto mail order house, have been willing to commit themselves to major reforms and form long-lasting working relationships with environmental organizations. Potential allies also exist in Germany's growing renewable energy and pollution control industries. Their associations are active, albeit not highly influential, in lobbying and public education. Finally, environmental organizations and business sometimes form alliances of convenience, such as cooperative efforts with the insurance industry to combat global warming. Clearly then, there are ample opportunities for both conflict and cooperation between environmentalists and business.

Business can also be a source of financial support for environmental organizations;[4] however, it is far from an ideal funding source. The level of donations available is dependent on the state of the economy, and business has, in general, preferred donating to sports and cultural activities over environmental causes. Some businesses even rule out gifts to environmental organizations altogether. Businesses that do donate often do so more out of a desire to polish their image than out of commitment to the environmental cause, and many firms have shifted from outright donations to "sponsoring" arrangements. Here businesses provide funds out of their public relations or advertising budgets and see their donations as one side of an exchange relationship in which environmental

---

[3] For specific examples of cooperative relationships between industries or firms and environmental organizations see Chapters 9 and 12.

[4] The following sources provide useful information about business firms as donors to environmental causes: Zilessen and Rahmel (1991); Haßler (1993); Schreiner and Wörnle (1994); Drews (1995); Grüßer (1995); Rat von Sachverständigen für Umweltfragen (1996); Luthe (1997); Urselmann (1998); Haibach (1998b, 2006), and Bauske (2003).

organizations are expected to provide benefits to the donor firm in return for its financial support.

Businesses with this orientation prefer to support projects that are appealing to the public and to have their contributions publicized to the maximum extent possible, and they are generally inclined to support only less confrontational environmental organizations. Even when sincerely interested in helping the environment, they are also leery of pressures from partner environmental organizations to modify their practices. Dalton (1994), for example, found that the most confrontational European environmental groups received the least financial support from business and government, and a survey of German businesses showed that they sought partners primarily among environmental organizations that were well-known and seen as responsible by their clients, customers, and stockholders (Halcour, 1995). Daimler-Benz, for example, listed the following criteria for choosing environmental projects to sponsor: clear goals that are broadly accepted in society, professional management, opportunities to use Daimler-Benz's know-how, and the understandability of the project to the general public and political decision-makers (Luthe, 1997). Commerzbank produced a similar list. It seeks partners that have the following characteristics: a positive image that can contribute to the bank's reputation, members who are potential bank customers, members throughout Germany, technical competence, and willingness to work with the bank. And even the ecological pioneer Otto mail order firm avoids supporting highly confrontational organizations (Zillessen and Rahmel, 1991). These criteria favor environmental groups that are inclined to work within the system. It is thus not surprising that respondents in Halcour's (1995) survey of German businesses rated the WWF most positively and Greenpeace at the bottom. Indeed, one WWF leader I interviewed readily acknowledged that this probably occurs because the WWF's goals are less problematic for business.

Agriculture

Another important force in the social context of the environmental organizations is the agriculture industry and the politically influential German Farmers' Association. This lobbying group has over a million members, but it is dominated by large farm operators. It has a long history of opposing efforts to reduce the considerable environmental damage caused by agricultural practices like overuse of fertilizers and herbicides and poor land management. It has also resisted efforts to promote organic agriculture (von Alemann, 1989; Bammerlin, 1998; Jänicke, Kunig, and Stitzel, 1999; Rogall, 2003).

Relationships between the organic foods industry and environmental groups are predictably better (Oswald von Nell Breuning Institut, 1996; Rat von Sachverständigen für Umweltfragen, 1996; Berliner Zeitung, 2000a, 2000b), and the Agrarian Alliance, an association of organic farming, environmental, and animal protection groups, works to promote agricultural reform and sustainable agriculture (Bammerlin, 1998; AgrarBündnis, 2006).

Labor Unions

Labor union membership in Germany ranks in the low to middle range for industrialized nations, but the percentage is about twice that of the US.[5] As in many other industrial nations, union membership has been in slow decline. German labor today is organized into nine large national unions, a number that has been falling due to mergers. Each union covers an industry or several related industries and negotiates wages and benefits with the employers' associations on the national level. All but the union of public officials are members of the German Federation of Labor (DGB), which represents them in public debate but does not exert tight control over them. The unions have historically been closely allied with a major political party, the Social Democrats. They have considerable influence within it; however, they maintain formal independence and are willing to work with other parties. Their large memberships, high level of organization, and financial resources make them a significant economic and political force, and they enjoy considerable access to government. A recent study found that 14 percent of the members of the *Bundestag* were union members (Rogall, 2003).

Alliances with labor are politically attractive for environmentalists, and unions became more open to cooperating with environmentalists during the 1980s and 1990s (see Chapter 7). The DGB and its constituent unions today all affirm their commitment to environmental goals and sustainable development and publish numerous statements and position papers on the environment. Unions and environmental groups have common interests in ensuring that industrial production processes do not threaten human health, and they can easily agree on the desirability of improving environmental standards in less developed countries. Like all citizens, union members would benefit from a cleaner environment, and advocating for a popular cause like environmentalism has the potential to burnish the unions' public image.

Still, in practice, many obstacles to lasting and broad-based alliances remain. Environmental protection has not traditionally been a core concern of unions, and there is a considerable cultural and educational gap between the average union member and the average environmentalist. At the policy level, arguments for limiting consumption to achieve sustainability can threaten the historical modus vivendi, in which unions and management vied for their share of the spoils from ever-increasing production. New environmental controls are still sometimes viewed as threatening prosperity and jobs, especially by workers in ecologically problematic industries. The unions' understanding of sustainable development inclines more toward ecological modernization than to reducing consumption, and some unions are much more committed to environmental issues than others.

---

[5] For basic statistics about German labor unions and comparisons to other countries see Visser (2006) and DGB (2007). Useful treatments of the relationship between labor unions and environmentalists include von Alemann (1989), Oswald von Nell Breuning Institut (1996), S. Krüger (1998, 2000), Reutter (2001), and Rogall (2003).

Permanent working relationships between labor and environmentalism have thus proved elusive, especially in the harsh climate of the late 1990s and the early years of the new century.

In short, environmentalists can reasonably hope to find influential collaborators among labor unions at times and for specific issues, but there are many obstacles to across-the-board, permanent alliances.

Political Institutions

The German political system is also an important part of the social context in which environmental organizations operate.[6] Direct government activity, such as road building, river channelization or dredging, designation of national parks and wildlife reserves, sewage treatment and waste disposal, and establishment of nature reserves, has major impacts on the environment. Government's more general goals, such as economic growth or sustainable development, can also have enormous environmental effects. Also consequential for the environment is government regulation of business and consumer activity, including industrial emissions and consumer packaging, emissions standards for autos, and tax rates for various energy sources. The priorities of the national and *Länder* governments and the political parties also play a major role in determining which environmental issues are "in play" at any given time. Finally, government is an occasional source of funding for the environmental organizations' projects. Consequently, government is a major target for influence attempts by environmental organizations.

Formal Political Structures

Germany has a decentralized federal system in which only a few policy domains are the sole responsibility of the national government or the sixteen *Länder*, and environmental policy is one of the many areas of shared responsibility.[7] In most areas of environmental policy, the *Länder* can legislate only if the federal government has not put a law in place or if *Land* law goes beyond national legislation without contradicting it. In other areas, including water protection and nature protection, the national government is authorized by the constitution only to set very broad guidelines through national legislation; the *Länder* then pass specific implementing legislation.

All national legislation is initiated by the more powerful of the legislature's two houses, the directly elected Bundestag, while the views and interests of the *Länder* are represented by their appointed representatives to the Bundesrat. The consent of

---

[6]See Katzenstein (1987), Hey and Brendle (1994), Bammerlin (1998), and Dryzek et al. (2003).

[7]For descriptions of the formal structure and functioning of the German political system see references cited in note 6 and the following: Bulmer (1989); von Alemann (1989); Smith (1989); Wilhelm (1994); Koopmans (1995); Jänicke and Weidner (1996); Pehle (1997, 1998); Jänicke, Kunig, and Stitzel (1999); Wurzel (2002); Schreuers (2002); and Rogall (2003).

the Bundesrat is required for legislation that directly affects the *Länder,* including almost all environmental legislation. A standing conference committee attempts to reconcile the views of the two houses when there are differences of opinion. Primary responsibility for some environmentally relevant issues, such as sewage treatment and garbage collection, is delegated to local governments, and they also have a role in implementing legislation from higher levels of government.

Drafts of national legislation are typically developed by the relevant ministries in a long process of consultation with interest groups and other relevant ministries at the national and *Land* levels. The legislators rarely have the staff or expertise to evaluate drafts of legislation themselves, making them dependent on the ministries and interest groups for information. The most important ministry for environmental matters is the Federal Ministry for Environmental and Nature Protection and Nuclear Reactor Safety, but decisions impacting the environment are made in many different ministries, including agriculture, transportation, etc., not monopolized by the environmental ministry. Many of these ministries have their own environmental sections.

Some ministries, such as the ones responsible for transportation, construction and housing, agriculture, and economics and technology, have traditionally been so tightly allied with business or agriculture that they often oppose or seek to undermine environmental proposals. The economics ministry, for example, often supports business claims that environmental legislation will hinder German's international competitiveness. As a relatively new ministry with a fragmented constituency, the environmental ministry is in a weak position in negotiations with these powerful ministries. The situation with the long recalcitrant agriculture ministry was, however, transformed—at least briefly—at the beginning of the new century. Following Germany's outbreak of mad cow disease, it was combined with offices for consumer protection and placed under the leadership of a Green Party minister (Berliner Zeitung, 2001e; Lange, 2004).

The environmental ministry does have access to considerable environmental expertise. Its Federal Environmental Office and the Federal Office for Nature Protection are large, semi-autonomous bureaus staffed by experts. They function in an advisory role and have considerable funds to support research and public education. Additional expert advice is available from the Council of Environmental Experts, an independent government-funded advisory group comprised of six leading environmental scientists and supported by a substantial staff. The council conducts research and issues regular reports on the state of the German environment and the effectiveness of environmental policy. Similar but less influential bodies advise about sustainable development and global environmental issues. Finally, the Bundestag has at times appointed ad hoc commissions of legislators and experts to prepare comprehensive reports about topics such as air pollution policy and energy policy.

National-level legislation often leaves the development of implementing regulations for environmental laws to the bureaucracy. The relevant *Länder* ministries are much larger and better-developed than those at the federal level, so the federal government generally relies on them to implement even national-level legislation. These arrangements open the possibility of inadequate implementation of impressive-sounding legislation, and there are frequent charges that this is the case.

To function effectively, this system requires extensive input from and consultation among various levels of government and ministries, as well as a relatively high degree of consensus among them, and there is a well-developed communication network regarding environmental legislation and regulations. One example is the Council of Environmental Ministers, comprised of ministers from the *Länder* and the federal government. It meets periodically to discuss environmental issues and pending legislation, and has affiliated working groups comprised of representatives from the federal and *Länder* agencies responsible for all the major areas of environmental policy (Konferenz der Umweltminister des Bundes und der Länder, 2007). It also meets bianually with representatives of the major environmental organizations and the DNR (DNR Deutschland Rundbrief, 2002c).

This complex system of divided, and sometimes overlapping, responsibilities and jurisdictions provides environmentalists with multiple avenues of influence at the national, *Land,* and local levels through lobbying of legislators and the relevant ministries. However, the system is so complex, and the expertise of the frequent opponents of environmentalists is so great, that effective political action in such a context is difficult to carry out without a large professional staff and significant financial resources. The system of dispersed responsibilities also favors the development of *Land* and local subunits of national environmental organizations.

Environmental organizations can also seek to improve environmental conditions by monitoring or providing input into administrative decisions or—in limited circumstances—by filing complaints against inappropriate decisions.[8] German law establishes elaborate procedures for issuing permits to potentially polluting facilities, for land use and transportation planning, and for approval of facilities that might damage protected areas or species. However, environmental organizations that wish to affect the outcome of such proceedings face many barriers.

In general, German environmental law on citizen input into such decisions and the right to challenge them in court is based on existing law about damage from industrial emissions, which conceived injury in terms of damage to private property. Protection of the general public in government decisions about such facilities was entrusted solely to the authorities, so there are no provisions for input from environmental groups or lawsuits on behalf of the public interest. Only parties that can show direct damage can file lawsuits challenging government decisions, so environmental organizations have standing in the courts only when they own or are able to acquire property that would be directly affected.

Moreover, until 1994 Germany had no counterpart to the US Freedom of Information Act, so access to documents and records from administrative deci-

---

[8] In addition to references cited in notes 6 and 7, the following sources address environmental organizations' rights to be consulted, appeal, and file lawsuits: Gebers et al. (1996); Jannasch (2001); Blume, Schmidt, and Zschiesche (2001); and Stein (2003).

sions relevant to the environment was limited to those legally entitled to participate. New legislation required by the European Union and passed in 1994 increased access to such information, but the EU has criticized the German legislation as inadequate, and there have been complaints that agencies continue their noncooperative approach to requests for information.

There is one important exception to these generalizations. Recognizing that nature cannot speak for itself in administrative hearings, German law since 1976 has allowed officially recognized nature protection organizations the right to file position statements, access information, and participate in hearings regarding new nature protection laws, planning documents, and government-initiated projects that may have impact on nature or the landscape. The large, national environmental organizations can be certified to participate, but so can many "nature using" groups, such as organizations of hunters or horseback riders. Local BIs, on the other hand, are excluded. Government is required to solicit and consider input from these organizations, but not necessarily to act on it, and until 2002 the groups entitled to participate had no right under national law to file lawsuits to challenge the decisions except on the grounds that they had not been consulted. Moreover, some significant decisions, such as nuclear plant construction, were not covered by the legislation.

Before 2002, laws in all but a few *Länder* did give recognized environmental groups limited rights to appeal decisions in areas such as road construction or exceptions to regulations about nature protection areas on substantive grounds, and the organizations were occasionally able to make good use of these. A revision of the federal nature protection law put in place in 2002 allowed such lawsuits nationwide.

These procedures have made it possible for environmental organizations to exert some influence through participating in hearings, submitting position statements, and filing complaints and lawsuits where these are allowed, and the recent liberalization of the laws may have opened the door a bit wider. However, careful monitoring of planned actions at the local level and large time investments are required. So, too, is considerable expertise, both about potential damage to the environment and about environmental law and administrative procedures. Environmental organizations still sometimes receive notice only a short time before hearings are scheduled—often at times inconvenient for volunteers—and background materials may be available only by going to government offices to inspect or copy them.

Political Parties

Environmental organizations can also attempt to influence environmental policy by influencing elections or by cooperating with parties and influencing their platforms.[9]

---

[9]This section is based, in part, on my own fairly extensive reading of the German press over the past decade. In addition to sources cited in note 6, other useful discussions of the parties and their

German political parties are well-funded and generously staffed organizations that play an influential, central, and continuous role in the German political system. In addition to membership dues, they receive both substantial state subsidies keyed to their proportion of the vote in the last election and donations from businesses—including occasionally scandal-provoking illegal contributions. Parties are expected to perform a key role in the political education of the public, and each party receives significant public funding for a party-affiliated foundation for political education and research. Only about 3 percent of voters are formal party members. Well-educated professionals, public employees, representatives of interest groups, and persons seeking political contacts or careers are especially likely to belong.

German voter turnout in national elections is high (79 percent in the 2002 national election) but has been declining (Rogall, 2003). Parties must receive at least 5 percent of the votes cast to be represented in the Bundestag, which elects the head of government, the chancellor. Since World War II, a single party has attained an absolute majority and formed the government on its own only once, in 1957. All other national governments have been coalitions. With the exception of "Grand Coalitions" between the two largest parties, which governed in the late 1960s and since 2005, all have involved one of the two major parties and one small party.

The German Social Democratic Party (SPD), the first of the major parties, was historically a socialist labor party closely allied with the unions, but it distanced itself from its socialist roots in 1959. This allowed it to broaden its base, attracting especially supporters from the professional middle class. Today it attempts to appeal to a broad public, and its supporters range from center to left, which leads at times to internal conflicts. The SPD has a history of intermittent concern with environmental issues, especially insofar as they affected the health of the working class, but environmental issues did not become key issues for the party until the 1970s. Today the party has a detailed environmental program, and its recent alliance with the Green Party to form a governing coalition forced it to take environmental issues even more seriously. Nevertheless, the SPD's strong labor union base can lead to ambivalence about environmental initiatives that might threaten jobs, and the SPD has been more likely to promote ecological modernization than radical reform.

The other major "party," the CDU/CSU, consists technically of two parties, the moderate to conservative Christian Democratic Union and its somewhat more conservative Bavarian sister party, the Christian Social Union. Both draw

---

environmental views include the following: Dyson (1982); Wey (1982); Padgett (1989); Smith (1989); Koopmans (1995); Beucher (1995), and Rogall (2003). Readers interested in the current environmental positions of the parties will find abundant material on their Internet sites: Christlich Demokratische Union Deutschlands (2007); SPD (2007); FDP (2007); CSU (2007); Bündnis 90/Die Grünen (2007); and Die Linke.PDS (2007).

support from business owners and managers, small proprietors, rural areas, and social conservatives. Like the SPD, the CDU/CSU attempts to appeal to a broad spectrum of voters. Its constituents range from moderate to very conservative, but it has longstanding commitments to social justice and the welfare state. Environmental problems have never ranked among its top concerns, although it has at times adopted relatively progressive stances on environmental issues. It continues to publicly espouse environmental protection, but there are few strong advocates of environmentalism in its ranks, and it opposes many environmental initiatives. There is also a cultural gap between the CDU/CSU and environmental groups. For example, one CDU legislator I interviewed complained at length about environmentalists who appeared inappropriately dressed at hearings and proposed what he saw as completely unrealistic measures.

The smaller parties active at the beginning of the twenty-first century included the Free Democratic Party (FDP) and the Party of Democratic Socialism (PDS). The latter, the successor to the GDR's ruling party, has significant strength only in the five *Länder* of the former GDR, but it allied itself with a new leftist party in the West for the 2005 election. The FDP represents primarily small business owners, self-employed professionals, and some white-collar employees. Its platform combines classic economic and social liberalism, but it is a more moderate party than its US counterpart, the Libertarians. In the late 1960s and early 1970s, the FDP, governing in coalition with the SPD, seized the initiative in environmental policy, but in the long run this emphasis proved difficult to reconcile with the party's ideology. As one FDP legislator and former member of the Greens with whom I spoke ruefully admitted, the party now has almost no profile in environmental politics.

The best known of Germany's smaller parties, the Greens,[10] is often wrongly understood abroad as simply an "environmental party." Environmental issues are indeed of great importance to the Greens, but the forces that wanted a party focused solely on environmental issues were defeated early in its history. Today the Greens pursue a broad agenda that includes environmental issues, women's issues, an end to compulsory military service, economic reforms, and greater inclusion of ethnic minorities. Although much reduced in comparison to earlier decades, tensions remain between the dominant faction, which advocates compromise as the price of having a say in the government, and fundamentalists, who brook little compromise. The result has been awkward intraparty struggles over such issues such as whether to support demonstrations against shipments of atomic wastes, shipments the party had already agreed to allow as part of the government's plan for withdrawal from nuclear energy production (Hunold, 2001a, 2001b; Berliner Zeitung, 2001c). The Greens have become almost exclu-

---

[10] The Greens have attracted more scholarly attention than the other parties. In addition to sources cited in notes 6 and 9, see Frankland (1995), Kleinert (1996), O'Neill (1997), Hoffman (2002), and Wiesenthal (2004).

sively a party of the former West Germany. They are not represented in any of the legislatures of the former East German *Länder*, and outside Berlin they have little representation in local government (Nölting, 2002).

The far right is represented in German politics today only by fringe parties with a history of factional infighting and schisms.[11] They have only occasionally attained the 5 percent of the vote required for representation in German legislatures, and then only at the *Land* level. When they address environmental issues, they usually attempt to coopt them for their own purposes. For example, they emphasize the environmental impact of Germany's dense population as a springboard for arguing against immigration. Very small right-leaning ecological political parties, such as the Ecological Democratic Party (ÖDP), which moderated its position after a schism in the mid 1990s, also continue to be active; however, none have attracted enough votes to be represented even in *Länder* parliaments.

The structure of the German political party system poses a difficult set of challenges for large, national environmental organizations. Attempting to influence elections through political contributions or endorsements of specific candidates—the strategy followed, for example, by the Sierra Club in the United States—is an unappealing strategy in Germany. Campaigns are almost entirely publicly financed, and as several of my interviewees explained, party discipline is so tight that it would make no sense to endorse individual candidates in hopes that they would break ranks with their party over environmental issues.

Alternatively, environmental organizations could commit themselves to continuing support of a single party that would agree to place environmental issues near the top of its agenda—at present the Green Party. Recent experience shows that such a party can garner enough votes to be represented in the Bundestag and become part of the governing coalition, but this strategy is not without its pitfalls. Organizations that tie their futures too closely to a single party risk losing their identity and independence. Indeed, the Greens are already a competitor for volunteers and—as BUND's press spokesman once noted in an interview (Die Zeit, 1998)—press attention. Outright endorsement of a single party could also drive away supporters who belong to other parties and bind the future of the environmental organizations too closely to the fate of a small party with a possibly precarious future.

A final strategy is to attempt to influence the actions of whatever government is in power through nonpartisan lobbying, petitions, public education, and protests. This approach offers the chance to exert at least some influence continuously and to attract and retain members from all parties. It sacrifices, however, the chance to take clear stands on elections, even when one party or coalition offers a much better environmental program than its opponents.

---

[11] Although they are neglected in most general discussions of German parties, a great deal has been written about right-wing parties. For sources that focus on their environmental views see the following: Wüst (1993); Biehl (1995); Statham (1997); Bergstedt (2002); Geden (1999); and Olsen (1999).

Political Traditions and Political Culture

The relationship between environmental organizations and the political system is also shaped by Germany's political traditions and culture.[12] Historically, Germany had a strong state bureaucracy that provided many services but also regulated many spheres of life. The postwar constitution limited the power of the centralized state bureaucracy by introducing a stronger national legislature, a decentralized, federal structure, and a stronger role for political parties; however, a powerful state bureaucracy remains an important assumption of German political culture. German officials typically view the state as representing the interests of the society as a whole, not simply as a mediator between competing interest groups.

This assumption, along with a desire to avoid the excesses of the past, has made the state disinclined to tolerate strong dissent from either end of the political spectrum. Far-right parties have several times been declared unconstitutional and abolished, and strong protests from the left during the period of confrontation were countered with repression. These traditions have been weakened somewhat as the generation shaped by the oppositional political culture of the 1960s and 1970s has assumed power, but they remain important, and they help to account for the formal and informal barriers that environmental organizations face in gaining access to information and input into administrative decisions.

German political culture and institutions display some aspects of neocorporatism, a system in which major interest groups, especially business, agriculture, and labor, are encouraged by government to organize as large national associations (see Chapter 2).[13] These associations enjoy privileged access to decision-making circles, where decisions are reached through compromise and consensus, and they are invited to cooperate in policy implementation. German business, agriculture, and labor have been organized into large national associations since the nineteenth century, and there is extensive formal and informal consultation between government and these groups both in framing legislation and writing and implementing regulations; however, such consultation is not mandated by law.

As might be expected in a society with strong corporatist elements and a well-developed welfare state, government financial support for nonprofit organizations in the arts, health care, and social welfare in Germany is among the highest in the world. Many welfare and health services are provided by third-sector organizations, including the health and welfare organizations of the Catholic and Evangelical churches, that are closely bound to government. With their enormous

[12] For discussions of various aspects of Germany's political culture and the environmental organizations see the sources in note 6 and the following: Dyson (1982); Smith (1989); Paterson (1989); Joppke (1993); Rat von Sachverständigen für Umweltfragen (1996); Oswald von Nell Breuning Institut (1996); Pehle (1997, 1998); Brand, Eder, and Poferl (1997); Lahusen (1998); Jänicke, Kunig, and Stitzel (1999); Stark (2001); Reutter (2001), and Wurzel (2002).

[13] In addition to sources in notes 6 and 12, discussions of German corporatism and the ties between interest groups, NGOs, and the government include Schmitter (1979), Lehmbruch (1982), (Anheier, 1992, 1998), Wilson (1992), and Salamon and Anheier (1998).

budgets and thousands of employees, these organizations play key roles in delivering social welfare services and are consulted in formulating welfare policy.

Such patterns also exist in the environmental field, but they are less well-developed. The German Association of Engineers, for example, has a key role in identifying the best available pollution-reducing technology, which many German laws mandate. And insofar as it has been willing to accept input from environmentalists, the federal government has often showed a preference for working with the German Nature Protection Ring (DNR), the major umbrella organization of environmental groups. On the whole, though, environmentalists have found it difficult to organize effective umbrella associations and speak with one voice (see below and Chapter 5).

Neocorporatist political cultures and structures typically resist the efforts of interest groups representing new constituencies to obtain a place at the table. Considerable historical evidence supports this contention. The German state has a long history of brushing aside environmental complaints from locally based citizens' groups, and the experiences of the environmental movement in the 1970s and 1980s are very much in accord with this claim. Researchers working within the rubric of social movement theories about political opportunity structures (see Chapter 2) have reached similar conclusions,[14] noting that the German political elite strongly prefers to work with accepted interest groups and actively resists the entry of new ones. However, they also point out that the German system is not altogether closed to new interests. Environmentalists have been able to exert some influence by working at the *Land* level, through the efforts of the Green Party in the Bundestag, and by working through the courts to block or slow down undesired projects. The establishment of an environmental ministry and the presence of the Greens in the governing coalition also suggest some movement toward a more inclusive approach.

Yet despite the institutionalization of environmentalism described in Chapter 7, many important decisions about the environment are made in discussions among various ministries and interest groups that remain relatively closed to environmental organizations. This not only limits the influence of the environmental organizations, it thrusts them into the role of critiquing either decisions already made or near-final drafts of proposed laws or regulations. The result is that relationships between environmentalists and the government ministries—including at times the environmental ministry—are marked by conflict and mutual suspicion. Government ministries thus often view environmental groups as obstacles to be overcome or annoyances to be marginalized or excluded rather than as partners, and they complain that different environmental groups often argue contradictory positions.

A related obstacle to full participation by environmental organizations is the high importance German environmental law and political culture accord to scientific and professional expertise and the resulting tendency to ignore input from

---

[14]See, for example, Kitschelt (1986), Koopmans (1995), Rucht (1996), and van der Heijden (1997, 2006).

groups that do not display deep expertise about the matter at hand. Although the environmental groups have increased their technical competence in recent years, they are not always in a position to compete with industry and government experts. Government bureaus and industry therefore frequently complain that the environmental groups lack expertise and do not understand administrative realities. Partly for this reason, experience working for an environmental organization has typically not been viewed as qualifying one for positions in the ministries, and there is little exchange of personnel.

In short, German political traditions and political culture create a difficult situation for environmental organizations. They find themselves confronted by powerful ministries that have a history of excludig newcomers and working closely with business, labor, and other traditional interest groups. Formal government structures established after the war provide the organizations with some opportunities for access and influence, and both the institutionalization of environmentalism that began in the 1980s and recent legislation have increased it. Nevertheless, as a relatively disunified group with relatively few resources for pressuring other actors, environmental organizations still frequently find themselves at the margins of decision making. That is, they are neither almost completely excluded—as they were during the period of confrontation—nor fully accepted. When they do try to work within the system, they are often cast in the role of junior partners with questionable intentions and expertise, whose presence at the conference table threatens to disrupt familiar working relationships and assumptions and produce conflict. This situation sets up a difficult choice between trying to operate within the system and confronting it as outsiders.

Legal Regulation of Environmental Organizations

German law predictably contains detailed regulations concerning the legal forms that environmental organizations may take.[15] The most commonly chosen organizational form is a registered *Verein* (association), a form often chosen by sports clubs and many other voluntary associations. To obtain this status an organization must fulfill legal requirements that include periodic membership meetings and election of a board of directors and officers. *Vereine* can also have other *Vereine* rather than natural persons as members, providing the potential to organize umbrella organizations of environmental organizations as *Vereine.* A nonmembership environmental organization may also organize as a *Stiftung* (foundation), set up to serve the public good and governed by a self-perpetuating board of directors. In order to receive tax-free donations these associations and foundations must be certified by the authorities as meeting relevant legal requirements and serving the public interest. The authorities have at times used the threat of withdrawing an organization's certification and making contributions to it taxable as a lever to discipline organizations they viewed as threatening.

[15] For overviews of German law in this area see von Alemann (1989), Heuser (1994), and Zimmer (1996).

Government as a Source of Financial Support

In comparison to government funding for other third-sector organizations, annual subsidies for environmental organizations have never been widespread or generous.[16] In recent years, they have been limited to small payments to the DNR, which received about €250,000, and a smaller amount for the Green League (DNR Deutschland Rundbrief, 2000a). In recent years many environmental organizations also benefited from free labor from participants in the civilian alternative to military service or government job creation programs; however, the latter have been cut back since the 1990s. Judges can allocate revenues from criminal fines to environmental organizations and sometimes do so, and in some *Länder* the organizations receive modest financial compensation for the work involved in preparing position statements on environmentally sensitive proposals.

Environmental organizations can also seek support for specific projects through grants from government agencies and from the German Federal Government Environmental Foundation, the Federal Environmental Office, the Federal Office for Nature Protection, and foundations administered by the various *Länder* and the Green Party. The amounts involved are not large. Direct grants from government agencies were under €3 million in 1980 (DNR Deutschland Rundbrief, 2000a); however, they can be a significant factor in the financing of some projects. The Federal Government Environmental Foundation made grants of €41.5 million in 2005, but most of the grants went to business firms for technological improvements rather than to environmental organizations (Deutsche Bundesstiftung Umwelt, 2006).

Obtaining government funding can be a seductive prospect for environmental organizations, but it has its drawbacks. Government subsidies are subject to unanticipated cutbacks during tough budgetary times, and grants for specific projects terminate with the project. Recipients may also find their programming shaped by what granting agencies are willing to fund, and in extreme cases government has used funding cutoffs to discipline or control environmental groups. Funding for the BBU, for example, was terminated when the CDU/FDP coalition took control of the government in 1982, and BUND's opposition to a Frankfurt airport expansion once precipitated threats to cut off its revenue from fines (Wolf, 1996). More recently, the Green League was ordered to rewrite or withdraw brochures that it had planned to distribute at a Youth Environmental Congress funded by the Federal Environmental Office (Bergstedt, 1998).

In any event, current low levels of government funding are unlikely to increase much in the near future. A stagnant economy and resulting budget pressures led to reductions in subsidies to many third-sector organizations in the late 1990s, and the budget cutbacks have since worsened, forcing many organizations to look elsewhere for revenue. As relative latecomers to the quest for large-scale government funding and frequent critics of government policy, environmental

---

[16] For discussions of government funding for environmental organizations see Luthe (1997), Anheier et al. (1997), Bergstedt (2002), Haibach (1998a), Urselmann (1998), and Bode (2003).

organizations are not well positioned to compete for increasingly scarce government subsidies.

Right-Wing Environmentalism

German environmental organizations differ from those elsewhere in having right-wing ecology as a prominent part of their past, and there continues to be a lively right-wing discourse about ecological issues in Germany.[17] Right-wing ecologists typically seek both to interject conservative ideologies into environmentalist thought and to use environmentalist themes to advance a conservative agenda. The activity centers around institutes for discussion and research, such as the Collegium Humanum, around organizations like the Federal Association for Ecology, which has close connections to the CDU and business, and around the World League for the Protection of Life, a pillar of the early anti–nuclear power movement that later drifted to the far right and lost its mass membership. These groups frequently host seminars and lecture series and put out a constantly mutating array of low-circulation magazines and newsletters.

Specific points of emphasis in right-wing ecology vary somewhat among various groups and authors, but there are some common themes: 1) the important role of heredity in shaping human behavior; 2) the uniqueness of human societies and the problems of ethnically mixed or multicultural societies, sometimes presented with undertones of the racial or national superiority of some societies; 3) the key role of population growth and immigration in producing environmental problems; 4) the need to subordinate individuals to the needs of the whole, often coupled with an emphasis on love of homeland; 5) use of the organic analogy, which sees societies as products of natural evolution of a people within a natural environment and interprets environmental problems as a threat to the entire *Volk;* 6) a propensity to deduce universally valid laws that apply to society from biological processes, normally coupled with claims that present-day societies fail to respect these laws; 7) a critical stance toward both socialism and consumerist materialism—the latter, ironically, shared with countercultural environmentalists; 8) a propensity to associate protection of the environment with a more generally conceived protection of "life," sometimes coupled with strong stands on social issues like abortion; 9) attribution of environmental and other social problems to alienation from society, deficiencies in individual morality, or lack of social responsibility rather than to economic or social structure; and 10) the need for strong leaders and a strong state to overcome environmental problems.

In recent years, conservative ecologists have focused attention especially on rapid population growth in less developed countries and on immigration. They typically portray population growth as the major cause of the problems of less-developed countries and describe excess population in such countries as threaten-

---

[17] Useful treatments of right-wing ecology include Jahn and Wehling (1990), Jaschke (1990), Wüst (1993), Ulbricht (1993), Biehl (1995), Statham (1997), Geden (1999), Olsen (1999), and Bergstedt (2002).

ing an already overpopulated Germany with waves of difficult-to-assimilate immigrants. Such arguments can easily harbor racism, although it usually remains latent rather than open.

Local citizen protests against the creation of nature protection areas (BUND-Magazin, 1998a; van der Heijden, 2005), reintroduction of predators (BUND, 1995), or windmill construction (Bammerlin, 1998; Hart, 2001) sometimes break out in Germany, but the German right has not been successful in harnessing this discontent, and German business has not been interested in funding such groups. Consequently, in Germany there is practically no organized populist anti-environmentalism of the sort frequently encountered in the US.

The European Union

An additional complicating factor in the context of German environmental organizations is the rapidly growing importance of European Union legislation and regulations.[18] Since the early 1970s, environmental policy has emerged as a major domain of EU activity, focusing on areas such as chemicals, air and water pollution, nature protection, and environmental impact assessment. Well over 300 EU directives and regulations now require member states to pass and enforce environmental legislation within parameters specified by the EU (van der Heijden, 2006), and member states' rights to pass legislation that exceeds EU standards is limited. During the closing years of the twentieth century and early years of the twenty-first, the EU standards exerted considerable pressure on environmental laggards. In a few instances, such as requirements to establish areas for the protection of flora and fauna, Germany too has experienced such pressure. Indeed, in one editorial, the editor of BUND's magazine expressed great pleasure at the prospect that EU guidelines would soon force recalcitrant German regional and local officials to take action to reduce particulate pollution (BUND Magazin, 2005c). However, EU regulations can also be a problem, as occured when strong resistance emerged in Germany to EU guidelines for patenting DNA sequences (DNR Deutschland Rundbrief, 2000l).

European Union environmental policy is made in the context of the overlapping and sometimes vaguely defined responsibilities of the European Commission (the EU's administrative bureaucracy), the Council of Environmental Ministers (composed of the environmental ministers from all the member nations), corresponding councils in other fields of EU activity, and the European Parliament. EU environmental regulations are initiated and drafted by the En-

---

[18] Major sources drawn on in this section include the following: Mazey and Richardson (1992); Hey (1994, 2004); Hey and Brendle (1994); Rat von Sachverständigen für Umweltfragen (1996); Brand, Eder, and Poferl (1997); Pehle (1997, 1998); Rucht (1997, 2001); Bammerlin (1998); Long (1998); Webster (1998); Jänicke, Kunig, and Stitzel (1999); Rootes (1999a, 2004a); Sbragia (2000); Schreuers (2002); van Tatenhove (2002); Roose (2002); Louis (2003); Rogall (2003); and van der Heijden (2006).

vironmental Directorate of the European Commission, often at the suggestion of one or more member states or the country that currently holds the rotating EU presidency, or in response to the requirements of the EU's governing treaty. Historically, the Environmental Directorate has been understaffed and politically weak, and environmental regulations must be drafted to avoid conflicts with the EU's core principle of ensuring free markets.

The Council of Ministers holds the major legislative authority in the EU structure, but the historically weak European Parliament has gained influence through recent revisions of the EU's governing treaties. Formal rules, customs, and the practicalities of keeping the member states "on board" dictate that Council of Ministers' decisions require near unanimity, often requiring long negotiations to arrive at lowest-common-denominator solutions. The parliament, by contrast, has a strong environmental committee and has been a strong advocate for environmental concerns.

The European Commission also oversees the member states' implementation of environmental legislation; however, its enforcement sections are weak and seriously understaffed. If the commission believes a member state is not fulfilling its obligations, it can institute action in the European Court of Justice to force compliance. The right of environmental organizations to file such complaints has been the subject of conflicting legal decisions and remains unclear; however, they can, at a minimum, bring instances of noncompliance to the attention of the Environmental Directorate. And in fact, environmental complaints have constituted a large part of the Court of Justice's workload.

The Environmental Directorate has historically sought out contacts with environmental organizations that could provide it with expertise and legitimize its efforts; however, it has not been very responsive to input from purely *national* environmental organizations, preferring input from networks of organizations from multiple nations. Other directorates are less open to input from environmental groups.

In response to the Environmental Directorate's preference for working with international networks of environmental organizations, as well as their own need to exchange information, coordinate their efforts, magnify their influence, and reduce the associated costs, twenty-five national-level environmental organizations formed the European Environmental Bureau (EEB) in 1974. Since its beginning, the EEB has received part of its financial support from the Environmental Directorate and enjoyed substantial access, and it remains today the most environmental important network. At present, it has 143 members from thirty-one countries. The EEB works in numerous environmental policy areas, and it has working groups comprised of EEB staff and representatives from the member organizations for twelve of these areas. It publishes a bi-weekly newsletter for its members and a quarterly magazine for outside circulation (European Environmental Bureau, 2007). Its governing board is composed of one member from each EU nation, plus Norway. As of the late 1990s, there were only eleven staff members (Webster, 1998), and its current Internet contact page lists fifteen.

The EEB has not proved a completely satisfactory vehicle for influencing EU policy. It has been criticized for taking an overly timid approach, not sharing enough information with member groups, and following its own agenda rather than those of its members. Greenpeace withdrew from the EEB in 1991, citing its ineffectiveness, and the EEB's other German members were reported to be inactive and dissatisfied. The situation has evidently improved since, but the EEB still has a great many passive members and a very small staff, and it now faces competition from more specialized environmental lobby groups.

Greenpeace, the WWF, and Friends of the Earth, whose German affiliate is BUND, established offices in Brussels during the mid to late 1980s. By the end of the 1990s, the Friends of the Earth office represented twenty-eight chapters in current or potential EU nations (Webster, 1998). It supplies its member groups with information about EU environmental issues, coordinates the European efforts of the national chapters, and engages in lobbying on its own. As of 2005, it had about twenty employees (BUNDMagazin, 2005l). In the late 1990s, Greenpeace's small Brussels office had only two employees (Webster, 1998). They were charged with keeping Greenpeace International and the various European chapters of Greenpeace informed about European developments and representing Greenpeace positions to the EU. Direct lobbying by Greenpeace and cooperative work with the other Brussels organizations has been limited because of the small staff, Greenpeace's preference for spectacular actions, and its desire to set its own agenda rather than respond to EU debates. The WWF's European office represented chapters in twelve European countries and had eight staffers (Webster, 1998). It supplies these chapters with information and lobbies the EU, focusing on advocacy for environmentally sound international development.

Several additional networks of European environmental organizations in specific policy domains, such as Bird Life International, whose German affiliate is NABU, the Climate Action Network Europe, and the European Federation for Transport and Environment, opened Brussels offices in the early 1990s. The Climate Action Network Europe, with sixty members groups at the end of the 1990s (Webster, 1998), is a branch of the international organization with the same name. Bird Life International, which then represented twenty-nine European groups (Webster, 1998), is a very well-funded and organized network that gathers data, provides its European affiliates with a great deal of information, coordinates their efforts, and lobbies the EU.

The networks listed above, plus the World Conservation Union, constituted the loosely organized "Group of Eight" whose members confer regularly to exchange information and coordinate their efforts. Their total Brussels staff remains quite small, so even working together it is impossible for them to be effective in all areas of EU environmental policy.

National groups in Germany and elsewhere have been reluctant to divert resources from home to support the networks, limiting their budgets and staffing. Consequently, the networks must focus their attention on only a few issues among many possibilities, and they are more often reactive to ongoing policy debates and requests for input than proactive. In general, they have emphasized lobbying, submitting reports or position statements to the European Commission, issuing press releases, and holding press conferences. They have not, for several reasons, attempted to mobilize a mass public for protest. The highly technical nature of much EU policy-making makes it hard to mobilize activists, publicizing such events in so many nations is problematic, coordinating protesters from so many different nations and organizations would be difficult, and travel distances are large. Moreover, most of the networks are financially dependent on funds from the EU to support their lobbying efforts—although Greenpeace, in line with its general policy, rejects such support—and the EU funds some European-wide projects conducted by the networks. The EU also supports nature protection projects of some national environmental organizations, but it does not provide funding for them to lobby in Brussels.

The Brussels environmental networks face many challenges. They are confronted by well organized and funded business lobbies, employing by one recent estimate between 15 and 20 thousand lobbyists (Rogall, 2003). The European Commission's Environmental Directorate has been open to environmental concerns, but environmental outcomes are affected by many other directorates, many of which are much more responsive to other constituencies. Explaining the arcane functioning of the EU bureaucracy and the complexities of European environmental policy to a Europe-wide public is intrinsically difficult, and there are few Europe-wide media and little good press reporting about EU environmental policy making. Consequently, there is as yet no general European "public" to which environmental organizations can address press releases, fundraising appeals, or mobilization efforts. Finally, differences among the cultures, central objectives, degree of institutionalization, and typical operating styles of organizations in the different countries can make cooperation difficult. The centrality of opposition to nuclear power in German environmentalism, for example, is almost unique, and delegates from countries with weak environmental regulations may be more satisfied with EU rules than groups from other countries where national regulations are stronger.

German environmental organizations that wish to be active at the EU level can also work through the DNR. In 1991, the DNR set up a coordinating office to mediate between its member organizations and the EU. Its executive secretary was a longtime member of the EEB's board, and another DNR staffer serves today. The DNR publishes a respected monthly newsletter on EU environmental affairs, which has about 400 subscribers. These steps have greatly improved the flow of information to the German environmental organizations (Deutscher Naturschutzring, 2002; Roose, 2002; Röscheisen, 2006).

The EU structure provides multiple points of access, so national environmental organizations, including those in Germany, can choose among several possible avenues for influencing the EU. They can directly lobby the European Commission and European Parliament members. This approach is associated with potentially large expenditures of time and money to acquire information and expertise, establish a regular presence in Brussels, and develop new styles of operation. Second, they use the ordinary measures at their disposal to influence German representatives in the European Parliament or the German government so that its representatives in the Council of Ministers will push for measures that they prefer. As the largest EU country and one of the driving forces in the EU, Germany is in an especially good position to do this. Third, they can work through existing European networks like the EEB, the European WWF office, or Bird Life International. This has the obvious advantages of cost effectiveness and addressing the EU with one voice; however, it requires investing some money to support the networks and a great deal of time in reaching compromise positions that do not perfectly reflect the national organizations' priorities and do little to reinforce their reputations at home.

The Media

The ability of environmental groups to advance their agenda rests, to a considerable extent, on whether the general public and political elites define the environmental conditions as problematic. Whether this occurs is, in turn, heavily dependent on the media.[19] Scientific research results and the experiences of individuals or local communities cannot influence the general public or politicians unless they become known to them. And even if known, the "facts" are rarely unambiguous or striking enough to force politicians and the public into action. Instead, "facts" about the environment become the subject of a public discourse carried out primarily in the media. Here environmental groups, scientists, business, and other interest groups attempt to persuade the public and political elites of the validity of their claims about the severity of environmental problems and the best solutions for them. Which information and points of view reach the public depends on the resources of the various interest groups, their skill in presenting their case in a way that attracts media interest, and the practices of the media in selecting and presenting environmental news.

Favorable press attention is especially important to environmental groups. Because they have relatively few opportunities to exert direct influence on decision makers, they must often resort to indirect influence by mobilizing public opinion. Lacking the financial resources to mount major advertising campaigns of their own, they are necessarily dependent on media coverage of environmental issues and the environmental organizations' interpretations of them in order to get their point of view across. Moreover, when environmental organizations are visible, positively portrayed, and sought out for expert comment in the media,

---

[19] For discussions of the role of media in shaping environmental discourse, public attitudes, and the political agenda see Kielbowicz and Scheer (1986), Leonhard (1986), de Haan (1995), Hannigan (1995), Brand (1995, 1999a), Oswald von Nell Breuning Insitut (1996), Brand, Eder, and Poferl (1997), Jänicke, Kunig, and Stitzel (1999), Braun (2003), and Rogall (2003).

their public stature and chance of attracting and retaining members and donors increases.

Survey results suggest that German citizens rely heavily on the press, especially local newspapers and television news, for information about environmental issues. Radio, nationally circulated newspapers, weekly newspapers and news magazines, specialty magazines, television documentaries and films, informational brochures, materials from environmental organizations, and the Internet are also used, but with less frequency. Most of these sources enjoy relatively high credibility, but almost half the public believes that there is too little environmental coverage, and about a fifth evaluate the quality of environmental reporting as deficient or unsatisfactory (European Opinion Research Group, 2002; Bundesministerium für Umwelt, 2004; Directorate General Environment, 2005).

Print Media

Newspaper readership in Germany is relatively high, and about 50 percent higher than in the US (World Bank, 2000).[20] In addition to numerous locally oriented dailies, several major newspapers oriented to serious readers circulate nationally, ranging across the political spectrum from the *Tageszeitung,* with its roots in the left and the counterculture, to the conservative, business-oriented *Die Welt.* Germany also has a very large and highly sensationalistic tabloid press. There are also two nationally circulated news magazines, the venerable, left oriented *Der Spiegel,* and the newer and more conservative *Focus.* Many of the counterculturally oriented weekly newspapers, which covered environmental issues heavily, have ceased publication or become more oriented to lifestyle issues, but Berlin's *Rabe Ralf* continues the older tradition of political reporting.

Environmental news is not always covered well. Environmental issues often cut across several established "beats," and relatively few newspapers have a separate environmental desk. Only a few reporters specialize in environmental reporting or have strong competence in science, and many environmental stories are farmed out to freelance reporters. Newspapers therefore rely heavily on stories from wire services and news agencies, press releases, and staged events, such as press conferences and demonstrations. Operating with limited resources and under time pressure, daily papers rarely undertake investigative reporting or extensive background research about environmental stories, and they are not well suited to presenting technically complex data or covering environmental stories that develop slowly and are not punctuated by crisis. On the other hand, waves of "follow the leader" stories about the same topic are common, and crises, such as the Chernobyl disaster, or media events like the 1992 UN Rio Conference

[20] Information about the print media presented here is derived from firsthand experience during several periods of residence in Germany and the following sources: Leonhard (1986); Voss (1990); Hömberg (1993); Rossman (1993); de Haan (1995); Brand (1995); Rat von Sachverständigen für Umweltfragen (1996); Bammerlin (1998); Jänicke, Kunig, and Stitzel (1999); bfp Analyse (2000b); Biermann and Böttger (2000); Bergstedt (2002); Rogall (2003); Braun (2003); and Röscheisen (2006).

can produce a high volume of stories in a short period. Sensational reporting is always the order of the day in the tabloids, but even the serious press is far from immune to tendencies to prefer stories about confrontation or dramatic events, stories with dramatic visual images, or stories with direct relevance to readers' interests or pocketbooks.

Several nationally circulated magazines cater to readers with environmental interests, but they reach a relatively small public. An inventory of these in 1989 found thirty-seven, with a total circulation of 4 million, but the number has probably declined since (Jänicke, Kunig, and Stitzel, 1999). Among the most widely read are the general environmental magazines *Geo* and *Natur + Kosmos,* while *Öko-Test* features evaluations of products from an ecological standpoint. *Natur + Kosmos* includes some coverage of the environmental organizations and their activities, but its audience is small. There are also publications read almost exclusively by insiders, such as the environmental ministry's magazines *Umwelt* and *Wölfe,* the DNR's newsletters about German and EU environmental politics, and three low-circulation newsletters, *Politische Ökologie*, *Umwelt Kommunale Ökologische Briefe,* and *Punkt.um.*

Television

The German television market is divided between the state-subsidized networks, which were established after the war, and private networks, which were first allowed in 1984.[21] The latter emphasize entertainment programming, and their appearance was associated with a rise in TV viewing. Critics charge that the private networks offer sparser, less serious news coverage and that competition from the private networks has degraded the quality of public information offerings on the public networks. Television and radio do offer some coverage of environmental themes through regular news reporting, documentaries, and "nature shows," but TV news may be even less likely than print media to pursue stories that develop over long periods. I could find no systematic investigation of the amount and types of environmental coverage; however, my own fairly extensive viewing indicates that, with the exception of nature shows, environmental themes receive little attention on German television.

Strategies for Effective Media Work

The environmental organizations thus face a situation in which representing their point of view in the media and maintaining a positive media image are of crucial importance, but the barriers to success are great.[22] Media attention to environmental issues has declined (see Chapter 7), and neither print nor broadcast media

---

[21] In addition to my own viewing of German TV, this section draws on Hömberg (1993), bfp Analyse (2000b), and Rogall (2003).

[22] For discussions of the challenge of dealing with the media see Dölle and Lüginbühl (1985), Kielbowicz and Scheer (1986), Hannigan (1995), de Haan (1995), Brand, Eder, and Poferl (1997), and Bammerlin (1998).

now devote much regular coverage to them. Moreover, the environmental organizations must compete with the well-financed public relations efforts of business.

Organizations that can deliver media content that is timely, well-prepared, interest-catching, and full of striking images are therefore at a considerable advantage. The temptation to sensationalize environmental issues to attract attention is therefore ever present; however, the German public today is less receptive to apocalyptic claims than in previous years (see Chapter 7), and maintaining accuracy is also important. As one environmental journalist I interviewed pointed out, Greenpeace's misstatement of the amount of oil and toxics remaining on the Brent Spar oil platform in the North Sea, whose sinking it opposed, damaged not only its public image, but also its long-term credibility with journalists (see also bfp Analyse, 2000d). She contrasted this with the high-quality information that BUND had supplied for several of her stories. Environmental organizations must also be conscious of the fact that the media's fondness for dramatic stories extends to reporting on schisms or scandals within social movements.

Success with the media also requires careful monitoring of public opinion, the media, and the political process; timing reports, demonstrations, and press releases to coincide with ongoing events; successfully projecting the organization's own concerns into ongoing debates; speaking with one voice; and making background information and expert spokespersons available on short notice. All of this entails sustained effort by well-financed and professional media operations.

Other Environmental Organizations

The context in which the four major national environmental organizations operates includes, in the first place, the other three. Although they have somewhat different emphases, competencies, and strategies, there is also much overlap in goals and strategies, as well as in potential supporters. A mail survey of readers of the NABU member magazine, for example, showed that 34 percent of respondents also belonged to Greenpeace, 22 percent to BUND, and 18 percent to the WWF (Naturschutz Heute, 1994), and a mail survey of BUND members showed that 59 percent had donated money—other than their BUND dues—for environmental protection (Bodenstein et al., 1998). Consequently, the four largest environmental organizations are simultaneously potential allies and competitors.

The social context in which these four organizations operate also includes many other types of environmental groups. They too are not only potential allies and sources of information and ideas, but also competitors for members, activists, donations, and press attention.

Citizens' Initiatives and Networks of Citizens' Initiatives

BIs are not as visible a part of the German environmental scene as they were during the confrontations of the 1970s and 1980s, but they continue to play a significant role. Many are single-purpose organizations that appear and disappear within a short time, but others endure for years. It is extremely difficult to determine their number, although Rucht and Roose (2001a, 2001b) estimated that there were over 9,000 "small environmental groups," including several thousand local chapters of national or regional organizations in Germany.

Some BIs belong to broader networks of such initiatives. Two of these, the BBU and the Green League, provide both the most opportunities for collaboration and the greatest competition for the large national organizations because they operate on a supra-regional scale and focus on a variety of different issues. By mutual agreement, the BBU operates only in the former West Germany, and the Green League only in the East. Although the BBU has lost most of the influence and membership it enjoyed in the period of confrontation, it continues to operate with a limited program and very small staff.[23] The Green League[24] has proved somewhat more resilient; however, recent years have seen a declining number of member groups and the shuttering of some regional offices. Its member groups retain their own names, enjoy great autonomy, and have independent budgets. They engage in a variety of environmental and public education projects, such as maintaining a network of environmental libraries, working on sustainable development projects for cities, lobbying against construction of environmentally destructive public works projects, holding youth camps, monitoring water quality, and assisting with creating green spaces in apartment courtyards.

The Green League is certified as an environmental organization under the provisions of the German Nature Protection Law and regularly submits position statements under the law's provisions. Its business office regularly issues press releases, and it occasionally undertakes protest actions. It also issues a monthly newsletter and other publications and has a close relationship with Berlin's monthly environmentalist newspaper. The budget of the central organization is very small in comparison to the large national organizations. Direct government subsidies and grants for specific projects from government and government-funded foundations have historically furnished most of the organization's budget, a problematic situation in a time of budget cuts. The Green League is a member of the DNR and the EEB.

There are also national networks of BIs focused on single issues, such as traffic and transportation policy, waste disposal, and genetically altered goods—a more contested issue in Germany than in the US. The oldest of these is "Turnaround," which began life as the BBU's working group for traffic problems but became independent in 1983. It offers assistance to BIs, conducts public information campaigns, lobbies, and publishes a magazine and other informational materials (Bergstedt, 2002; Umkehr e.V., 2007).

---

[23] See Bergstedt (2002), Markham (2005b), Bundesverband Bürgerinitiativen Umweltschutz (no date, 2005, 2007).

[24] For information about the Green League see Weinbach (1998), Grüne Liga (1999, 2003, 2005), Rink (2001), Rucht and Roose (2001b), Grüne Liga Berlin (2001), Bergstedt (2002), Nölting (2002), and Rink and Gerber (no date).

The BIs and their umbrella groups represent an alternative approach to environmental action, an approach emphasizing volunteer activism and local action that is deeply rooted in the history of German environmentalism and still appeals to many potential activists. Consequently, although not as prominent as they once were, they remain competitors for activist involvement, appealing especially to activists who are skeptical of the more tightly organized national environmental organizations.

Other National Environmental Organizations

An intermediate approach between those of the large, formalized national environmental organizations and loose networks of BIs is represented by Robin Wood,[25] a Greenpeace spinoff formed originally by activists who wanted to combat threats to the German forests via a grassroots, democratic organization. Forests have remained an important focus of its attention; however, as the 1980s wave of concern about forest death in Germany waned, Robin Wood expanded its focus to include related problems, such as air pollution, destruction of tropical rain forests, traffic and automobile dependence, energy, and paper recycling.

As its founders intended, Robin Wood operates as a relatively small, decentralized national organization of grassroots activists organized in local groups with a great deal of autonomy. A board of directors elected by an assembly of delegates from local groups is responsible for coordination, fundraising, and review of proposed actions to be sure they do not involve too much legal risk. Robin Wood was able to expand its number of donors rapidly in the late 1980s and early 1990s through mass mailings, but their implementation led to bitter confrontations at annual meetings and the withdrawal of some groups (Scholz, 1989). It is careful to distinguish occasional donors from members, who are required to participate actively in meetings and actions.

Robin Wood focuses on increasing public awareness of environmental problems primarily by means of attention-grabbing demonstrations and protests, which are often illegal but never violent. It also seeks to educate the public through its quarterly magazine, as well as books and other publications, and it calls on the public to put pressure on corporations that contribute to pollution through actions like paying their electric bills in ways that are inconvenient and expensive for the power companies. It sometimes cooperates with other environmental groups, but it does not undertake scientific research or practical nature protection activities, and it does not purchase nature protection areas.

As of 2003, Robin Wood's web site claimed 2,300 members in twenty groups (Robin Wood, 2003a), but in 2007 only thirteen local groups and no membership total were listed (Robin Wood, 2007a). The most recent budget report available shows that the organization had receipts of just under €900,000 in 2004

---

[25] Information about Robin Wood used in this section was derived from the following sources: Scholz (1989); Rucht (1989); Zillessen and Rahmel (1991); Blühdorn (1995); Bergstedt (2002); and Robin Wood (2003a, 2007c).

and operated at a loss in 2003 and 2004 (Robin Wood, 2007b). Except for advertisements in its magazine, it accepts no funds from business or government. The business office is in Bremen. Although Robin Wood has clearly weakened in recent years, it continues to offer a home for environmentalists who prefer a more confrontational approach than that of BUND, NABU, or the WWF and more democracy than Greenpeace.

Two organizations with a long history of nature protection activity continue to engage in environmental protection activities. The BH now bears the name League for Homeland and Environment. It continues to claim nature protection as a goal, with a particular focus on preserving historic and scenic landscapes (Fischer, 1994; Bund Heimat und Umwelt in Deutschland, 2007). Perhaps because of this somewhat retrograde emphasis, it gains little recognition of its environmental activities. It is rarely mentioned in recent discussions of major environmental organizations in Germany (Blühdorn, 1995; Oswald von Nell Breuning Insitut, 1996; Rat von Sachverständigen für Umweltfragen, 1996), and my review of newspaper stories in the late 1990s showed that it is almost never mentioned or asked for comment in connection with environmental issues. It may, however, provide an organizational home for citizens with environmental or nature protection inclinations who are too conservative to be comfortable in other environmental organizations.

The Friends of Nature, another historically important environmental organization, today claims almost 100,000 members in about 700 local groups, a slight decline since the beginning of the century. Its nature protection activities focus on inland waterways, environmentally friendly tourism, climate protection, and regional and organic food production, but its environmental objectives remain only one aspect of a multifaceted agenda that includes hiking, outings, and sports (Deutscher Naturschutzring, 2000; Naturfreunde Deutschlands, 2007). Like the League for Homeland and Environment, its environmental activities have not been visible enough to earn it recognition as a major environmental organization, especially in comparison to organizations with environmental goals as their sole concern, and it is rarely mentioned in overviews of important German environmental organizations or press reports on environmental issues. Still, it may compete with the four major environmental organizations for supporters whose interests lie on the border between nature and environmental protection and outdoor recreation.

There are also numerous single-purpose national organizations with environmental goals.[26] These include the Association for Protection of the North Sea Coast, the German Transportation Club, which promotes ecologically sound auto use, the bicycler's association, and the German Association for Landcare. The last-named organization has a select membership of top nature protection advo-

[26] Descriptions of these and similar organizations can be found in Zillessen and Rahmel (1991), Oswald von Nell Breuning Institut (1996), Chaney (1996), Rat von Sachverständigen für Umweltfragen (1996), Bammerlin (1998), Englehardt (2000), Deutscher Naturschutzring (2000), Bergstedt (2002), and Deutscher Verband für Landschaftspflege (2005).

cates, scientists, business leaders, and political leaders, and works to maintain the health of Germany's rural landscapes. Several organizations, including German Environmental Assistance, Euronatur, and Oro-Verde, which raise money and make grants, are also significant actors on the environmental scene. German Environmental Assistance, for example, raises funds from sponsoring arrangements with business and private donations and uses them to support grants to organizations engaging in nature protection and its own projects (Deutsche Umwelthilfe, 2005). There are also numerous regional environmental groups,[27] such as the Nature Protection Association of Lower Saxony and regional bird watching or ornithological societies that are independent of NABU.

The most prominent of the national single-purpose organizations is the German Forest Protection Association,[28] which pursues its objective mainly though public education and lobbying. Its membership contains many foresters and government employees, and it receives significant government subsidies. As of 2000, it had 20,000 members (Deutscher Naturschutzring, 2000). Although it is best described as a niche player in the German environmental scene, it does successfully compete for members and donors, press attention, and government subsidies within that niche, as do the other single-purpose organizations. On the other hand, it and the other single-purpose organizations can also become allies in specific projects or campaigns.

Organizations whose members use nature for recreation or that conduct nature study, such as the hunters', hikers', and anglers' associations and the German Alpine Association, have long claimed nature protection as a secondary goal. Their impressive memberships and their ability to mobilize constituencies not easily reached by environmental organizations make them potentially useful allies; however, their relatively conservative approach, narrower focus, and desire to make use of nature in ways that other environmentalists sometimes see as destructive makes them problematic allies and even occasional opponents (Rat von Sachverständigen für Umweltfragen, 1996; Röscheisen, 2006). For example, a bitter conflict broke out in Brandenburg in 2001 over whether lands formerly owned by the GDR should be turned over to Greenpeace or to the hunters' association, which envisioned a less restrictive use of them (Kirschey, 2001). Because of differences in goals, they probably compete with environmental organizations for supporters to only a limited extent.

Another important segment of Germany's environmental scene is its numerous environmental research institutes.[29] These include the Institute for Applied Eco-

---

[27] For descriptions of some of these organizations see Rucht (1991a), Oswald von Nell Breuning Institut (1996), Rat von Sachverständigen für Umweltfragen (1996), Englehardt (2000), Deutscher Naturschutzring (2000), and Bergstedt (2002).

[28] See Rucht (1991a), Dominick (1992), Deutscher Naturschutzring (2000), Bergstedt (2002), DNR Deutschland Rundbrief (2000d), and Schutzgemeinschaft Deutscher Wald (2003).

[29] Information about the history and recent programs of these institutes is available in Rucht (1991a), Jänicke and Weidner (1996), Brand (1999a), Rucht and Roose (1999), Dryzek et al. (2003), Rogall (2003), and Rink and Gerber (no date).

logy, the Eco-Institute, the Wuppertal Institute for Climate, Environment, and Energy, the Institute for Energy and Environmental Research, and the Independent Institute for Environmental Questions in East Germany. Most of these were founded in the late 1970s with the support of the environmental movement to counter industry and government experts and provide a more scientific research base for its work; however, they have since become more independent. By the end of the 1990s, organizations like the Eco-Institute were functioning as hybrids of environmental organizations and commercial research institutes; however, they continue to maintain their public information and fundraising operations and compete with environmental organizations for donations.

The German Nature Protection Ring

As the largest umbrella organization representing organized environmentalism in a nation with significant elements of corporatism, one would expect the DNR to be a key player in German environmentalism.[30] It is, in fact, a very important force in the social context of the four large environmental organizations, but a number of factors have kept it from attaining the central position that it might occupy.

The DNR can point with pride to its wide network of connections. It is regularly consulted by government agencies, which often prefer to work with a single umbrella organization in framing legislation, and its leaders and representatives are appointed to many commissions, study groups, and award selection panels, such as the national Advisory Council on Sustainability. During the sixteen years of CDU/FDP government that ended in 1998, its president met once to twice a year with the chancellor, and regular meetings between the president or other DNR representatives continued under Chancellor Schröder. Its representatives lobby regularly at the national and EU levels, presenting position statements and testifying at hearings on environmental legislation.

The DNR can also point to significant recent accomplishments. It has been especially active in promoting domestic German follow-ups to the 1992 UN Rio Conference and coordinating German participation in the various successor conferences, and the Forum for Environment and Development (see below) is housed in its offices. It also assumed the leading role in, and devoted a great deal of effort to, coordinating and publicizing the ultimately successful campaigns for an energy tax and the most recent revision of the national nature protection law.

The DNR is active in public education through its conferences on environmental problems, publications, and press conferences, and it provides a great deal of information to its member organizations through various newsletters. Its widely

---

[30]The most comprehensive source about the DNR today is a recent book by its longtime executive secretary (Röscheisen, 2006), but this section also draws on a number of other sources: Flasbarth (1993); Hey and Brendle (1994); Oswald von Nell Breuning Insitut (1996); Rat von Sachverständigen für Umweltfragen (1996); Wolf (1996); Pehle (1997); Bammerlin (1998); Bergstedt (2002); Dryzek et al. (2003); Rogall (2003). Useful reports published by the organization itself include Deutscher Naturschutzring (1998, 2000) and DRN Deutschland Rundbrief (2004b).

circulated *DNR Deutschland Rundbrief* (DNR Germany Newsletter), published since 1994—and since 2003 with financial support from the environmental ministry—provides the most comprehensive single source of information about German environmental politics, environmental issues, and the activities of environmental organizations. The DNR's monthly EU newsletter, issued since 1991, provides a similar service for EU environmental affairs, focusing on their relevance for Germany. Both are available in print and Internet versions. The low-circulation *DNR-Kurrier,* an internal newsletter for member groups, appears about a half-dozen times yearly.

Despite these accomplishments, the DNR has failed to emerge as the single effective voice for environmentalism. Numerous internal problems have limited both its external effectiveness and its attractiveness to potential members as a co-ordinating body or spokesman. Historically, the most significant of these internal problems was the long-running conflict between DNR member organizations that focused on nature protection and those for which nature protection was secondary to recreation. This problem was considerably reduced by the withdrawal of many of the latter groups during the late 1980s, and the DNR has become a more effective organization as a result. Still, conflicts between nature-protection and nature-using groups have continued to bob to the surface. Five years after the expulsion of the hunters' association, conflict broke out over the DNR's position on revision of Germany's hunting law, and the organization's final position paper had to note the dissenting views of the horseback riders, canoeists, the Forest Protection Association, and several other member groups. Another conflict occurred when complaints and threats of resignation by the Alpine Association forced the DNR to soften its position on the recreational use of mountain areas, and the Alpine Association and other recreation-oriented organizations later fell into conflict with the environmental organizations over the DNR's position on revision of Germany's Nature Protection Law.

Not all the recent internal conflicts, however, have been organized along this fault line. One conflict broke out over wind power. Member groups oriented to environmental protection favored it as a measure to conserve energy and reduce pollution, while some nature protection organizations worried about the resultant bird kills. There was also a dispute over whether an environmental certification for automobiles would be desirable or would simply encourage the growth of a fundamentally undesirable mode of transportation. A long-time DNR leader I interviewed told me that the internal conflicts have been much reduced in recent years, but he admitted that the need to reach consensus among so many diverse organizations before taking a public stance still limits what the organization can do. More critical outside observers point to conflicts between the DNR's more activist member organizations and its leadership, which several of my interviewees characterized as stodgy, out of date, and too close to conservative political parties and business interests. Indeed, an analysis of newspaper reports between 1988 and 1997 (Rucht and Roose, 2001b) showed that the DNR almost never assumed a leadership role in environmental protests. The retirement of the DNR's long-serving president in 2000 and his replacement by Hubert Weinzierl has reduced but not completely silenced this criticism.

The DNR has also been weakened by the fact that two of Germany's four largest environmental organizations are no longer members. Greenpeace belonged to the DNR for only two years, 1987 to 1989, and the WWF withdrew in 1990. Predictably, these organizations often resist DNR efforts to present itself as *the* voice of the environmental movement. When the government requested input on energy politics in 1999, Greenpeace unsuccessfully attempted to coordinate a combined position for the major environmental organizations without including the DNR, while the WWF protested not being consulted by the DNR. But even BUND and NABU, the major environmental organizations that belong to the DNR, have preferred maintaining their own direct lines of communication with the press and government over having the DNR speak for them. In one example, an effort by the DNR to organize regular annual forums among it, the four largest environmental organizations, and other environmental groups engaged in lobbying was met with little enthusiasm by Greenpeace and by a demand from NABU that the organizations involved rotate the post of convener. In the end, the forum met only five times. It may have been conflicts like these that motivated Hubert Weinzierl, who soon thereafter became DNR president, to complain that the small-mindedness, selfishness, and focus on maintaining their own visibility displayed by some member organizations hindered the DNR's effectiveness (DNR Deutschland Rundbrief, 2000g).

Since the departure of the hunters and related groups, NABU and BUND have had a stronger voice in DNR politics, but the results have not been altogether positive for the DNR. The two organizations sometimes compete over appointments of their leaders as DNR representatives to various bodies, and beginning in the 1990s, they began pressing the DNR to focus its attention on information gathering and distribution, membership service functions, and its linkages to the EU, not on staking out positions or initiating numerous campaigns on its own. They complained that efforts to do the latter were characterized by failure to get all member groups on board and keep them informed, poor coordination, and competition with the activities of member organizations, but many suspected that the desire to avoid having the DNR as a competitor was also at work. In any event, these suggestions were largely implemented in the new DNR constitution put in place in 1995 and have been reflected in subsequent DNR activities. In 2004, NABU and BUND, the organization's two largest financial supporters, successfully used these concerns and their own financial difficulties as the basis for a successful argument for reducing their dues, weakening the DNR financially.

As of 2000, the DNR claimed ninety-eight member organizations with a total of just over 5.2 million members (Deutscher Naturschutzring, 2000), and the organization's Internet site listed ninety-seven members in 2007 (Deutscher Naturschutzring, 2007). The DNR's board of directors is elected at an annual conference of representatives from the member organizations. It is responsible for most contacts with government and business leaders. The DNR maintains offices in Bonn and Berlin, with a total of about a dozen employees. It has a set of working groups for various topics, and a board member is assigned to coordinate with each. The DNR has long received a modest amount of financial support from the government to fund its basic operations, but its budget remains small, and most interviewees also noted that its effectiveness is limited by lack of financial resources. Lack of staff and funds also requires it to rely heavily on volunteers for lobbying and technical expertise.

Disenchanted with the growing dominance of the large environmental organizations in the DNR, a former activist recruited a group of smaller, largely regional nature protection organizations, none of which had been DNR members, and formed an alternative umbrella organization, the German Nature Protection Forum, in 2001 (Röscheisen, 2006). As of 2007, its Internet site listed thirteen member organizations and cited goals similar to the DNR's, but with an emphasis on meeting the needs of smaller environmental groups (NaturschutzForum Deutschland, 2007).

The DNR constitutes a troublesome conundrum for the large national environmental organizations. Some of the services it offers them, such as information about environmental politics in Germany and the EU and the opportunity to maintain contact with other environmental organizations, are quite valuable. Should the organizations wish, they could use it as a forum for exchanging information and coordinating their own actions. Yet the sheer number of organizations the DNR encompasses and the diversity of their views make using it this way cumbersome and problematic, and there is the possibility that allowing the DNR the coordinator role would increase its influence at their expense. Similar concerns exist about using the DNR as conduit to convey the views of environmentalists to government, industry, or labor. However desirable presenting a unified view might be, formulating consensus positions has proven difficult, time consuming, and full of difficult compromises, if indeed it is possible at all. Furthermore, the DNR has displayed a penchant for assuming the role of sole spokesman for organized environmentalism. This not only threatens the other large organizations' aspirations in this area, it opens the possibility that the DNR might become so deeply entwined in corporatistic decision making that it tries to force undesired compromises on the organizations. In short, there is good reason for them to be cautious.

In addition to the DNR, various other national-level umbrella organizations or working groups in Germany focus on promoting the environmental agenda in specific policy areas. The German Bird Protection Council, for example, works to coordinate the efforts of various bird protection groups (Deutscher Rat für Vogelschutz, 2006), and the Agricultural Alliance promotes organic and sustainable agriculture (AgrarBündnis, 2006). The Forum for Environment and Development, an umbrella organization of environmental and economic development groups sponsored by the environmental ministry and housed in the DNR offices, seeks to promote sustainable development in both developed and less developed countries. All the major national environmental organizations are members (Bammerlin, 1998; Schreuers, 2002; S. Krüger, 2000; Forum Umwelt und Entwicklung, 2006). A somewhat less establishment-oriented approach to the same issues is represented by German Watch. It monitors German and European politics, focusing on sustainable development in Germany, north-south trade issues, and climate change (Wollenweber, 1994; S. Krüger, 2000; German Watch, 2006).

Potential Members, Donors, and Activists

Historically, environmental organizations have depended heavily on the efforts of volunteer activists and the support of members and donors, and the availability of these forms of support is a significant factor in how they respond to the challenges of the twenty-first century.

Volunteering and Giving in Germany

Germany has strong political parties, labor unions, and organized interest groups, but it does not have especially strong traditions of volunteer work and private donations.[31] In part, this is a result of generous government subsidies for the arts, health and human services, and the churches, which are supported primarily through a state-collected church tax. For many years, this generous funding, now on the wane, made it easy for Germans to assume that private donations were not needed and that paid staff would attend to social problems. This does not mean that volunteer work and private donations are altogether absent. Recent studies rank Germany slightly above the European average—although well below the US—in both volunteer work and donations (Gensicke, Picot, and Geiss, 2005; Priller and Sommerfeld, 2005).

Obtaining donations and volunteers in Germany has been made more difficult since the mid 1990s by continuing economic stagnation, lack of growth in real wages, and high unemployment. Only crude estimates of the total volume of donations exist, but the available evidence suggests that the percentage of citizens who make charitable donations has not increased much since the early 1990s. And except for a jump in 2001, the year the euro was introduced, size of the average donation has remained stable even as inflation continued (Priller and Sommerfeld, 2005). Meanwhile, the number of organizations seeking donations has steadily increased. The total percentage of the population active in volunteer work has also remained about the same over the last decade (Gensicke, Picot, and Geiss, 2005).

---

[31] This section is based primarily on the following sources: von Alemann (1989); Anheier (1998); Anheier and Seibel (1999); Oswald von Nell Breuning Institut (1996); Schumacher (2003b); and Gensicke, Picot, and Geiss (2005). Articles and books focusing mainly on donations include Wallmeyer (1996), Luthe (1997), Urselmann (1998), Böker (2001), Felbinger (2005), Priller and Sommerfeld (2005), and Haibach (2006). Especially useful assessments of the situation in Eastern Germany include Behrens (1993) and Weinbach (1998).

Volunteering and Donating for Environmental Organizations

Recruiting volunteers and donations for environmental organizations has been rendered more difficult in recent years by changing perceptions of environmental problems.[32] The absolute percentage of Germans who view environmental problems as important remains high, but other issues, such as unemployment and the performance of the economy, now receive higher importance rankings, and many citizens perceive environmental conditions as improved (see Chapter 7). National surveys also show that knowledgeable and engaged environmentalists constitute a small proportion of the population, and research by the organizations shows that even some of their supporters are only weakly committed to the cause.

German citizens are more apt to volunteer their services to church, social service, and community recreation groups than to environmental organizations. A recent survey conducted for the Ministry for Families, Seniors, Women, and Youth, for example, showed that 2.5 percent of persons over age fourteen claimed to engage in voluntary activity for environmental or animal protection organizations, a percentage essentially unchanged from a similar 1999 survey. Eleven percent volunteered in sports groups, 7 percent in schools, and 6 percent each in churches, culture and music, and social services (Gensicke, Picot, and Geiss, 2005). A somewhat higher percentage of the population reports simple membership in environmental groups. In surveys sponsored by the environmental ministry between 1996 and 2004, between 4 and 9 percent of respondents claimed to belong to an environmental group, but there was no clear trend of increase or decrease (Bundesministerium für Umwelt, 2004; Kuckartz and Rheingans-Heintze, 2006). Surveys conducted for BUND by the Wickert Institute (1989) and for the WWF by the EMNID Institute (1993) found percentages in this same range, as did the 2002–2003 European Social Survey (Bozonnet, 2005) and a 2004 study for the Ministry for Families, Seniors, Women, and Youth (Gensicke, Picot, and Geiss, 2005). In surveys conducted for the environmental ministry in 1996 and 1998, 9 and 11 percent of respondents respectively reported that they had attended a meeting of an environmental organization, and fewer than 10 percent reported having ever complained to a government office about an environmental problem or taken part in a demonstration (Bundesministerium für Umwelt, 1998; Preisendörfer, 1999).

While there is no evidence of a decline in the number of citizens who volunteer their services to environmental groups, there appear to have been changes in

---

[32]This section draws on data from national surveys presented in the following research reports and summaries: Inglehart (1990); Rohrschneider (1991); Billig (1994); Fuchs and Rucht (1994); Dalton (1994); Wimmer and Wahl (1995); Preisendörfer and Franzen (1996); Bundesministerium für Umwelt (1996, 1998, 2000, 2002, 2004); Brand (1997); Preisendörfer (1999); Gruneberg and Kuckartz (2003); Priller and Sommerfeld (2005); Kuckartz and Rheingans-Heitze (2006); Gensicke, Picot, and Geiss (2005), and Bozonnet (2005), as well as the following reports of surveys conducted or sponsored by the organizations themselves: Naturschutz Heute (1994); Wallmeyer (1996); Bode (1996b); Bodenstein et al. (1998); Wickert Institute (1989); and EMNID Institute (1993).

the kinds of commitment volunteer activists are willing to make. A number of my interviewees, as well as numerous well-informed observers (e.g., Bammerlin, 1998; Schumacher, 2003a, 2003b; Mitlacher and Schulte, 2005; Kuckartz and Rheingans-Heintze, 2006), believe that it has become increasingly difficult to find volunteers who will commit themselves to working intensively for an organization over a period of years, and that declining social movement mobilization has reduced the number of highly motivated new volunteers. Conflicts with job demands, the increasing proportion of women in the labor force, concerns about personal security in a stagnating economy, and a growing desire for personally meaningful and interesting volunteer work have led more and more people to limit the time they are willing to devote to the organization to a few hours a week or to short-term projects of interest to them. A national survey conducted provided at least some support for this position; almost a fourth of those who reported doing volunteer work for environmental or animal protection groups said that they felt overburdened by their work (Gensicke, Picot, and Geiss, 2005). Similarly, a survey of BUND activists conducted in 2004 showed that the activists worked very long hours for the organization and found juggling volunteer work and family difficult (BUNDSchau, 2003a).

Financial contributions to environmental causes lag far behind gifts to churches, health and welfare organizations, and emergency relief (Haibach, 2006). In environmental ministry biennial surveys between 1996 and 2004, between 14 and 27 percent of respondents said that they had contributed money for some type of environmental protection, with the most recent survey reporting 15 percent (Bundesministerium für Umwelt, 1998, 2004; Preisendörfer, 1999; tabulations for 2000 and 2002 supplied by Prof. Udo Kuckartz; Kuckartz and Rheingans-Heintze, 2006). Results from the TNS Emnid Spending Monitor annual surveys from 1995 to 2004 also show minor year-to-year variation without a clear trend, but cluster around the 10 percent mark (Tnsemnid, 2007), the same percentage reported by the European Social Survey for 2002 (Bozonnet, 2005). The head of fundraising for one of the organizations told me that he estimated that there were about three million potential contributors for his organization in Germany, or about 5 percent of the adult population.

Persons who might become activists, donors, or even members of an environmental organization are far from representative of the entire population. According to the most recent environmental ministry survey, the highest rates of donation occur among those in late middle age (Kuckartz and Rheingans-Heintze, 2006). The 2004 Ministry for Families, Seniors, Women, and Youth survey of volunteering showed the same result: both rates of membership or other participation in environmental or animal protection activities and rates of active volunteering were highest among those ages 46 to 65, followed by those ages 31 to 45 (Gensicke, Picot, and Geiss, 2005). These results echo patterns from questions about environmental consciousness, in which persons in middle age have replaced young adults as the most environmentally conscious group (see Chapter 7).

For the environmental organizations, the strong representation of this cohort—many of whose members cut their teeth during the confrontational environmental movement of the 1970s and 1980s—among their potential supporters poses a significant challenge. Some, like one old-time activist I interviewed, continue to fight for countercultural ideologies and strategies of confrontation from within the organizations, while others, like a local Berlin activist I interviewed, are no longer active but retain their memberships and criticize from outside. Others, such as a former employee of one of the organizations of my acquaintance, have moderated their views over the years but still think of themselves as veterans of the movement. Although they recognize current-day realities, they retain an underlying skepticism of business and government; too much compromise makes them uncomfortable, and a broad-based democratic movement remains an ideal for them.

Surveys also typically show that members and supporters of environmental organizations are well-educated and economically comfortable—although the most recent studies show little relationship between income and more general environmental consciousness. Those most concerned about environmental problems and most likely to participate in environmental organizations also display a distinct occupational profile, often characterized as the "new middle class" (Eckersley, 1989; Kriesi and Giugni, 1996). They include many teachers, scientists, independent professionals, government employees, health and human services workers, and students. Notably less well-represented are members of the working class and corporate management and ownership. Recent surveys show higher levels of environmental consciousness among women, but women are only just now catching up with men in self-reported membership in environmental organizations.

Supporters of the environmental cause also tend to be politically to the left, to hold more critical attitudes toward society, and to be less oriented toward business values and material success and more oriented toward self-development. For some of them environmentalism assumes the tone of a moral crusade, but relatively few reject basic social institutions, and there are far more reformers than revolutionaries. Self-reports about motivations for joining or supporting environmental groups must be interpreted cautiously, but altruistic motives such as love of nature and desire to take responsibility for problems are ranked much higher than direct personal benefits such as gaining professional qualifications or developing social contacts.

The Catholic and Evangelical (Lutheran) churches claim the great majority of Germans as at least nominal members, although rates of active involvement are low. Both faiths have a long record of advocating for environmental protection (Rogall, 2003), and religion is associated in population surveys with greater environmental concern and engagement, as well as with donating to environmental organizations. Informed observers believe that supporters of environmental organizations are more likely than the general population to have ties to "alternative" organizations and cultural practices, including organic agriculture, stores offering "natural" products, alternative medicine, alternative schools, and "New Age" groups, which frequently embrace ecological themes or terminology (Wüst, 1993; Biehl, 1995; Geden, 1999; Bergstedt, 2002).

Summary and Conclusions

The goals and strategies pursued by the large, national environmental organizations at the beginning of the twenty-first century were, as the institutional perspective suggests, partly a result of their histories. Yet no organization in a changing society can remain completely wedded to historical patterns. As open systems, the organizations had to come to terms with the major social trends described in Chapter 7, for these trends affected their ability to obtain the resources they needed, maintain themselves, and move toward their goals. The complex and challenging social context the organizations faced as they entered the twenty-first century was, in part, the result of these trends. It created for them a series of dilemmas centered around building internal structures suited to their environment, acquiring needed resources, and setting organizational goals and developing strategies for reaching them. In the remainder of this summary, I briefly review this social context and provide an overview of the problems it posed for the organizations.

Germany's dense population, its industrialized economy and ecologically problematic industries, its lack of fuels other than coal, and its high consumption and increasingly auto-dependent economy create a multitude of environmental problems for environmental organizations to address: air and water pollution, greenhouse gas emissions, waste disposal, urban sprawl, loss of scenic landscapes, dependence on coal and nuclear energy, traffic problems, and endangered species and ecosystems. The organizations had to decide what combination of these problems to emphasize and how best to address them. They had to consider not only potential benefits for the environment, but also what mixture of goals and strategies would allow them to obtain the resources and legitimacy they needed to continue their work.

Germany's high unemployment rate, heavy reliance on exports, and high dependence on imported energy sources provide the basis for arguments that environmental measures will damage Germany's international competitiveness, cost the economy jobs, and increase its dependence on foreign oil. The economic stagnation of the late 1990s and early twenty-first century increased the persuasiveness of such arguments; therefore the organizations had to choose goals and strategies carefully so that they could move forward without undermining their own legitimacy or driving away their less committed activists or donors.

Germany is a leader in ecological modernization, and the economy includes some especially ecologically conscious firms and numerous producers of environmentally friendly products; however, even businesses with minimal ecological consciousness are anxious to be viewed as "green." They mount public relations efforts to tout their ecological awareness, and they may be willing to donate to noncontroversial environmental projects that will polish their image. Labor unions too no longer offer across-the-board condemnations of environmental controls as "job-killers," and they are sometimes willing to work with environmentalists. These conditions raised difficult questions for the environmental organizations about how closely to cooperate with business and labor and with which industries, firms, or unions to cooperate. The potential availability of donations from businesses, whose motivations rarely line up perfectly with those of the organizations, also posed difficult strategy choices.

The influential and well-organized German business sector frequently and effectively resists environmentalists' proposals through well-funded and effective lobbying and public education programs. Firms in ecologically problematic sectors are especially vigorous opponents of measures that affect them directly, and they are sometimes joined by their unions. Commonly used strategies include portraying environmental measures as damaging to Germany's international competitiveness, public relations programs that amount to "greenwashing," arguing for deferring action until interminable disputes among experts are resolved, offers of voluntary compliance as an option to legislation, and offers to negotiate with environmentalists on business' terms. How the organizations cope with these business strategies, whether and with whom they choose to negotiate, and how strongly they confront business laggards have implications not only for the environment, but also for the organizations' legitimacy, reputation, and access to resources of various types.

Effective environmental protection requires the ability to influence the political system. Germany's federal political structure, three branches of government, and multiparty system provide multiple points of potential access, but the complexity of the system with all its interdependent parts is daunting, and exerting political influence is no easy task for amateurs. The political system displays many elements of neocorporatism, providing business and labor with privileged access to the circles in which legislation and implementing regulations are formulated. The system's earlier stubborn resistance to the environmental movement has been replaced by a degree of acceptance. Yet even after the elaboration of a substantial framework of environmental law and the establishment of an environmental ministry, environmental organizations remain rather marginal participants in the system, and mutual suspicion lingers. Neither wholehearted participation in the system nor confrontational opposition from the outside proved to be a completely satisfactory adaptation to this situation, so the organizations found themselves negotiating a potential minefield as they tried to find the right mix of social movement and interest group strategies and decide which of the other possible assignments of civil society organizations they wished to undertake. Concretely, this implied making hard choices about how deeply to become involved in the political system, when and how to confront the authorities, and whether to solicit or accept government subsidies.

Additional complications resulted from the rise of the Green Party and its recent participation in the governing coalition. The party's successes have been a clear plus for the environmental organizations seeking to influence policy, but the Greens remain a small party, which limits its ability to produce results, and the environmental organizations have many members from many other parties. Outright endorsement of the Greens would therefore have many drawbacks, and tight party discipline precludes endorsing individual candidates. The remaining parties' orientations toward environmental issues range from mild support to rather reluctant acceptance, so a strategy of attempting to work with whatever coalition is in power is a viable, though at times unrewarding, option. Each of these choices offers rewards and benefits for the organizations, which must try to choose the most promising avenue.

Working with the administrative bureaucracy also poses many challenges for the environmental organizations. The tradition of a strong state bureaucracy with limited citizen input has been tempered by postwar reforms and by additional reforms in recent years, but it has not disappeared. Rights to acquire information and to intervene legally remain limited, gaining credibility within the system requires the ability to display factual expertise, and relations between environmental organizations and the bureaucracy remain strained. Efforts to influence administrative decisions are thus most likely to succeed if executed by organizations with substantial expertise, continuous involvement, and considerable resources of funds and time, all of which suggest professionalization of their staffs. Yet this approach is not without its costs, especially for organizations that wish to remain informally structured or keep their distance from government.

Right-wing environmentalism, once at the core of German environmentalism, is relegated today to the margins. Yet the old themes live on, and they bubble to the surface in the writings of far-right intellectuals and the platforms of right-wing political parties. Charges that the organizations remain open to seduction by the right continue to be raised from time to time, and they must tread exceptionally carefully in this terrain.

The expansion of the European Union's role in environmental policy making introduced an additional level of complexity into the environmental organizations' political context. The impact of the EU on German environmental politics and policy is substantial and growing, and it is a potential ally of the environmental organizations in some struggles. Yet finding ways to build contacts with the EU or influence it is no small assignment. The development of a European media and public opinion are in their infancy, so expert lobbying, rather than public education or mobilizing for demonstrations, is the order of the day for those that wish to influence the arcane European policy-making process. The EU bureaucracy has not been very receptive to individual national organizations' attempts to influence it. And even if the organizations wished to pursue this approach, developing expertise about EU environmental politics and maintaining a lobbying office in Brussels would require major commitments of time and money. Alternatively, the organizations can attempt to affect EU policy by influencing the German government's votes in the Council of Ministers or lobbying Germany's representatives to the relatively weak European Parliament, or they

can work through one or more of the numerous international networks of European environmental groups. The time and financial resources required by the latter approach are smaller, but substantial willingness to compromise is required, and the organizations gain little visibility at home in return for their efforts.

With limited resources to finance their own advertising campaigns, environmental organizations are necessarily very dependent on favorable media coverage of their issues and viewpoints to influence public opinion and maintain visibility and a positive image. Specialized environmental publications do exist, but they have a very limited circulation. Local and nationally circulated newspapers and television reach a wide audience, but they devote ever fewer resources to investigative reporting on environmental issues and are not well equipped to cover complex, slowly developing stories. Environmental organizations that can meet the media's needs for timely and accurate stories with dramatic events and images can still attract media attention, but success in the media does not come cheap, and the game is a difficult one for amateurs. It requires detailed knowledge of the taste and habits of the media, the skill to provide the right story at the right time, and finding the right balance between dramatic confrontation and sober analysis.

The large environmental organizations face a great deal of competition from one another and from other environmental groups in their efforts to gain influence, be visible and respected, and attract members and donations. Potential members, activists, and donors have their choice of supporting BIs and their umbrella organizations, general environmental organizations at the national or regional level, and single-purpose organizations. For individual environmental groups, this places a premium on visibility, legitimacy, and a reputation for effectiveness. Consequently, gains in effectiveness realized by cooperating with other environmental groups can be offset by the possible loss of visibility or unique profile, and the large national organizations face the challenge of finding the right balance.

This dilemma has proved especially acute in the case of the DNR. Germany's quasi-corporatist system offers the environmental organizations many incentives to channel their efforts through an umbrella organization like the DNR. However, the DNR's efforts to be inclusive have led to a degree of diversity among its member organizations that has made it difficult to reach consensus or operate effectively—even after the departure of the most discordant voices during the 1980s. These problems, as well as the absence of the WWF and Greenpeace from its membership roll, undermine the DNR's claim to be *the* spokesman for the environmental organizations. BUND and NABU have obtained considerable influence within the DNR, but—evidently fearing that it would become a competitor—they have used their power to channel it toward functioning as a service bureau rather than staking out positions and developing campaigns of its own. Nevertheless, in a semicorporatist system the DNR has its uses, and decisions about when and how to work with and through it remain difficult ones for the environmental organizations.

Although the German public continues to rate environmental concerns as important, economic difficulties, unemployment, and other issues now rank at the top of the public agenda, exacerbating the problems the organizations face gaining media attention, political influence, and supporters. The pool of potential supporters is substantial, but it represents only a small percentage of the population, and it is concentrated among the better-educated, persons in typical "new class" occupations, and increasingly among middle-aged veterans of the confrontations of the 1970s and 1980s. In this environment, finding the right mix of organizational goals and effective strategies for attracting members, activists, and donors is no simple matter. Committed activists, in particular, have become a scarce commodity. Confrontational strategies that attract some supporters may repel others, so organizations that can use carefully targeted and designed appeals for support are at a considerable advantage.

How the organizations developed successful strategies for coping with these challenges is the subject of the remainder of this book. The next chapter profiles the four largest organizations, their goals, and their strategies. The following three examine how the organizations have coped with seven key dilemmas centered respectively around organizational structure, resource acquisition, and selection of goals and operational strategies.

Chapter 9

Major Environmental Organizations in Germany

Four Profiles

By the last decade of the twentieth century, four large organizations, the German Nature Protection League (NABU), the Worldwide Fund for Nature (WWF), the German League for Environment and Nature Protection (BUND), and Greenpeace, each with over 250,000 supporters, had come to be generally recognized as the most important German environmental organizations (Rucht, 1989; Blühdorn, 1995; Bergstedt, 2002; Foljanty-Jost, 2004; Brand, in press). The precipitous decline of the BBU, the demise of the GNU, the self-imposed limitation of the Green League to the territory of the former East Germany, the stalled growth of Robin Wood, and the continued marginal role of organizations like the Friends of Nature and League for Homeland and Environment mean that these four organizations face little direct competition as general-purpose environmental organizations. Environmental organizations with specialized goals and regional environmental organizations also exist, but they have memberships and general impact much smaller than those of the "big four."

The following three chapters describe in detail the key strategy dilemmas the four organizations had to cope with in order to attain and hold their dominant positions. This chapter, mainly descriptive in content, contains profiles of the four organizations. For each I provide a very brief recap of its history, an overview of its major activities and emphases, a description of its organizational structure, and data about its membership and finances. The profiles focus on the characteristics of the four organizations as they entered the twenty-first century. I have extended the membership and budget statistics to include most recent available data at the time the manuscript was completed and noted a few recent key developments in programs, leadership, or structure, but none of the organizations have experienced major changes since the beginning of the century.

The German Nature Protection League (NABU)

The German Nature Protection League was founded in 1899 as the League for Bird Protection (BfV). Dominated by educated, middle-class leaders, it focused on acquisition of ecologically sensitive areas for bird habitats, practical efforts to protect birds, public education, and behind-the-scenes lobbying. It remained relatively resistant to the seductions of prewar right-wing ecology, but it did not actively contest being placed under government control during the National Socialist era and benefited from its willingness to cooperate with the Nazis by being declared the official bird protection organization for all Germany, and from its receipt of government subsidies. After the war, the organization reverted to its earlier structure, goals, and style of operations; however, it was much weakened by association with the Nazi regime, membership losses, and wartime destruction of its facilities and required many years to recover.

In the polarized atmosphere of the late 1970s and early 1980s, the BfV gradually transformed itself into a more general environmental organization with a broader set of goals, an increased level of political involvement, and a more confrontational approach; however, nature protection remained foremost among its goals. The transformation was marked by sharp internal conflicts—since largely resolved—between older, more conservative members and younger members, who favored a more activist action repertoire and broader agenda. The BfV grew rapidly in the 1980s, and in 1990, it incorporated numerous local groups from the German Democratic Republic's state-sponsored nature protection organization and adopted its present name and structure.

Goals, Programs and Emphases

In its most recent goal statement, issued in 2000, NABU identifies climate, energy, transportation and land use policy, maintenance of biodiversity and ecosystems, sustainable use of land and natural resources, strengthening environmental consciousness and ethics, and the effects of environment on health as among its key concerns (NABU, 2000a; see also NABU, 1999). Its Internet site highlights the first three of these and adds protection of wild animals, protection of ecosystems and biodiversity internationally, waste reduction, and sustainable development in Germany and worldwide to the list (NABU, 2007). Almost all of my interviewees agreed that NABU's strong suits are protection of specific species and ecosystems and agricultural policy, and these goals, which reflect the organization's historical focus on nature protection, receive much emphasis in its goals statement. NABU has also recently extended its reach to selectively support nature protection projects in Eastern Europe, Russia, and Africa (e.g., Naturschutz Heute, 1999c; NABU, 2000b; Naturschutz Heute, 2003b). Many of the organization's local groups remain heavily involved in concrete nature protection efforts, such as caring for nature reserves and guarding bird nesting sites. These activities offer members the opportunity to be directly involved in protecting nature and to enjoy participation in a social network (Cornelsen, 1991; Bergstedt, 2002).

NABU also continues the BfV's longtime emphasis on acquiring nature reserves. It maintains a network of over 5,000 nature reserves and continues to acquire new ones. Over fifty of these, including its recently constructed facility at Blumberger Mühle near Berlin, double as nature educational centers. Others, such as the Michael Otto Institute (formerly the Institute for Bird Protection), engage in scientific research, and some of NABU's members engage in amateur research, such as nationwide bird counts (NABU, 2001, 2004a, 2005b; Naturschutz Heute, 2000c, 2000d, 2005a, 2005b, 2005f). From 1977 until 2003, NABU also operated a major nature protection and volunteer training center (Naturschutz Heute, 2003c). Since 1971 NABU has named a "bird of the year," often selecting a species that illustrates its current concerns (e.g., Naturschutz Heute, 2000b). Since 1993, it has named a "Dino of the Year" to point up environmental scandals (e.g., Naturschutz Heute, 2000a).

NABU engages in lobbying at the national, *Land,* and local levels. It has had a Berlin office for several years and is in the process of moving its headquarters from the former West German capital, Bonn, to Berlin to increase its access to the national government. Local chapters still contain many relatively apolitical members whose primary focus is local nature protection, and several of my interviewees, including those from the organization's national and *Land* headquarters offices, pointed out that even today political education, political activity, and lobbying are often pursued more enthusiastically at the national level than the local (see also NABU, 2000a; Bergstedt, 2002). They insisted, however, that local chapters and the national office have learned to coexist peacefully and respect and support one another's different contributions. A poll of environmental journalists in the late 1990s found that NABU's political effectiveness was rated lower than that of BUND and Greenpeace (Naturschutz Heute, 1998e); however, several of my interviewees from various political parties said it can be effective in its areas of specialty. NABU is recognized as a nature protection organization under the German Nature Protection Law, which entitles it to submit position statements and file complaints about specific projects that might have a negative impact on nature (Rogall, 2003).

NABU is also active in public education. In addition to education centers at its nature reserves, the national office publishes many pamphlets and reports on environmental topics, and some of its local groups are also active in this area (Cornelsen, 1991; Fritzler, 1997; Bergstedt, 1998; NABU, 1999). NABU's member magazine appears quarterly in an edition of almost 200,000 copies (Rogall, 2003). It includes articles about natural history, environmental problems and politics, and NABU projects and political positions, as well as articles offering advice on ecological lifestyle topics such as natural food and clothing, nature photography, and green investing. Also included are listings of activities at NABU nature centers, tips on where to buy maps and booklets on ecological lifestyles, classified ads for vacation cottages, and book reviews of nature guides and books on nature photography, national parks, and more serious topics. Each issue includes a section for readers from NAJU, NABU's affiliate for young people. Finally, there are occasional articles about the business firms with which NABU has cooperative arrangements. Each issue also contains an advertising section for items for sale from the NABU Nature Shop, including books, models of birds, binoculars, bird nesting boxes, and clothing with the NABU logo, as well as a good many ads from private firms marketing to persons interested in ecological lifestyles and outdoor activities. NABU also maintains a very extensive, professionally constructed Internet presence.

Organizational Structure

NABU and its semi-autonomous *Länder* chapters are all legally organized as *Vereine* (see Chapter 8). NABU has chapters in every *Land* except Bavaria, where there is an independent bird protection organization that formed an alliance with NABU in 2001 (NABU, 2001). One highly-placed interviewee told me that, although NABU had wanted to merge with the Bavarian sister organization for many years before 2001, the latter's president had stood in the way. One of NABU's strengths is its strong network of approximately 1,500 local groups throughout Germany (Bergstedt, 2002; Rogall, 2003; NABU, 2007). Delegates from the local groups meet annually to elect boards of directors for their *Land* chapters, and delegates from the *Land* chapters meet annually to elect the national board of directors (Naturschutz Heute, 1999b). The national board is filled with volunteer workers, leaders from the *Land* chapters, and national office staffers. Business leaders and government officials are rarely elected (NABU, 1999, 2000b, 2001, 2004a, 2007), although a few are included on the organization's larger advisory board (NABU, 2001). As of the mid 1990s NABU had approximately eighty full-time employees (Bergstedt, 2002). Today it has about seventy-five (Helge May, e-mail correspondence).

The president draws a full-time salary and acts as CEO. NABU's president from 1992 to 2003 was Jochen Flasbarth, the former leader of the affiliated youth group, which during the 1980s had pushed for a broader and more political agenda. NABU enjoys a reputation for stable leadership and effective administration, and several of my interviewees commented on the skill with which Flasbarth was managing the organization's gradual transition from a nature protection organization to a more activist environmental organization. Flasbarth resigned in 2003 to accept a position as director of the Nature Protection Division of the environmental ministry and was replaced by Olaf Tschimpke, the chair of the board of directors of the Lower Saxony chapter (DNR Deutschland Rundbrief, 2003e).

When the BfV merged with elements of the GDR's Society for Nature and Environment (GNU), it adopted the GNU's and BUND's practice of constituting working groups comprised of volunteer experts from around the country to gather information, prepare reports, and assist with advocacy work concerning specific environmental problems. NABU also has a national advisory board of scientific experts (Naturschutz Heute, 1999b; Rogall, 2003). Its affiliated youth organization, NAJU, is semi-independent but heavily subsidized by the parent organization (NABU, 2000b, 2005a). In 2000, NABU established an affiliated foundation, primarily for the purpose of acquiring and preserving state-owned land in the former GDR (NABU, 2000b, NABU-Stiftung, 2002, 2005). NABU is a member of the DNR, Bird Life International, and the EEB (NABU, 2000b, 2007).

Membership and Finances

NABU emerged from its merger with elements of the GNU with active chapters in every *Land* except Bavaria. During the second half of the 1990s, it experienced steady growth and entered the new century with just over 287,000 members, an increase of 37 percent since 1995. In 2000, NABU began reporting membership totals including its new partnership—not yet a full merger—with Bavaria's bird protection organizations, which had 66,000 members. Rapid growth continued until 2002 but then stalled out.

NABU's membership is geographically unbalanced. In 2000 the organization averaged 6.4 members per 1,000 inhabitants in Germany's five southernmost *Länder* (including its Bavarian affiliate). In the *Länder* of the former GDR (including Berlin), it had only 2.4 members per 1,000 population. In the remainder of Germany, it averaged 4.8 members per 1,000 inhabitants (NABU, 2001). Still, NABU's representation in the East has been on the increase (NABU, 1999, 2003), and several leaders I interviewed reminded me that membership totals for the *Länder* of the former GDR can be misleading because a much higher proportion of members there are active.

Figure 9.1. Membership Trends, 1995–2004

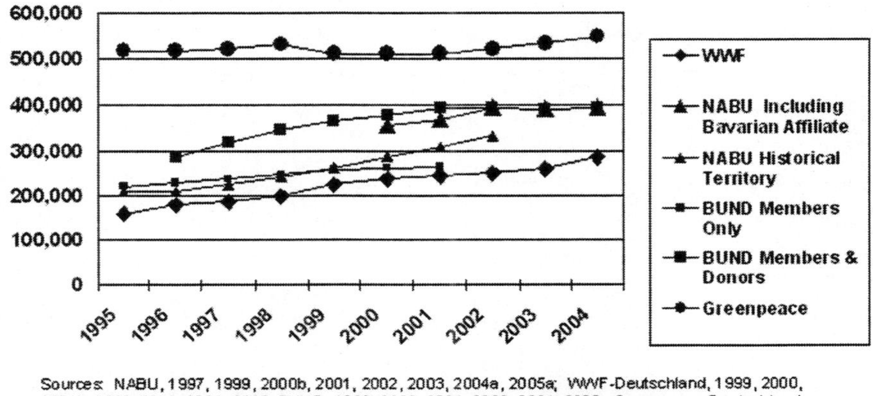

Sources: NABU, 1997, 1999, 2000b, 2001, 2002, 2003, 2004a, 2005a; WWF-Deutschland, 1999, 2000, 2001b, 2002, 2003, 2004, 2005; BUND, 1998, 2000, 2001, 2002, 2004, 2005; Greenpeace Deutschland, 1996, 1997, 1999, 2000, 2002, 2003, 2004, 2005a; Greenpeace Nachrichten, 2001a, 2005c.

According to my interviews, only about 10 percent of NABU members are active in the sense of attending meetings and taking on volunteer tasks (see also Cornelsen, 1991; NABU, 2004a; Felbinger, 2005). A survey in the mid 1990s indicated that the organization's members are typical of environmentally concerned citizens in Germany, except for being about two-thirds male and less urban (Naturschutz Heute, 1994; see also Bammerlin, 1998). An interviewee from outside the organization who knows NABU well characterized its members as "Loden Coat Greens;" another simply said that NABU members were less likely than BUND members to be "Müsli eaters," and a highly placed NABU leader speculated that many live in suburbs and rural areas. As of 2000, 43 percent of NABU members were women, but they constituted only 6 percent of local group leaders, 20 percent of the representatives to the national assembly, and 10 percent of the board of directors. This underrepresentation was of enough concern for the national leadership to sponsor a study of it (Naturschutz Heute, 2000m), and in a 2001 interview the president of a *Land* chapter volunteered that males remained a clear majority in his chapter and that this was an issue of some concern.

NABU's total revenue in 2000 was €15,732,000, an increase of 32 percent over 1995, and by 2004 it had reached €19,361,000. In 2005, 66 percent of the organization's receipts came from member dues, and an additional 3 percent from inheritances. Donations from business and individuals are combined in NABU's financial statements; together they constituted 18 percent of revenues, with about 90 percent of this derived from individual gifts (Felbinger, 2005). However, over the years, it has accepted donations or entered sponsoring arrangements with many business firms, including IBM, Volkswagen, Lufthansa, which has long supported a joint WWF/NABU project to save cranes, and a porcelain firm, which makes plates representing each year's bird of the year (Zillessen and

Figure 9.2. Gross Revenue Trends, 1995–2004

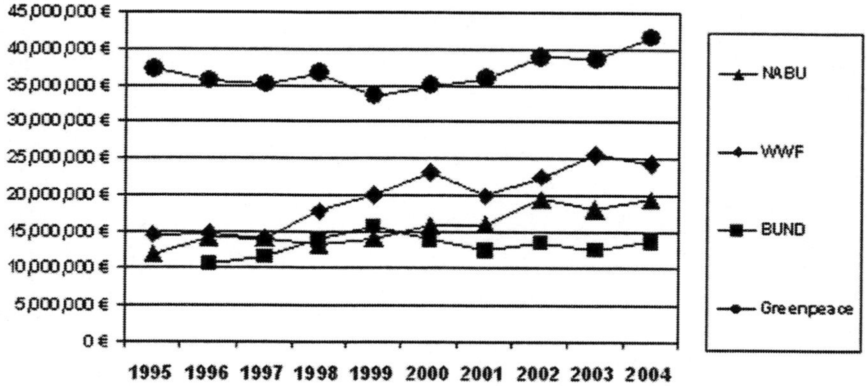

Sources: NABU, 1997, 1999, 2000b, 2001, 2002, 2003, 2004a, 2005a; WWF-Deutschland, 1999, 2000, 2001b, 2002, 2003, 2004, 2005; BUND, 1998, 2000, 2001, 2002, 2004, 2005; Greenpeace Deutschland, 1996, 1997, 1999, 2000, 2002, 2003, 2004, 2005a.

Rahmel, 1991; Bergstedt, 2002). Government subsidies of NABU projects accounted for only 5 percent of revenues. Smaller amounts of income came from fines transferred to NABU, earnings on investments, and sales of materials and services (NABU, 2005a).

NABU has not sought to set aside substantial financial reserves; consequently, its year-by-year expenditures closely approximate its revenues. Unfortunately, expenditures are not reported by type of activity, such as public information, lobbying, acquisition and maintenance of nature reserves, and the like. The narrative portion of the 2000 annual report states that NABU spent 5 percent of its budget on administration and 24 percent on member services and new member acquisition (NABU, 2001), and the 2004 report lists the same percentage for administrative expenses (NABU, 2005a).

The Worldwide Fund for Nature (WWF)

The Worldwide Fund for Nature was founded in 1961 in Britain by international scientific, business, and political elites to raise funds to protect wildlife worldwide. The German branch was established in 1963. For almost two decades, WWF-Germany was a small, elite organization that functioned mainly to raise funds for WWF-International. With its business-friendly approach and relatively noncontroversial mission, the WWF was little involved in the confrontations of the late 1970s and 1980s; however, during the 1980s it was reinvigorated by new leadership. It added staff, professionalized its operations, and increased its emphasis on projects in Germany. Somewhat later, it broadened its mission to include selected environmental issues. These changes, coupled with high public interest in environmental issues during the 1980s, led to impressive expansion in the size and breadth of its donor base and in its budget.

Goals, Programs, and Emphases

WWF-Germany is very closely intertwined with WWF-International, and its magazine and annual reports (e.g., WWF-Journal, 1998d; WWF-Deutschland, 2001b; Wünschmann, 2003) do not always clearly distinguish between the goals and activities of the two. WWF goals, as articulated in recent overviews, include maintaining biological diversity, protecting ecosystems that are necessary to sustain wildlife populations, sustainable use of resources, providing economically viable and sustainable options for local citizens who might otherwise destroy wildlife or ecosystems, and fighting problems—such as pollution, excessive consumption, destructive resource extraction, unsound agricultural practices, and climate change—that threaten wildlife worldwide. Substantive concerns emphasized in recent years include climate and atmosphere, oceans and coastal areas, rivers and wetlands, forests, agriculture, and tourism (WWF-Deutschland, 1999, 2001b, 2004, 2007; WWF-Journal, 2001; Wünschmann, 2003). The WWF also continues its traditional efforts to protect "megafauna," such as elephants and tigers, through creation of wildlife reserves, training of rangers, and the like, and it features this work prominently in its self-presentations (Bergstedt, 2002).

WWF-Germany's premier activity remains fundraising for WWF projects, both in Germany and abroad. In 1999, it was supporting 31 international projects and 35 in Germany; in 2003 the corresponding numbers were 48 and 30 (WWF-Journal, 2000g; WWF-Deutschland, 2005). Since 1981, the organization also has a major commitment to the WWF's international emphasis on combating trafficking in scarce wildlife, especially at Frankfurt's international airport (Cornelsen, 1991; WWF-Journal, 2002f), and it has heavily promoted the Forest Stewardship Council's program of certifying products from forests managed according to ecologically sound practices and a parallel Marine Stewardship Council program for sustainable fisheries (Bauske, 2003; WWF-Deutschland, 2005). WWF-Germany also purchases land in wetlands, coastal areas, and other ecologically sensitive areas and operates several nature reserves and research stations—such as its Floodplain Institute—that are integrated into these emphases (WWF-Deutschland, 1999, 2002, no date; WWF-Journal 2000h).

WWF-Germany is also active in research, public education, and lobbying, mainly in the areas of climate and air pollution, coastal areas and the North Sea, and rivers and floodplains. It issues brochures on a variety of topics, stages exhibits, holds press conferences and issues press releases, and prepares materials for schools (Heuser, 1994; Fritzler, 1997; WWF-Deutschland, 1999, 2001b, no date; Groth, 2003). It also occasionally funds applied research intended to influence public opinion or politics, such as a 2002 study of the potential for reductions in CO2 emissions by switching to renewable energy in Germany's power industry (WWF-Magazin, 2003a). The WWF's twelve-person lobbying staff followed the national government to Berlin in 2003. However, some of my interviewees reported that the WWF's lobbying effort was less successful than those of BUND or NABU. Almost to a person, they described the WWF as a relatively conservative organization with close ties to business and a relatively non-confrontational approach (see also WWF-Deutschland, 2001b; Bergstedt, 2002; WWF-Journal, 2002e; Kohl, 2003). The Berlin office is also responsible for coordinating with the WWF's Brussels office and its efforts in Central and Eastern Europe (WWF-Magazin, 2003b; WWF-Deutschland, 2003; von Treuenfels, 2003; WWF-Magazin, 2005).

The WWF's quarterly member magazine is more professionally produced than its NABU or BUND counterparts. It is printed on higher-quality paper, and each issue includes many pages of professional-quality nature photography. In addition to short articles about WWF activities and successes, successful collaborations with business and government, and contemporary environmental issues, each issue contains a series of longer articles, most organized around a specific theme, such as forests, the Arctic, oceans, or climate. Most articles are written by WWF staff. Also included are book reviews, a short section on items available from the WWF's mail order house, a page for "Young Pandas," and a short section describing the activities of its local groups. There are few advertisements.

The WWF maintains an elaborate, professionally produced Internet site, which reported over half a million hits in 2000 and 700,000 in 2003 (WWF-Deutschland, 2001b, 2004). The site also contains opportunities to donate and occasional opportunities to complete an online petition (WWF-Journal, 2002b). By 2003 the WWF had almost 30,000 subscribers to its e-mail newsletter (WWF-Deutschland, 2004). It has also made successful use of e-mail to mobilize members to e-mail government officials protesting decisions it believes are environmentally unsound (e.g., WWF-Journal, 2000d).

Organizational Structure

With almost five million supporters and 4,000 employees, the WWF is the largest international environmental organization. It has independent branches in thirty nations and offices in over twenty more. In 2005, its total income worldwide was almost €390,000,000, and it supports almost 2,000 projects in over 100 countries worldwide (WWF-International, 2006a, 2006b). The WWF national organizations are independent organizations, but they cooperate closely with the international organization by sending representatives and experts to work with it and by supporting it and its projects financially (WWF-Deutschland, no date). WWF-Germany is the fourth largest national branch (WWF-Journal, 2002f).

WWF-Germany is organized as a foundation with a self-perpetuating board of directors. Its annual report does not list its Board members' occupations, but in 2004 only 20 percent were women (WWF-Deutschland, 2005). The board is advised by a Council of Scientific Experts. The thirty-member, self-perpetuating board of directors has several standing committees, the most important of which is the Executive Committee. The president, who is employed-full time, chairs the Executive Committee and serves as the chief executive officer. Routine administration is in the hands of the executive secretary. Individual supporters participate primarily by making donations, reading the member magazine, or occasionally participating in a petition drive. They have no say in the WWF's policies or management (Blühdorn, 1995; Bergstedt, 2002; Zidek, 2003). As of 2004, the WWF had 112 employees, the equivalent of eighty-nine full-time workers (WWF-Deutschland, 2005). Well over half are women, and about a third work part-time (Zidek, 2003). The headquarters and business office is in Frankfurt, but many of the professional employees are located at various satellite offices.

The WWF has, from time to time, experimented with informal regional groups that give its supporters the chance to meet together, make presentations at local environmental events, or staff information stands (WWF-Journal, 2000i; Bergstedt, 2002; Felbinger, 2005). As of 2003, there were thirteen such groups (WWF-Deutschland, 2004). However, the WWF has no formally established chapters at the *Land* or local level and—according to an interview with its president in 2000—no plans to establish them. The WWF's youth group was dissolved in 1996 (Fritzler, 1997), but it has a separate membership category, newsletter, and web page for "Young Pandas," who totaled 8,200 in 2004 (WWF-Journal, 1998f; WWF-Deutschland, 2001a, 2005). It also offers supporters the opportunity to participate in various nature observation or education outings it sponsors (Bouman, 2003).

The WWF was a member of the DNR during the 1990s, but it dropped its membership in 2000 (Röscheisen, 2006). WWF-International maintains an office in Brussels to represent its interests at the EU, and this office coordinates its work closely with the Berlin office (Webster, 1998; Roose, 2002; WWF-Deutschland, 2004).

Membership and Finances

As a foundation, the WWF has no true members. Persons who obligate themselves to an annual contribution of at least €40 annually (WWF-Deutschland, 2007) receive the organization's quarterly magazine. For many years they were usually described as "supporters" (*Förderer*), though the term "members" (*Mitglieder*) sometimes appeared. The organization's web site now refers to them as members; however, here and elsewhere, I use the more technically accurate term, supporters.

The number of WWF-Germany supporters grew by a striking 47 percent during the late 1990s, making it the fastest-growing of the four organizations. Nevertheless, with 236,000 supporters in 2000, it remained the smallest. It grew at a much slower rate during the first years of the new century, but membership jumped by 10 percent in 2004.

A monthly contribution of €30 to the WWF earns one the status of "patron" (WWF-Journal, 2002c, WWF-Deutschland, 2005). The number of patrons grew from approximately 2,800 in 2000 to 4,500 in 2003. Their contributions protect species like the Siberian tiger and European brown bear, as well as national parks in Germany and Indonesia (WWF-Deutschland, 2001b, 2004). A program begun in 2000 designates those who contribute at least €1000 annually as "Global 2000 Protectors" (WWF-Deutschland, 2001b). By 2004 this program had 365 participants (WWF-Deutschland, 2005).

I was unable to locate any quantitative data about the geographic dispersion or characteristics of WWF-Germany's supporters; however, an article by a WWF fundraiser describes the typical supporter as above average in income, very well-educated, and over thirty-five. Women donors outnumber men, and supporters are more likely to live in southern Germany and outside central cities (Bouman, 2003). Interviewees who know the organization well characterized WWF supporters as people who have money to give, love nature, are often politically inactive, and are rarely political or environmental radicals. One interviewee described the organization as "rather elitist," while another described its supporters somewhat more pointedly as "worthies." WWF-Germany's total revenues increased by 59 percent between 1995 and 2000, providing it with by far the fastest income growth of any of the organizations. Its 2000 income of €25.5 million ranked it far ahead of BUND or NABU, even though WWF had significantly fewer supporters. The organization's receipts experienced several sharp swings during the first years of the new century, but its 2004 income was 5 percent above that of 2000.

As of 2004, WWF-Germany had an endowment of about €9.1 million, the interest from which is used to pay its administrative costs, as well as other reserves of almost €9 million. Income from the endowment provided 2 percent of its income in 2004. The largest share of the organization's revenues (60 percent) came from donations from individuals, and an additional 11 percent was derived from inheritances. Thirteen percent of the budget consisted of government grants for its projects. Despite the organization's reputation for being close to business, only 7 percent of its 2004 receipts came from corporate contributions, a decline from 12 percent in 2000. Included were donations from many of Germany's largest corporations. Another 2 percent came from a subsidiary that licenses businesses to use its name or panda logo (WWF-Deutschland, 2005). For example, the WWF received a percentage of revenue from sales of wildlife stamps developed in cooperation with various national governments and a Swiss stamp manufacturer. It licensed the right to use its panda logo to a toothpaste manufacturer that made a major donation, to a collection of stuffed animals, and to a nature-themed umbrella. It also received license fees for use of its name on a book of photos of Russian wildlife for which its president wrote the introduction (WWF-Journal, 1998c, 1998e; WWF-Deutschland, 2004). The WWF also received €1.8 million as a percentage of the sales of a well-known beer brand for protection of a rain forest in the Central African Republic (WWF-Journal, 2002d; WWF-Magazin 2003c). Other revenues came from its affiliated mail order business, whose Panda-Catalogue offered over 3,000 items and served over 500,000 customers yearly in 1999 (WWF-Journal, 1999e). The organization even offers a Panda Investment Fund in cooperation with the Deutsche Bank (WWF-Journal, 1998a; WWF-Magazin, 2004).

WWF-Deutschland aims to allocate about 60 percent of its project expenditures to international projects and 40 percent to domestic projects (Wunschmann et al., 2003). As of 2004, it reported spending only 6 percent of its budget on administration and 13 percent on services to its supporters and recruitment of new supporters (WWF-Deutschland, 2005).

The German League for Environment and Nature Protection (BUND)

The German League for Environment and Nature Protection (BUND) was an outgrowth of the Bavarian League for Nature Protection (BN). The BN was founded in 1913 to support the Bavarian government's nature protection efforts and remained closely linked to the government for many years through leadership overlaps and coordination of activities. The BN was placed under government supervision during the Nazi era, but after the war it resumed its traditional roles: advising the government about nature protection issues, public education, and behind-the-scenes lobbying. During the late 1960s, discontent within the BN led to the election of new leaders, who strengthened it, broadened its mission to include environmental issues, and pushed for a more activist approach.

In 1975 leaders of the BN and similar groups elsewhere, joined by prominent environmentalists, formed BUND. They wanted a national voice for environmental concerns that would be more effective than the traditional nature protection organizations, the DNR, or the BBU. BUND's early years were marked by problems in finding a political identity and establishing itself at the national level; however, by the end of the 1970s it had established itself as a key social movement organization for the more moderate wing of the burgeoning environmental movement, and it entered a period of rapid growth that continued throughout the 1980s.

Goals, Programs, and Emphases

BUND's areas of interest cover virtually the entire spectrum of nature and environmental protection (BUND, 1988, 2007b; Wolf, 1996; BUND/Misereor, 1996; Fritzler, 1997). Its current statement of guiding principles emphasizes working for a sustainable economic and social system via a volunteer-based, democratic organization that works at all levels, from the local to the international (BUND 2007a). Areas of emphasis listed on its Internet site include agriculture, chemicals, energy, EU environmental politics, globalization, climate, sustainability, nature protection, transportation, and water (BUND, 2007b).

BUND works vigorously to influence Germany's political agenda through lobbying and public education at the local, national, and regional levels (Rogall, 2003; Felbinger, 2005). Its commitment to effective lobbying was exemplified by its being the only large environmental organization to immediately move its headquarters office to the new capital in Berlin after reunification (BUND, 2000, 2001). BUND's approach to government is more confrontational than that of NABU or the WWF, and in my interviews, politicians from across the political spectrum frequently complained that it was too rigid and uncompromising—especially with regard to the nuclear power issue. Nevertheless, BUND's agenda is reformist, not revolutionary (Kazcor, 1989; Blühdorn, 1995; BUND/Misereor, 1996; Bergstedt, 2002).

BUND stresses the solid factual grounding of its lobbying efforts, and several of my interviewees noted that it is respected for the high quality of its research reports and position statements. Many of these are produced with the assistance of its network of working groups, which cover areas such as solid waste, agriculture, transportation, genetic technology, and sustainable development (Bergstedt, 2002; Rogall, 2003; Maxim and Degenhardt, 2003; Degenhardt, 2003). BUND has also contracted with environmental research institutes to prepare research reports on topics such as increased fuel taxes (Oswald von Nell Breuning Institut, 1996), the effects of increased use of nonrecyclable beverage containers (Die Zeit, 1995h), and plans to deepen the Elbe to promote shipping (DNR Deutschland Rundbrief, 2001i). Its book-length report on how Germany might develop an ecologically sound and successful economy (BUND/Misereor, 1996; see also Roth, 1992)—cosponsored with a Catholic social welfare organization and prepared by the Wuppertal Institute—was widely circulated and received much positive comment (e.g., Der Spiegel, 1995b; Brand, 1999a; BUNDMagazin, 2000b). BUND has continued to emphasize the perspective of this report in its publications (e.g., BUNDMagazin, 2000a, 2000b; BUND, 2001) and in a follow-up volume (BUND/Misereor, 2002). In line with this approach, the organization was an early and strong supporter of ecological tax reform, which raises taxes on fuel consumption and uses the resulting revenues to reduce employment taxes (e.g., BUND, 2001).

BUND also emphasizes environmental education and alerting the public to environmental problems. In the late 1990s, for example, it conducted a comprehensive evaluation of the waste avoidance and disposal practices of German cities, gave each city a grade, and publicized the results widely. It also evaluated computers according to how environmentally friendly they were (BUND, 1998). The national organization, *Länder* branches, and local groups answer inquiries from citizens about environmental issues, sponsor nature study trips, stage lectures, symposia, and exhibitions, sponsor environmental festivals, and publish brochures and fact sheets (Cornelsen, 1991; Hey and Brendle, 1994; BUNDMagazin 2000b; Bergstedt, 2002). For example, it recently published guides to ecological home construction (BUNDschau, 2004a) and local nature protection measures (BUNDschau, 2004b). It also occasionally undertakes demonstrations and other actions to attract public attention. Recent examples include having an activist clad as Santa Claus thank train riders for protecting the environment (BUND, 2001), dumping thousands of empty beer cans in front of a brewery (BUNDMagazin, 2000f), setting up a fake beach scene in Düsseldorf to dramatize the threat of rising sea levels due to global warming (BUNDschau, 2003c), and setting up a giant dinosaur at a Bonn climate conference to symbolize the world's current fossil fuel dependence (BUNDschau, 2004c).

From the beginning, BUND emphasized not only environmental issues, such as traffic and nuclear power, but also traditional nature protection (Rauprich, 1985), and a recent statement by its president makes clear that this dual emphasis will continue (BUNDMagazin, 2004c). BUND owns and maintains nature reserves throughout Germany and operates five environmental education centers. It has also been heavily involved in efforts to acquire open land along the former boundary with East Germany. Local chapters engage in practical nature protection projects, such as maintenance of ponds and traditional organic orchards, tree plantings, stream renaturalization, reintroduction of formerly represented species into the wild, protection of nesting areas, and inventory and protection of bat nesting areas. The national organization sometimes undertakes national nature protection campaigns, such as a 1997 tree planting effort (Cornelsen, 1991; Blühdorn, 1995; BUND, 1998; BUNDMagazin, 1998b, 2000b, 2004b).

BUND's quarterly magazine, today called *BUNDMagazin,* has been published under various titles since 1977 (BUNDMagazin, 2000b). It contains articles about environmental problems in Germany and abroad, environmental policy, and BUND projects and activities at the national, regional, and local levels. Also published are numerous consumer-oriented articles on topics such as energy-saving appliances, balcony gardening, ecologically sound home repair, environmentally oriented kindergartens, and natural foods, as well as listings of BUND-sponsored excursions and hikes, museum exhibits, and book reviews. Advertisements for merchandise available through BUND's mail order service, as well as ads for insurance, pension funds, a rail discount card, green electric power, and other services with the BUND seal of approval also appear. There are also occasional poems and stories, classified ads for vacation apartments, and even a small section of singles ads. In 2003, the magazine underwent a minor redesign in response to a reader survey (BUND, 2004). The new format places more emphasis on BUND's affiliation with Friends of the Earth, employs a less academic writing style, and incorporates more pictures and more modern layouts. BUND also publishes a quarterly newsletter for leaders and activists. In 2004 it was changed from print to on-line format; at that time it had a circulation of 3,400 (BUNDschau, 2004e).

Organizational Structure

BUND is formally organized as a *Verein,* with an organizational structure similar to that of NABU. It is a federation of chapters in each of the sixteen German *Länder,* which enjoy considerable autonomy. Below the *Länder* chapters are approximately 2,200 local groups (BUND, 2007c). The *Länder* chapters send representatives to an annual delegate assembly, which in turn elects a board of directors and a scientific advisory board. The latter is comprised primarily of the chairs of working groups focused on specific environmental issues. The sixteen *Länder* chapters also appoint representatives to an advisory committee that represents their interests at the national level (Felbinger, 2005; BUND, 2007e). In recent years, BUND's board of directors has been comprised primarily of BUND leaders from the national or *Land* level. Those with paid employment have typically held professional jobs. There are no representatives of business or government. In 2007, seven of nine members were women (BUND, 2001, 2002, 2007d). Like NABU, BUND has an affiliated youth organization.

BUND's long-term president, Hubert Weinzierl, who presided over the transformation of BUND's predecessor organization in Bavaria into a more activist environmental organization, stepped down in 1998. He was succeeded by longtime vice president Angelika Zahrnt. BUND's president, unlike the presidents of NABU and the WWF, is unpaid and not on-site full-time. Day-to-day administration is in the hands of the executive secretary. As of the late 1990s, there were about seventy permanent employees (Bergstedt, 1998).

BUND has twenty working groups at the national level and over a hundred at the regional level (BUND, 2007f). The working groups are full of experts in fields like forestry, city planning, law, and natural science who produce reports and position statements in their areas of interest (see also Cornelsen, 1991; Hey and Brendle, 1994; Wolf, 1996; Bergstedt, 2002; Rogall, 2003). BUND is recognized under the provisions of the Nature Protection Law as an environmental organization entitled to submit position statements about local projects that might damage nature.

Since 1989, BUND has been the German affiliate and largest member of Friends of the Earth (BUNDMagazin, 2000b, 2000c), an international federation of over seventy national environmental organizations, which claims about 1.5 million members worldwide. Friends of the Earth originated in the mid 1970s as a more radical offshoot of the US Sierra Club and sister organizations in Britain, Sweden, and France. It has emerged in recent years as one of the strongest opponents of globalization and industrialized agriculture and a strong advocate of grassroots democracy and locally based economies, biodiversity, and environmental justice between the global North and global South (Jordan and Maloney, 1997; Friends of the Earth International 2001, 2005). BUND's affiliation with Friends of the Earth and its work on the EU level have received increasing emphasis in recent publications, and it has established a working group devoted to international environmental issues (BUNDMagazin, 2000d, 2003b; BUND, 2000, 2002). BUND is also a member of the DNR and the EEB.

Membership and Finances

The BUND experienced steady but unspectacular growth during the late 1990s. It entered the twenty-first century with 260,000 members, 19 percent more than in 1995. Nevertheless, its relatively slow growth during this period cost it its membership advantage over NABU, even before the latter organization began counting members from its Bavarian affiliate, and BUND's lead over the WWF diminished from 59,000 to 24,000.

Since the beginning of the century, BUND's membership has stagnated. Perhaps partly in response, it has begun to report not only the number of dues-paying members, but also the number of persons who authorize a regular donation to it from their bank account even though they choose not to become members (BUND, 2000). The latter number continued to increase impressively until 2001; however, since 2001, the combined total of members and donors has hardly changed.

BUND's members and donors are even more geographically concentrated than NABU's. Although there are active chapters in every *Land*, as of 2004 almost two-thirds of the organization's members and donors lived in Germany's two southern *Länder*, Bavaria and Baden-Württemberg. Supporters in the territory of the former GDR, including Berlin, comprised only 7 percent of BUND's members and donors, although this region holds about 18 percent of the German population. Well over half of these supporters are in Berlin, and the organization has not been able to sustain strong growth in the East (BUND, 2005).

Both my interviewees and other informed observers report that the profile of BUND members is similar to that of supporters of other environmental organizations, but with an especially strong representation of well-educated government employees, scientists, educators, and university students (Cornelsen, 1991; Bodenstein et al., 1998; Bergstedt, 2002; BUNDMagazin, 2002a). Data from BUND's Bavarian chapter in the mid 1980s (Weiger, 1987b) showed that only 4 percent of the members were farmers and 7 percent "workers," while 15 percent were teachers or professors and 30 percent were "officials." A 1997 membership survey showed that men made up 55 percent of the membership and that the membership was heavily concentrated in the age range 30–59, but the response rate was poor (BUND, 1998; see also BUND, 1995) and the percentage of women members has probably increased since the survey was completed. As in the case of NABU, my interviewees estimate that no more than 10 percent of BUND members are active (see also Wolf, 1996). A recent survey of activists found that the majority were men, married, and had been members for over fifteen years; however, the response rate for the mail survey was only 31 percent and Bavaria was not included (BUNDschau, 2003a; see also Bammerlin, 1998).

Representatives and organizations of the German right were prominent among BUND's founders. Most withdrew after it became clear that they would be unable to push through their agenda; however, a few remained (Biehl, 1995; Geden, 1999; Olsen, 1999; Bergstedt, 2002). Adherents of right-wing parties or groups continue to surface, albeit infrequently, in leadership positions or in policy debates, sometimes causing the organization embarrassment (Geden, 1999; Bergstedt, 2002).

BUND's revenues reached their high point of €15.5 million in 1999 and declined 11 percent the next year. Nevertheless, 2000 receipts remained 32 percent above those of 1996. Since 2000, revenues have varied from year to year, but the 2004 total was almost identical to 2000's. Eighty-nine percent of the organization's income in 2000 and 78 percent in 2004 came from member dues and gifts from individuals. In the same years, 2 and 5 percent, respectively, were derived from government grants, usually for specific projects. Less than 1 percent came from business in each year (BUND, 2001, 2005), and BUND's 1997 annual report listed only eight businesses that donated more than €5,000 (BUND, 1998). In 2004, the organization reported spending about 5 percent of its budget on solicitation of new members and donations and 6 percent on administration. Like NABU, it has minimal reserves (BUND, 2005).

Greenpeace Germany

Greenpeace Germany is the German branch of Greenpeace International. Greenpeace International originated in Canadian-based protests against US nuclear testing, but quickly broadened its agenda to environmental issues and spread to the US and Europe. Greenpeace Germany was established in 1980, and its spectacular protests quickly earned it a great deal of media attention. Riding on the coattails of favorable media attention, rapid growth of public concern about the environment, and the success of Greenpeace International, the German branch achieved almost exponential growth in its number of supporters, budget, and paid staff during the 1980s and emerged as the largest and best-financed German environmental organization.

Goals, Programs, and Emphases

Greenpeace Germany is so tightly intertwined with its international partner that the emphases and activities of the two are not easy to distinguish. Thematically, Greenpeace has focused on a limited number of areas, including nuclear power and energy, forests, oceans, the chemical and oil industries, climate change, and genetically altered food. It engages in a wide variety of activities including demonstrations and protests, lobbying, research about environmental problems, and public education (Rucht, 1995a; Greenpeace Deutschland, 1996, 2007; Wapner, 1996; Flechner, 1999).

Greenpeace's trademark is the staging of attention-grabbing actions to point up environmental abuses.[1] These have included sailing ships into nuclear testing zones, driving rubber rafts between whalers and whales, and climbing smokestacks of polluting industries to hang banners. The actions are professionally planned, rehearsed, and executed media events, with the place and time calculated to gain media attention, raise public consciousness about specific environmental issues, and put pressure on businesses or government to change their practices. In general, Greenpeace actions focus on environmental problems that are easy to understand and report in the press, and its protests usually provide striking, action-filled visual images, such as the 10-foot-high alarm clock and ticking Geiger counter that activists placed in front of the federal office that approves the transport of nuclear wastes (Berliner Morgenpost, 1998), the wall of building insulation it set up in front of the construction ministry to highlight the need for greater efforts to fight global warming (DNR Deutschland Rundbrief, 2000b), or the freight container of genetically altered corn it dumped in front of the German health ministry (DNR Deutschland Rundbrief, 2001k). Such actions present Greenpeace as a trustworthy, independent, and morally indignant David risking life and limb to challenge a selfish and callous Goliath. At times Greenpeace also conducts attention-catching research designed to highlight environmental problems, such as taking samples of polluted water from rivers (Eitner, 1996) or using radar to count the number of ships risking shipwreck by violating navigational rules in narrow ocean channels (DNR Deutschland Rundbrief, 2001j).

Successful implementation of Greenpeace's strategy requires that it capture media attention, and Greenpeace has worked hard and successfully at this.[2] Its large, highly professionalized press office makes use of professional media agencies and photographers and is well informed about the needs and culture of the

---

[1] A great deal has been written about Greenpeace's use of spectacular actions. Sources used in this section include the following: Rauprich (1985); Reiss (1988); Kunz (1989); Eyerman and Jamison (1989); Der Spiegel (1991b); Reichert and Schmied (1995); Rucht (1995b); C. Krüger (1996, 2000a); Böttger (1996, 2000); Eitner (1996); Wapner (1996); Brand, Eder, and Poferl (1997); Biermann and Böttger (2000); and Vandamme (2000).

[2] Greenpeace's media operation has also attracted considerable comment. In addition to sources cited in note 1, see Rossman (1993), Hansen (1993), and Radow and Krüger (1996).

media. It works hard to cultivate good relationships with the media by holding regular press conferences, promptly answering media inquiries, providing contact persons who are easy to reach, and issuing regular news releases. Its spectacular actions provide good copy and striking visual images for both print media and television, and it has the technology and organizational skills to ensure that the press can be present or receive live feeds from the scene. It can also deliver striking images of environmental abuses, such as slaughtered seals and forest clearcutting.

Even Greenpeace campaigns that do not involve spectacular actions—such as its campaigns against accidents at Hoechst Chemical Plants (Brand, Eder, and Poferl, 1997) and the sinking of Shell Oil's Brent Spar oil storage platform (Vandamme, 2000), its crusade against potentially hazardous softening agents in children's mouth toys (Die Zeit, 1999b), its campaigns against genetically altered foods (Die Zeit, 1997c), and its use of the BSE-infected cows as a springboard for a campaign to eliminate industrialized agriculture (Berliner Zeitung, 2000d)—often seem to have been chosen with media impact in mind. Complex problems that require continuing attention over an extended period, such as the environmental effects of Shell's operations in Nigeria (Vandamme, 2000), are targeted less often.

This does not mean that every Greenpeace campaign is a success or garners media and public attention. An effort to block construction of an oil pipeline in the North Sea in the early 1990s ended in failure due to the complexity of the issue and lack of media attention (Ramthun, 2000), a multi-year campaign for fuel-efficient cars has barely dented the consciousness of the German public, and its efforts to promote Germany's ecology tax were little more successful (Biermann and Böttger, 2000; bfp Analyse, 2000a, 2000b). Greenpeace pursued these campaigns doggedly for years, but others, such as a campaign to reduce car use (Rucht, 1995b), were abandoned when they stirred resistance and failed to take off.

Overall, however, Greenpeace has enjoyed very good press. During the 1980s, many major media outlets covered it positively and ran its advertisements free of charge (Reiss, 1988). Both Greenpeace's own research (Radow and Krüger, 1996) and two studies of media coverage in the 1990s (Rossman, 1993; Müller-Henning, 2000; bfp Analyse, 2000b) showed that the organization was widely and positively covered in the press—although its viewpoints were often presented in conjunction with those of its opponents or other groups (Müller-Henning, 2000). Between 1992 and 1998, it averaged over 13,000 annual mentions in the print media and over 3,000 on TV in the sample of media monitored by bfp Analyse (2000b).

Public opinion polls have long indicated that the German public generally reacts positively to Greenpeace's approach (Der Spiegel, 1991a; Rucht, 1994; Blühdorn, 1995). About 90 percent of Germans recognize its name (bfp Analyse, 2000a), and in a 1989 survey, it was identified by 72 percent of respondents as the environmental organization they most trusted, ranking it far ahead of all others (bfp Analyse, 2000a). Greenpeace is also viewed with more trust than almost any other institution or person by adolescents (Krüger and Müller-Henning, 2000).

Nevertheless, during the 1990s Greenpeace ran into media criticism that it chooses only easily winnable battles that it can use to burnish its image (e.g., Die Zeit, 1991; Der Spiegel, 1991a, 1991b, 1998) and that its research is sometimes shoddy. It apparently overestimated the amount of radiation emitted by a shipment of nuclear waste (Frankfurter Allgemeine Zeitung, 1997b), and it was forced to apologize after mismeasuring, perhaps intentionally, the amount of oil remaining in the Brent Spar oil platform (Die Zeit, 1996e; Greenpeace Magazin, 1996d; bfp Analyse, 2000d).

Whether in response to these criticisms or the declining environmental movement mobilization described in Chapter 7, Greenpeace has increased its emphasis on activities other than spectacular actions (Bode, 1996a, 1996b; Der Spiegel, 1997b; Die Zeit, 1997a; Rogall, 2003). Yet the remarkable success of its Brent Spar campaign, in which it stirred sufficient public outrage to head off the sinking of an oil storage platform in the North Sea and attracted massive press attention, suggests that spectacular actions remain a viable approach (Bennie, 1998; Vandamme, 2000; Ramthun, 2000), and they remain a key part of the organization's strategy and culture (Krüger, 1996). Indeed, their impact has been so great that many of my interviewees described Greenpeace almost exclusively in these terms, omitting its other activities.

Greenpeace's lobbying activities (Steenbock, 1996a, 1996b) address all levels of government. It attempts to persuade local governments in Germany to adopt ordinances requiring environmentally friendly building materials (Greenpeace Deutschland, 1995), employs small staffs of lobbyists in Berlin (Greenpeace Nachrichten, 2005b) and Brussels (Webster, 1998; Roose, 2002), and actively lobbies the UN and International Whaling Commission to end all whaling (Wapner, 1996). It sometimes supports its lobbying with petition campaigns, such as one that collected a million signatures urging the German government to pressure Canada to stop the slaughter of baby seals for their pelts (Eitner, 1996), and it has organized effective consumer boycotts to apply pressure to alleged environmental offenders. Examples include boycotting Shell for planning to sink the Brent Spar oil platform, a campaign in which consumers mailed spray cans powered by ozone layer-destroying propellants back to the manufacturers, and a boycott of Norwegian products to protest whaling (Reiss, 1988; Die Zeit, 1995a).

Greenpeace proudly proclaims that it does more than point out problems; it also suggests workable solutions. It has contracted with ecological research institutes to produce studies of policy issues, including fuel efficiency and transportation policy (Die Zeit, 1996c), energy taxes (Greenpeace Deutschland, 1995), and insulation standards (Frankfurter Allgemeine Zeitung, 1999j; Greenpeace Nachrichten, 2000a). It has worked with manufacturers to develop demonstration models of a fuel-efficient car (Die Zeit, 1995g; Frankfurter Allgemeine Zeitung, 1997c), a refrigerator without coolants that damage the ozone layer (Die Zeit, 1996b), an ecologically optimal bicycle (Zillessen and Rahmel, 1991), and a bicycle recycling program (Greenpeace Nachrichten, 2000b). It even published a replica of Germany's best-known news magazine on chlorine-free paper to demonstrate the viability of the technology (Greenpeace Magazin, 1996c). Greenpeace has also developed suggested standards for clean production in the chemical industry and for sustainable forestry (Bode, 1996a), and it set up a supplier of green power (Frankfurter Allgemeine Zeitung, 1999i; Greenpeace Nachrichten, 2001b).

Historically, Greenpeace did not purchase or operate nature reserves or undertake local nature protection efforts. However, since 1999, Greenpeace Germany has operated a separate foundation that undertakes some nature protection projects, such as preservation of heritage species of domestic animals (Greenpeace Nachrichten, 2004b; Greenpeace Deutschland, 2007b; Greenpeace Deutschland, no date). Greenpeace has not sought recognition as a nature protection organization that could participate in hearings or file complaints (Rogall, 2003).

Greenpeace Germany publishes a nationally circulated magazine, which is sold by subscription and at newsstands, and an eight-page tabloid-style newsletter, which is circulated free of charge to its donors (Pietschmann, 1996). The professionally produced magazine contains general interest articles about environmental scandals and environmental politics. As of the mid 1990s, over a third of its circulation of 130,000 was to persons who do not donate to Greenpeace (Pietschmann, 1996). It enjoyed a much higher circulation in the 1980s than it does today (Zillessen and Rahmel, 1991). Like its annual reports, Greenpeace's newsletter for supporters is full of striking images and articles about environmental scandals and Greenpeace actions. Neither the articles nor the numerous capsule descriptions of Greenpeace actions distinguish clearly between the activities of Greenpeace Germany and Greenpeace International. Resistance to Greenpeace actions and criticism of Greenpeace by business and government often receive prominent attention. General interest articles about nature and articles about ecological lifestyles like those found in the WWF, BUND, and NABU magazines are largely absent, but there is an insert describing clothing and other items available for purchase from Greenpeace's mail order house.

Greenpeace was the first of the organizations to develop an elaborate Internet site, which is full of dramatic action scenes and attention-catching phrases, such as "Taten statt Warten!" (Don't wait; act!) (Markham, 1999). Greenpeace also distributes pamphlets, stages exhibits, and sets up information stands in public places to distribute information. It even briefly had its own national TV show (Die Zeit, 1997b; Krüger, 2000b).

Organizational Structure

Greenpeace International claims about 2.8 million donors and an income of €162 million worldwide. It has offices in forty-one nations, although only twenty-seven are full-fledged national chapters (Greenpeace International 2006a, 2006b). It is governed by a council of representatives from nations with Greenpeace chapters (Greenpeace Deutschland, 2002). Greenpeace International grew very rapidly in the 1980s, but many national chapters experienced sharp declines in number of supporters during the 1990s (Greenpeace Deutschland, 1995). Because Greenpeace Germany experienced a smaller decline, it emerged as the largest and most influential chapter. Since the beginning of the century, it has contributed about a third of Greenpeace International's total budget (Greenpeace Deutschland, 2000, 2005a), and it has about a fifth of all Greenpeace members worldwide (Greenpeace Nachrichten, 2005d). This has given the German chapter a strong voice in Greenpeace circles, and the last two executive secretaries of Greenpeace International have been Germans.

Like the other national chapters, Greenpeace Germany is bound to Greenpeace International by a contract, which entitles it to use the Greenpeace name in return for subordinating itself to the latter's requirements. These include nonviolence, nonpartisanship, independence from business and government influence, and participation in international campaigns. Major protests must be cleared with the international headquarters, which also controls the allocation of ships and equipment and operates the international press office. Greenpeace Germany's Hamburg headquarters harbors the international organization's fleet of ships and has a warehouse full of equipment for use in protest activities (Der Spiegel, 1991a; Kaczor, 1992; Rucht, 1995b; Günther, 1996; Flechner, 1999). Its publications (e.g., Greenpeace Magazin, 1996a; Greenpeace Deutschland, 2000) emphasize the importance of an international approach, which Greenpeace leaders argue is necessary to successfully address the most important environmental problems (Bode, 1996b; Greenpeace Magazin, 1996b).

Greenpeace seeks to enroll supporters who will obligate themselves to an annual contribution, and it sometimes blurs the distinction between supporters and members; however, in reality, the supporters have no direct voice in decisions. Greenpeace Germany is organized as a *Verein,* but there are only forty voting members, including ten each from the organization's paid staff, Greenpeace International, and local activists. The voting members elect a board of directors, which in turn appoints the executive secretary and staff. Day-to-day operations are in the hands of the executive secretary, who acts as Greenpeace Germany's primary spokesperson (Greenpeace Deutschland, 2005a, 2005b). The employees include a mixture of founding activists and more recent hires recruited for their professional and management expertise. As of 2005, Greenpeace Germany had 180 employees (Greenpeace Deutschland, 2005b), up from 130 in the late 1990s (Frankfurter Allgemeine Zeitung, 1999h).

Greenpeace Germany has an unusually hardworking network of local activists—a total of approximately 2,400 organized in over eighty local groups, mainly in urban areas (Greenpeace Deutschland, 2005a, 2005b). Most local activists are young, and there is a great deal of turnover. Local groups play an important role in fundraising and public information campaigns by staffing information stands in public places, and they sometimes collect signatures for petitions or assist with Greenpeace actions. Greenpeace Germany is highly centralized, and local groups receive most of their marching orders from the Hamburg headquarters, to which they are bound by contract (Rucht, 1995b; Flechner, 1999; Greenpeace Nachrichten, 2000c; Rogall, 2003; Felbinger, 2005). Greenpeace justifies this rather corporate model of operation in terms of its need for efficiency and speedy decision-making, its large size, and its international scope (Die Zeit, 1995f; Rucht, 1995b; Flechner, 1999; Felbinger, 2003). Greenpeace also maintains separate "Greenteams" for youth and "Greenpeace 50 plus" groups for older adults (Greenpeace Deutschland, 1995; Greenpeace Nachrichten, 1998).

Membership and Finances

Greenpeace is by far the largest of the four organizations covered here, and it proudly reports the number of persons who contribute regular financial support. The number of Greenpeace supporters increased very rapidly in the 1980s, peaking at about 750,000 (Der Spiegel, 1991a; Rucht, 1993a); however, the early 1990s saw a sharp decline to 517,000 supporters in 1995. Between 1995 and 2002 the number of supporters fluctuated erratically without a clear trend, but an upward trend, reaching 548,000 supporters, emerged in 2003 and 2004.

Greenpeace's internal research indicated that, as of the mid 1990s, two-thirds of its supporters had a college-preparatory high school degree, 51 percent were male, and that only 8,500 lived in the former East Germany (Wallmeyer, 1996); unfortunately, no information about the design and quality of the underlying research is available. Several of my interviewees commented that Greenpeace supporters are more diverse in social class than those of the other organizations. One journalist added the interesting, albeit somewhat impressionistic, comment that Greenpeace's action-oriented approach is especially appealing to Germans because they live relatively regimented lives.

Like its membership, Greenpeace's revenues fluctuated during the late 1990s. Its income of just over €35 million in the first year of the new century was 6 percent below its 1995 receipts, and its expenditures consistently exceeded its revenues in the late 1990s; however, since 2000 revenues have risen to just over €41.5 million, and receipts now exceed its expenditures. The great majority of Greenpeace's support comes from donations from regular supporters, supplemented by inheritances and occasional donations from others. The organization's annual reports combine these income sources with smaller amounts of revenue from fines allocated to it by judges and in-kind donations. In 2004, these sources together provided 95 percent of Greenpeace's revenue, with donations of under €100 providing 71 percent. Only 3 percent came from donations larger than €500, and Greenpeace screens larger donations for conflicts of interest. Greenpeace also earns interest on its almost €39 million in financial reserves, by far the largest reserve held by any of the groups. In 2004, these earnings funded 2 percent of its budget (Greenpeace Deutschland, 2002, 2005a). The affiliated foundation had reached an endowment of €1.8 million by 2004 (Greenpeace Nachrichten, 2004b; Greenpeace Deutschland, no date).

Greenpeace neither accepts advertisements for its publications nor seeks financial support from business, although it has accepted occasional business donations that it believes do not threaten its independence or reputation (Wallmeyer,

1996). It does receive some income from its affiliated mail order house, which sells Greenpeace publications, clothing and merchandise with the Greenpeace logo, and various "green" products, including recycled paper (Bruns, 1996; Wallmeyer, 1996; Felbinger, 2005). Greenpeace receives no direct financial support from government.

During the 1990s, Greenpeace Germany devoted about a third of its budget to international campaigns and a third to campaigns in Germany (Heuser, 1994; Greenpeace Nachrichten, 2002b). Recent budget reports do not provide a breakdown of international versus domestic programs, but in 2004, 7 percent of the budget was devoted to administrative expenses, 6 percent to direct costs of acquiring donations, and 19 percent to "communication," which included €2.5 million for member service (Greenpeace Deutschland, 2005a).

Conclusions

The four large environmental organizations profiled in this chapter, NABU, the WWF, BUND, and Greenpeace, entered the twenty-first century as the dominant actors in German environmentalism on the national level. Their success evidences their ability to adapt successfully to the ongoing social changes described in Chapter 7. This was no small feat, for the changes markedly restructured the social context in which they operate and created the very complex operating environment described in Chapter 8.

As open systems, the organizations had to decide how to structure themselves internally, how to attract the resources they needed to pursue their work and retain their legitimacy, and how to deal with key actors in their environment. Only by finding adequate answers to these questions could they survive and move toward their goals. The fact that all four have adopted relatively similar internal structures, methods of obtaining support, and strategies for dealing with other organizations suggests that their social context made some strategic choices considerably more attractive than others. It also evidently favored convergence around a common pattern, rather than efforts to occupy highly specialized niches. It appears, for example, that the current social context favors a strategy that mixes working within the system with occasional protest, not strategies tilted strongly toward accomodation or confrontation. Greenpeace and BUND arrived at this strategy by moving away from past strategies that had been more oriented to protest. The WWF, by contrast, moved toward a more activist stance and somewhat more frequent use of protest.

The outcome has not, however, been a completely uniform set of structures and strategies. Organizations are also institutions with histories and traditions that can be hard to change. Moreover, they differ in their constituencies, and their leaders sometimes respond differently to similar situations. Consequently, the four organizations take approaches that are perhaps best described as variations on common themes. All four organizations display, for example, tendencies toward centralization of power in the hands of national office staff; however, at BUND and NABU, formally democratic structures, democratic traditions, and heavy reliance on volunteers have significantly restrained this tendency, while their absence at the WWF has allowed more centralization. Still, in comparison to the full range of organizational forms and strategies that have existed historically and at present, the four organizations are more similar than diverse. Certainly, none of them look very much like the BBU, with its emphasis on confrontation and grassroots democracy, nor the almost entirely volunteer-staffed nature protection organizations of the prewar era.

The organizations' similarities—and the differences among them—manifest themselves in seven key areas, which I examine further in subsequent chapters.

First, in comparison to their own histories and to other environmental organizations in Germany today, all four organizations are large and well-funded. All underwent rapid expansion during the 1980s, and they have continued to grow, albeit at different rates and with occasional interruptions. Their budgets are small in comparison to those of business and labor associations, but they are large in comparison to their own histories or the budgets of other national environmental organizations. Their public statements, their rush to attract support in the East after reunification, and the amount of money and energy they spend soliciting supporters and donations make clear that maintaining and expanding their number of supporters and their budgets remain key goals for all four.

Second, all of the organizations display levels of professionalization far higher than anything observed among their predecessors. Staff members organize professional membership recruitment and fundraising efforts, lobby the Bundestag and government agencies, plan and organize major campaigns, and conduct professional public relations work. With professionalization has come considerable concentration of influence in their national offices. Nevertheless, all the organizations but the WWF also rely to a significant extent on volunteers. Widespread citizen involvement remains a core tenant of NABU's and BUND's ideologies and a key to successful execution of their strategies. Greenpeace, the organization with the largest staff of professionals, also continues to rely on the young volunteers in its local groups to accomplish some important tasks and reinforce its image as a citizens' crusade. And even the WWF encourages the formation of local groups to support its work. This formula—heavy reliance on professionals combined with continuing use of volunteers—has proven much better adapted to the current milieu than the minimal professionalization and reliance on grassroots activism that characterized the confrontational environmental movement.

Third, all of the organizations obtain the bulk of their financial support from large numbers of individual donors and seek to increase their financial support mainly by attracting even more such supporters, including many who have no strong inclination toward volunteer activism. This method of financing their operations was a major innovation, almost unknown in their histories, and it helps to explain their rapid growth. All but Greenpeace also seek out and obtain some support from business and government; however, even the WWF, which pursues business and government support most avidly, obtains most of its support from small donations.

Fourth, unlike many organizations of the past—where nature protection was only one goal among many and was sometimes subordinated to objectives such as protecting historical architecture or building working-class consciousness—NABU, BUND, the WWF, and Greenpeace have converged on a common definition and framing of the problems they call "environmental." All focus on protecting ecosystems from the burdens of human activity as the key to protecting biodiversity and the bases of human life. They do sometimes address other issues, such as BUND's focus on inequities between the global North and South and Greenpeace's traditional concerns about war and weaponry, but they do so within the context of their primary emphasis on environmental concerns.

Fifth, although they differ in how broad a spectrum of issues they address, all of the organizations pursue a wide variety of environmental problems. These include not only problems derived from Germany's tradition of nature protection, but also issues, such as traffic, air pollution, and possible resource exhaustion, that grew to prominence during the environmental movement of the 1970s and 1980s, and newer themes, such as genetically modified organisms. Not every organization pursues each of these issues with equal fervor, suggesting some tendency toward specialization, but each has a broad palette of goals, and new issues have generally been added to existing goals rather than supplanting them.

Sixth, despite some significant differences in emphases, all of the organizations pursue their goals through a wide variety of strategies that combine confrontation with working within the system. All engage in lobbying and public education, all seek visibility for their issues in the media, and all occasionally gather signatures on petitions. All acquire and operate nature protection areas, although Greenpeace's involvement in this activity is recent and marginal. All four also stage protest events. The degree of emphasis they place on protest varies considerably, but none of them function as social movement organizations in the sense that the BBU or BUND did for the environmental and anti-nuclear movements a quarter-century ago. On the other hand, none can be accurately described merely as interest groups participating in a corporatist system or as civil society organizations rendering a contribution to society without challenging it.

Finally, all of the organizations emphasize maintaining their own distinct identities and programs. While they do at times cooperate, their cooperation remains episodic and limited. The propensity for cooperative work also varies among the organizations, with Greenpeace ranked well below the other organizations in this respect. BUND and NABU belong to the DNR and EEB, but they jealously guard their autonomy and become concerned when these umbrella organizations show signs of displacing them in the media or in public consciousness. In other words, the organizations appear to have converged on a compromise between competition and cooperation.

The fact that the structures and strategies of the four organizations appear as variations on common themes is not the outcome of easy or obvious choices.

Instead, many of their decisions about how to structure themselves, obtain resources, and pursue their goals are the result of difficult struggles to balance competing alternatives with a mixture of advantages and disadvantages. To take a single example, all four have moved toward professionalized fundraising from a mass public. Yet the temptation to turn instead to business or government for funds remains seductive, and old-time activists complain that this turn to "checkbook environmentalism" undermines traditions of civic involvement and volunteer activism and makes the organizations unwilling to challenge the establishment.

Choices like these are true dilemmas. There are no easy solutions to them, coping with them evokes conflict and consumes a great deal of energy, and they are often resolved with uneasy compromise rather than a clear choice among alternatives. The next three chapters explore seven such dilemmas in detail. Chapter 10 looks at dilemmas of structure, focusing especially on how much professionalization and centralization the organizations choose to employ. Chapter 11 examines dilemmas of resource acquisition, and in particular choices about how much to rely on financial support from government and business vs. from a mass of individual donors. Finally, Chapter 12 looks at three dilemmas in relations with the outside world: cooperation versus confrontation with government and business, cooperation versus competition with other environmental organizations, and how many different objectives to pursue.

Chapter 10

DILEMMAS OF INTERNAL STRUCTURE
Professionalization and Centralization

Numerous dimensions of organizational structure have been examined in the literature about organizations (Price and Mueller, 1986; Hall, 2002); however, not all of these are equally salient for understanding how environmental organizations cope with pressures from their external environments, how they function internally, and how effective they are in trying to advance their objectives. The literature reviewed in Chapter 2 identifies two dimensions of internal structure, professionalization and centralization, as especially problematic, and my research supports this hypothesis. As the organizations entered the twenty-first century, questions about how much to rely on paid professionals versus volunteers and whether to opt for centralized or decentralized organizational structures were dilemmas very much alive at all four organizations, for these key structural decisions have major effects on their effectiveness. Ongoing changes in their contexts have led to increasing pressures for professionalization and centralization, but counter pressures exist as well, and the choices involved are by no means easy (Felbinger, 2003).

Professionals vs. Volunteers

The history of German environmental organizations contains many examples of national environmental groups that operated with minimal levels of professionalization, and present-day environmental groups vary considerably in this respect. Loose networks of local environmental groups, such as Robin Wood and the

Green League, resemble historical organizations, such as the BfV, in that they have few paid employees and rely very heavily on volunteers. At the other end of the continuum, WWF-Germany relies almost exclusively on paid staff. Other organizations, including NABU, BUND, and Greenpeace, lie between these extremes.

Advantages of Professionalization

Small committed groups of activists with strong interpersonal ties and clear, shared objectives have sometimes been able to mount sustained and relatively effective environmental protection efforts, especially at the local level;[1] however, this approach has many potential difficulties. The commitment of volunteer activists often rises and falls in response to competing role obligations and their satisfaction with the organization's goals and their volunteer experiences. Consequently, it is frequently easier for organizations to mobilize the sustained effort needed to effectively complete complex, long-term projects by relying on professional staffs.

Reliance on amateurs is especially problematic when an organization's core tasks require high levels of expertise. The greater expertise and more sustained effort that professionals bring to their work is especially beneficial for tasks like effectively lobbying legislators, preparing reports and position statements on technical matters, negotiating with experts from government or business, participating effectively in hearings and legislative deliberations, and monitoring the progress of legislation through the parliamentary process and later implementation. Reliance on professionals is also advantageous for many other technical tasks, including preparation of attractive, interesting member publications, cultivation of good relations with the media, providing the media with timely and reliable information that attracts public attention, fundraising from the general public, and preparation of grant proposals.

Environmental organizations whose approach to these tasks is amateurish not only find it more difficult to perform them well; they may also lose the confidence of the public and the cooperation and respect of key organizations on which they depend for assistance, financial support, and legitimacy. Other interest groups in Germany, including business, unions, and agriculture, have been steadily professionalizing in recent years, and German environmental law and regulatory procedures set a very high premium on scientific and professional expertise and tend to denigrate input from interest groups that do not display technical expertise about the matter at hand (see Chapter 8). As a result, when environmental groups display lack of professionalization, their negotiating part-

[1] In addition to general discussions of professionalization of interest groups and social movements cited in Chapter 2, the following sources include useful discussions of the advantages and disadvantages of professionalization of environmental organizations in particular: Mitchell (1989); Mitchell, Mertig, and Dunlap (1992); Hey and Brendle (1994); Ehlers (1995); Rat von Sachverständigen für Umweltfragen (1996); Oswald von Nell Breuning Institut (1996); Brand, Eder, and Poferl (1997); Jordan and Maloney (1997); Rucht, Blattert, and Rink (1997); Bammerlin (1998); Lahusen (1998); Jänicke, Kunig, and Stitzel (1999); Vandamme (2000); Bergstedt (2002); Dryzek et al. (2003), and Bosso (2005).

ners in business or government are apt to complain about their lapses and refuse to cooperate. Environmental organizations staffed primarily by amateurs may also find themselves overmatched in confrontations with more professionalized opponents. In the area of environmental communication, for example, environmental organizations must compete with highly professionalized and effective media campaigns from their rivals in business.

There is evidence that expenditures on professional staff do pay off in terms of increased contacts and acceptance. Dalton's (1994) study of European environmental organizations showed that organizations with larger budgets had more contacts with government ministries, greater participation in government commissions, and more media contacts. NABU's effort to forge a cooperative relationship with a major German bank, on the other hand, foundered when the bank complained that its staff was not professional enough and turned instead to the WWF (Bergstedt, 2002). And even at the environmental ministry, professional staff frequently complain that the environmental groups often lack the needed expertise to contribute effectively to policy making.

Similar dynamics apply in environmental organizations' relationships with the press and public. Several reporters and politicians I interviewed complained about press releases or position statements from environmental organizations that were amateurish or not adequately researched, and the BBU's inability to provide timely press releases is sometimes cited as one of the reasons for its decline (Ehlers, 1995). More recently, Greenpeace came under damaging press and public criticism when it was found to have mismeasured the amount of oil remaining in an oil platform that Shell planned to sink in the North Sea (see Chapter 9).

Another advantage of professionalization is that it contributes to an image of being modern and effective. Professionalization of activities such as lobbying, fundraising, environmental research, and public relations has become increasingly institutionalized as standard practice in recent years—not only in environmental organizations, but in other interest groups and nonprofit organizations as well—so organizations with few professionals are almost automatically suspect. Professionalization spreads from organization to organization by exchange of staff, professional societies, and professional publications, and its growing pervasiveness makes eschewing professionalization both unthinkable and unworkable.

Disadvantages of Professionalization

On the other hand, professionalization also has drawbacks for environmental organizations, especially against the backdrop of their efforts to present themselves as citizens' crusades. Highly professionalized environmental organizations are subject to criticism from the media, from volunteer environmental activists—including those inside the organization—and from less professionalized environmental groups. The critics accuse them of being too bureaucratic, too unresponsive to members' wishes, too willing to compromise core organizational values, and too much influenced by overpaid professional staffs filled with careerists, who allegedly prefer working within the system to challenging it.

Greenpeace, for example, has thus been plagued by criticism[2] that it operates from its 10,000-square-foot headquarters more like the corporations it criticizes than the David attacking Goliath it claims to be. Critics note that its highly professionalized media operation swallows a large chunk of its budget and that the volunteer activist leaders of its early days have been displaced by professionals. Indeed, several recent Greenpeace executive secretaries seem to have been chosen more on the basis of management skills developed in business than of their commitment to the environment (Kazcor, 1992; Frankfurter Rundschau, 1999). The critics also point out that Greenpeace's highly professionalized fundraising operation has allowed it to accumulate a huge financial reserve more appropriate for a business than a putative citizen's crusade (Die Zeit, 1995d; Der Spiegel, 1998). Even Greenpeace's trademark protest actions sometimes come under criticism for their professionalization. The critics note that they are carried out like professional military operations by small groups of highly trained activists. Spontaneous involvement of ordinary citizens is neither sought nor desired. It is clear that the professionalization issue is a sensitive one for Greenpeace leaders, who frequently feel called on to defend it (e.g., Wallmeyer, 1996; Flechner, 1999). Indeed, the materials the organization supplied in response to written questions I submitted by mail contained a defense of the salaries paid to Greenpeace employees, even though I had posed no questions about this.

Another well-known set of difficulties with professionalization involves tensions between paid professionals and volunteers. These can stem from three major sources. First, professional employees may value the work of other professionals above that of volunteers, leading volunteers to feel devalued. Second, professionals are understandably interested not only in the organization's goals, but also in their own professional development and career advancement. They thus often seek control over decisions and projects, and they may attempt to relegate volunteers to mundane tasks, generating resentment. Third, professionals often have more time to devote to the organization than volunteer leaders, and they typically have more information. If they attempt to use these resources to increase their own power, power struggles may erupt between professionals and governing boards or elected leaders.

BUND, with its traditionally heavy reliance on volunteers, has been particularly plagued by such problems. At the national level, there have been repeated power struggles and disagreements over goals between the paid professionals and BUND's scientific working groups, which are comprised of volunteer experts (Weiger, 1987b; Wolf, 1996). Activists from the left have also criticized professionals as too willing to compromise. In one particularly well-known case, a BUND staff expert in genetic technology first argued for less rigid opposition to genetic technology, and then left the organization to work for industry (Die Zeit, 1995i). Conflicts have also erupted between professional staff and leaders

[2] For reviews of these criticisms see Ehlers (1995), Klein (1996), bfp Analyse (2000e), and Vandamme (2000).

of local groups, who complained that professional staff were usurping their functions (Weiger, 1987b; Wolf, 1996). Wolf chronicles a series of such conflicts in Lower Saxony that resulted in the resignations of some volunteer activists. And in North Rhine-Westphalia, harsh criticism of professionalization from activists escalated into heated conflicts at annual delegate's meetings in the early 1990s and forced a rollback of the size of the paid staff (Maxim and Degenhardt, 2003; Degenhardt, 2003).

Conflicts between professionals and volunteers have also broken out at Greenpeace, arguably the most professionalized of the organizations (Kunz, 1989; Leif, 1993). In the early 1990s, for example, a group of long-time employees, whose Greenpeace involvement began before professionalization became the order of the day, hotly contested a plan to greatly increase management salaries, which the leadership claimed was necessary to attract qualified experts (Die Zeit, 1991). In a similar incident several years later, Greenpeace's executive secretary was dismissed—according to newspaper reports, at least in part because of complaints from volunteer activists that his high salary and lifestyle more closely approximated a business executive than a dedicated environmentalist (Frankfurter Rundschau, 1999; Frankfurter Allgemeine Zeitung, 1999a).

A NABU *Land* president whom I interviewed cited examples of similar conflicts, this time between professional staff in the *Länder* offices and volunteer governing boards. Indeed, even the much less professionalized BBU experienced conflicts between professional staff and volunteer leaders in the 1980s (Sternstein, 1981; Kazcor, 1986).

Trends toward Professionalization

Despite these disadvantages, all four of the organizations discussed here have moved steadily toward greater professionalization.[3] Indeed, by one estimate the number of employees of national environmental organizations increased by 70 percent between 1988 and 1997 (Rucht and Roose, 2001a). After its 1980 founding in Germany, Greenpeace quickly transformed itself from a volunteer-dominated organization into the exemplar of professionalized protest. And the WWF, once a small organization whose board of directors carried out much of its fundraising on a volunteer basis, greatly expanded the size of its professional staff during the 1980s as it broadened its mission and sought to broaden its donor support base (see Chapter 5). BUND and NABU have followed the same path, albeit with

---

[3]Virtually every description of trends in the development of the four environmental organizations notes their growing professionalization since the 1970s. The description here is based on the following sources: Leonhard (1986); Weiger (1987b); Schenkluhn (1990); Cornelsen (1991); Blühdorn (1995); Oswald von Nell Breuning Institut (1996); Rat von Sachverständigen für Umweltfragen (1996); Wolf (1996); Brand, Eder, and Poferl (1997); Bammerlin (1998); Brand (1999a, 1999b); Diani and Donati (1999); Rucht and Roose (2001c); Bergstedt (2002); and Dryzek et al. (2003). Discussions that focus only on Greenpeace include Radow and Krüger (1996), Der Spiegel (1998), bfp Analyse (2000a), and Vandamme (2000). Wollenweber (1994) focuses on its lobbying operations.

some time lag and a good bit more ambivalence—first at the national and then at the regional level. At NABU, the number of paid employees in the headquarters office increased between 1988 and 1997 from five to fifty-three (Rucht and Roose, 2001b). In NABU and BUND, especially, the result has been steadily decreasing reliance on volunteers to accomplish key tasks and decreasing influence for volunteer leaders and governing boards.

Increasing professionalization has manifested itself in almost every aspect of organizational activity, beginning with scientific experts and later expanding to include professionals in communications, fundraising, and economics. All four organizations moved to professionalize their lobbying at the national level in the 1980s. Greenpeace was first to develop a professionalized media office in 1986 (bfp Analyse, 2000a), and it substantially expanded its staffing during the 1990s. The NABU press office, which did not exist until the end of the 1980s, had reached five full-time employees by 1990 (interview data). Greenpeace and the WWF professionalized their fundraising during the 1980s, and BUND and NABU hired their first professional fundraisers at the end of that decade. By 2000, NABU and BUND had full-time professional staffs working exclusively on fundraising, and the WWF fundraising staff numbered fourteen (WWF-Deutschland, 2000).

BUND and NABU have also added professional staff in their regional offices in the individual *Länder.* Wolf (1996) documented the rapid growth of paid staff in BUND's chapter in Lower Saxony during the 1980s—from zero to thirty in less than ten years—and the growing reliance of the elected board on the staff. Similar growth occurred in the chapter in Bavaria during the 1970s and early 1980s, including the establishment of regional offices and the employment of paid staff by some of Bavaria's larger local chapters. Administrative costs, which had been negligible before the war, had already reached 44 percent of the Bavarian chapter's budget by 1979, and by 1987 there were twenty-seven permanent employees and over 100 temporary workers provided through government employment programs or the civilian alternative to the draft (Hoplitschek, 1984; Weiger, 1987b). Similar rapid professionalization occurred in BUND's chapter in North Rhine-Westphalia during the early 1980s; the paid staff grew from zero in the late 1970s to thirty-one by 1990. This trend was, however, at least temporarily reversed in the early 1990s (Maxim and Degenhardt, 2003).

At least four major factors appear to have been key in moving the environmental organizations toward professionalization. First, the institutionalization of environmental concern that began in the 1980s created enormous pressures on the organizations to professionalize. As the German parliament passed one new environmental law after the other, the complexity of German environmental law and regulations increased almost exponentially. The consultations and hearings in which the organizations were invited or entitled to participate became correspondingly more technical, and the environmental ministry and relevant sections of other ministries grew larger and hired more experts. More recently, the growing importance and complexity of EU legislation and regulations has added another level of complexity for the organizations to juggle. Business and labor responded by adding fleets of technical experts and professional lobbyists to their staffs, requiring the environmental organizations to hold their own against highly professionalized opponents. In short, the range of environmental issues under discussion, the number of opportunities for participation, the professionalization of other key actors, and the level of expertise expected of the organizations all escalated simultaneously. These changes, combined with the traditional emphasis on expertise in German administration, greatly increased the incentives for professionalization.

The institutionalization of environmentalism has also taken place in the economic sector. As a result, business and labor have become more willing to cooperate and negotiate with environmental organizations; however, like negotiations with government, discussions of ecological modernization and pollution control strategies can quickly turn technical, and business has considerable expertise at its disposal. Environmental organizations that cannot provide experts with technical data and power point presentations of their own run the risk of being perceived by their business negotiating partners as lacking expertise and thereby losing credibility.

Second, the changing nature of the organizations' relationships with the press has placed a premium on professionalization of their media and public education efforts. Several factors described in chapters 7 and 8 have contributed to this development. Environmental protest is no longer a novelty, and dramatic mass demonstrations like those of the 1970s and 1980s seldom occur. Environmental problems are no longer at the forefront of every citizen's consciousness, and they have become more technically complex, taking on the character of disputes among experts. Consequently, press attention to environmental issues has declined. Furthermore, environmental organizations must now compete with the well-staffed and highly professionalized media relations and public relations departments of major businesses. This working environment places a premium on large and professional public education departments.

Third, changes within the organizations themselves have promoted professionalization. The rapid growth of their membership rolls and budgets described in chapters 7 and 9 has made it increasingly difficult for volunteers to manage the necessary administrative work. To cope, the organizations need accountants, lawyers, and trained business managers. Increased use of direct mail and other new member recruitment and fundraising techniques, described in detail in the next chapter, have generated pressure to professionalize fundraising departments as well. The institutionalization of environmentalism has also led to creation of grant programs from government and sponsoring programs from business, but preparation of successful proposals for such funds is a time-consuming and technical activity.

Finally—with the relative priority of environmental issues in decline, social movement mobilization waning, and other social changes increasing the stress and time pressures that many potential volunteers feel—it has become ever more difficult to find volunteer activists who will make the kind of lifelong commitment to work intensively for an organization that was once much more common (see also Chapter 8). The local BUND group in Berlin became so concerned about this issue that it reemphasized volunteer training and began to recruit volunteers for short-term projects with limited time commitments (Schumacher, 2003a, 2003b), while Greenpeace held a workshop for local activists to try to develop new ways to attract volunteer activists (Greenpeace Nachrichten, 1999d).

Trends toward professionalization also tend to become self-sustaining. Newly hired professionals press for professionalization of additional tasks, and professional career ladders have developed that provide career paths within environmental organizations, between environmental organizations, and between environmental organizations and other nonprofits. Several of the fundraisers I interviewed, for example, had previously raised money for other environmental organizations or other nonprofit organizations. As this process continues, more and more activities come under the purview of professionals. Media experts from NABU's headquarters even released a handbook for local chapters, prescribing a uniform appearance for the organization's logo, stationery, and publications (Naturschutz Heute, 2001a), and WWF-International has a similar set of regulations (WWF-Deutschland, 1999). In such an environment, organizations that do not display an image of professionalism may lose credibility with negotiating partners and the public. Professionalization is also associated with increased emphasis on activities that professionals do best, such as public education and lobbying within the system, as opposed to amateur efforts like local nature protection projects or confrontational mass demonstrations. As these activities gain prominence, the need for professional staff increases further.

Centralization vs. Democracy

The second strategic dilemma facing large national environmental organizations is the tension between centralization of control in the hands of a few leaders in the central office versus decentralization of influence and grassroots democracy. Discussions of this dilemma—often framed in terms of Michels' hypothesis that a drift toward oligarchy in such organizations results from the desire of their leaders to retain the perquisites of office—have a long history in both the theoretical and research literatures on political organizations, labor unions, and voluntary associations (see Chapter 2). Discussions of centralization of power in environmental organizations, on the other hand, frequently cite additional factors, especially professionalization, as the major causes of centralization.[4]

Professionalization increases tendencies toward centralization in several ways. First, professional staff typically have more time to devote to an organization

[4]This section draws on the discussions of the causes of centralization of influence in environmental organizations in the following: Mayer-Tasch (1985); Wiesenthal (1993); Dalton (1994); Dowie (1995); Rat von Sachverständigen für Umweltfragen (1996); Wolf (1996); Jordan and Maloney (1997); Faber (1998); Bammerlin (1998); Flechner (1999); Felbinger (2003); and Röscheisen (2006).

than volunteers or unpaid governing boards, and they are less apt to be distracted by competing obligations. Consequently, influence tends to accumulate in their hands. Moreover, professionals, who view themselves as qualified experts, often find it inconvenient and frustrating to work in highly democratic structures, where things happen slowly and their decisions are subject to review from all quarters. By increasing their influence, they can avoid these difficulties and enhance their job security, influence, and prestige.

Centralization of influence has benefits, not just for professionals, but also for other stakeholders. In contrast to the characteristically slow and disorderly decision-making of decentralized, grassroots organizations, centralized structures promise quick decision making and better coordination. Clearly identified leaders and lines of authority can help to stabilize an organization, and they provide a way to cope with internal divisions that might otherwise lead to endless bickering or paralysis. The media often prefer to work with organizations with a clearly defined leader or a leadership group that speaks with a single voice, and negotiating partners in government and business prefer to work with organizations that have clear lines of authority and can make decisions without lengthy discussion or internal strife. For all these reasons, supporters, volunteers, and staff may all prefer some centralization of influence. To the extent that centralization results in more effective action, it might even lead to stronger member and staff identification with the organization and its goals.

Tendencies toward Centralization in German Environmental Organizations

Well into the postwar period, German nature protection organizations, with their volunteer boards, very limited budgets, and almost complete absence of professional staff, remained decentralized simply because they lacked the resources to plan and coordinate major operations from their central office or to exert tight control over their widely dispersed local chapters. Charismatic, highly respected leaders, such as the BfV's Lina Hähnle, could exert a modicum of influence, but their power was personal, not structural.

The substantial increases in organizational budgets, central office staff, and professionalization that followed the organizations' rapid membership gains during the period of confrontation in the late 1970s and 1980s, as well as their continued growth since then, have made greater centralization of power a viable and appealing option. In this new situation, BUND, the BfV, and Greenpeace all moved toward greater centralization.

BUND was founded, in part, as an alternative to the frequently chaotic BBU and was from the beginning a more centralized organization (see Chapter 5). Moreover, not many years elapsed before BUND began to display some of the symptoms of Michels' famous iron law of oligarchy. Its formal structure, described in Chapter 9, is that of a representative democracy. Yet despite the democratic forms, BUND elections are often uncontested, elected leaders at both the national and regional levels frequently remain in office for long periods, some paid staff members have held elected offices, some persons hold more than one office, and leaders who are already in office have played a major role in the selection of new leaders (Hoplitschek, 1984; Sternstein, 1981; Wolf, 1996; Bergstedt, 1998; Röscheisen, 2006).

Despite the concentration of office holding in a few hands, some aspects of BUND's structure have hindered centralization of power. Its federalistic structure provides the *Länder* branches with a great deal of autonomy, and during the 1990s three department heads in the national office reported directly to the board of directors, which attempted to coordinate operations without an executive secretary (Hey and Brendle, 1994; Wolf, 1996; Geden, 1999; Felbinger, 2003, 2005). The federal structure has the advantages of expediting work with the *Länder* governments and keeping the organization in touch with its grassroots support (Felbinger, 2003). Nevertheless, during my interviews in 2000, several outsiders familiar with BUND suggested that the organization's relatively slow growth during the 1990s was due in part to the resulting lack of clear direction, internal conflicts, and rapid staff turnover. Advocates of increased centralization within the organization, including longtime leader Gerhard Thielke (BUND, 1995), also complained about lack of coordination, too much regional autonomy, and lack of clear lines of authority in the headquarters office (Wolf, 1996; Felbinger, 2003).

These complaints did eventually lead to the appointment of a single executive Secretary in 1999 (BUND, 2000; Felbinger, 2003), but several of my interviewees argued that the reforms did not go far enough. They believe that BUND should follow NABU's lead by placing its volunteer board chair on salary so that she could act as a full-time CEO and rein in the power of its regional leaders. One interviewee who was intimately familiar with the organization commented that the power of the regional leaders still made it almost impossible for the national office to accomplish anything, leading to frustration and rapid turnover in the national staff. Several other interviewees, including one high-ranking BUND leader, reported that local groups sometimes simply decline to become involved in campaigns initiated in the national office because they prefer to focus on local nature protection work. Indeed, even the movement toward centralization of power that has occurred has not always been warmly received at the regional level. One *Land* president commented that term limits for national leaders might be a good thing, as it is undesirable to allow concentration of power in the hands of a few people.

Despite this critique, BUND's *Länder* chapters have also experienced increasing centralization of power. Hoplitschek (1984) documented a series of changes in BUND's largest state chapter in Bavaria that progressively reduced the influence of rank-and-file members and the large advisory board and concentrated power in the hands of a small board of directors and paid business managers, and Wolf (1996) described similar developments in Lower Saxony a decade later. Nevertheless, with their relatively small staffs, *Land* chapters of BUND necessarily remain more dependent than the national office on activists to carry out tasks such as lobbying and research.

NABU has also moved down the path to centralization. Between 1965 and 1974, it had a decentralized, federal structure in which members belonged directly only to a *Land* chapter, and the national organization had only a small staff to coordinate the activities of the *Land* organizations. In 1974, it adopted a structure in which members belonged simultaneously to the *Land* and national organizations. This change was followed by rapid growth of the national office staff and the number of tasks the national organization undertook (Hanemann and Simon, 1987; May, 1999). Like BUND, NABU is a representative democracy, with elections that follow democratic forms (Cornelsen, 1991). Nevertheless, also as at BUND, there is limited leadership turnover, and many elections are uncontested (Bergstedt, 1998). The current national president of NABU, for example, held the paid position of executive secretary of NABU's branch in Lower Saxony before becoming the elected chairman of the board there (DNR Deutschland Rundbrief, 2003e).

Some interviewees suggested that having its president on salary and always in house increased NABU's effectiveness, and most rated NABU's central leadership and administration as more effective than BUND's, but some informants were of the opinion that the central office nevertheless needed greater influence. Like BUND, NABU has a federal structure, and its *Land* chapters enjoy, if anything, even greater autonomy (Bergstedt, 2002). Consequently, national leaders must rely on persuasion rather than direct orders to see their campaign goals implemented at the local level. With this in mind, one national office staffer with whom I spoke referred somewhat ruefully to the power of the local "princes" and joked that NABU has to design its national campaigns to last several years because it takes that long to get the *Land* and local groups to fall in line. Indeed, the organization's new president used his column in NABU's 2004 annual report to comment on the problems of knitting an organization with hundreds of local groups into a coherent and effective whole (NABU 2004a).

Even though it was founded by activists with a background in local environmental protest, Greenpeace evolved very quickly after its founding into a much more centralized organization than either BUND or NABU. The organization has only forty voting members, including many representatives from the paid staff and Greenpeace International, its campaigns are tightly coordinated with Greenpeace International, and local groups of Greenpeace activists receive most of their instructions from the German headquarters, to which they are bound by contract. Greenpeace justifies this centralized model of governance in terms of the need for efficiency and speedy decision-making and the difficulty of coordinating a large international organization without centralized control.

As a foundation without regional or local chapters or members with voting rights, the WWF has always been a highly centralized organization. It does not have to confront the question of how to divide influence among the headquarters office, regional and local groups, and grassroots supporters, since its regional groups are intended only to give supporters the opportunity to meet together informally and support the organization. In an interview with me, the WWF's president mentioned the same advantages of centralized decision-making cited by Greenpeace and added that having regional or local chapters precipitates too much internal conflict.

Problems Resulting from Centralization

Although increasing centralization has been the order of the day among the environmental organizations, the results have been far from unequivocally positive. Indeed, the trend has generated considerable criticism and internal conflict.

In the first place, as noted in Chapter 8, Germany's federal political structure and the resultant overlapping responsibilities for dealing with environmental issues make having strong regional and local chapters advantageous. While decentralized organizations can work at multiple levels and tailor their programs to regional and local conditions, centralized organizations are apt to be less sensitive to regional issues. Critics of environmental organizations also argue that centralized governance structures, especially when accompanied by professionalization, lead to a conservative approach to environmental issues that overemphasizes working within the system (Dowie, 1995; Bergstedt, 2002). At least one piece of comparative empirical research supports this argument. Dalton's (1994) study of European environmental groups found that organizations with strong leadership control were more likely to have contacts with government ministries and participate in government commissions, but less likely to engage in protest activities. The reverse was true of organizations with high membership input.

The most important drawback of centralization, however, has been its effects on activists' motivation and its potential to stir up conflict with activists who prefer grassroots democracy. Literature on the reasons for participation in voluntary organizations, summarized in Chapter 2, as well as literature specifically about environmental organizations (e.g., Dowie, 1995; Rucht, Blattert, and Rink, 1997; Bammerlin, 1998) suggests that the desire to have a role in shaping the organization is one significant reason for activists' involvement. Commitment to grassroots democracy, opposition to large bureaucratic structures, and a desire to focus on local affairs played a very prominent role in the BIs of the 1970s and 1980s, where many of today's older activists cut their teeth, and East Germany's city ecology and church-related groups also had a strong commitment to grassroots democracy.

While the number and influence of BIs has faded, the worldview of a good many of today's potential activists was shaped by the movement, and a number of my interviewees noted that resistance to centralization among environmental activists is far from dead (see also Diani and Donati, 1999; Bammerlin, 1998; Bergstedt, 1998, 2002). For example, one local environmental activist with whom I spoke expressed great interest in my research until he learned that its focus was the large national organizations, which he then curtly dismissed, saying "Oh, that's just bureaucracy." Survey research also suggests considerable resistance to centralization. In a 2004 national survey, almost a third of volunteers in environmental and animal protection said that their opportunities to participate in decisions were either insufficient or only partly sufficient (Gensicke, Picot, and Geiss, 2005).

Greenpeace has proved especially vulnerable to internal disputes over its lack of democracy and secretive decision-making—probably because the needs of the activists it attracts via its spectacular actions are incompatible with its highly centralized structure. Within two years of its founding, activists in local Greenpeace groups were complaining about the headquarters' expectation that they function simply as street-corner fundraisers, about the centralization of power in the hands of a governing board with many members who also served on the staff, and about the organization's refusal to let them pursue their interest in the effects of acid rain on forests. Some of them soon withdrew to start their own purposefully decentralized organization, Robin Wood; however, somewhat ironically, it too later experienced complaints about excessive centralization of power at the national headquarters (Streich 1986; Reiss, 1988; Kunz, 1989; Scholz, 1989; Rucht, 1991a; Kazcor, 1992; Bergstedt, 2002).

Greenpeace responded to the complaints by permitting its local groups to elect a coordinator to represent their interests in the headquarters office (Eitner, 1996), but neither this step nor the withdrawal of the most disaffected activists resolved the problems. In the mid 1980s, a group of dissatisfied employees from the national office approached Greenpeace's leadership with requests that it share more information and allow the employees more input. Dissatisfied with the outcome of these discussions, they exercised their right under German law to form a works council, which the management is legally obligated to consult in making certain personnel and work scheduling decisions (Eitner, 1996; Löhndorf and von Lühman, 1996).

Criticisms of Greenpeace's centralized decision-making surfaced again in the early 1990s, when some headquarters staff members strongly criticized a plan to increase the number of levels of hierarchy from four to eleven (Die Zeit, 1991), and a consultant's report once again revealed concerns about lack of democracy among activists in the organization's local groups (Leif, 1993; Flechner, 1999). Some of the local groups, for example, resisted the headquarters' plan to develop a demonstration model of a fuel-efficient car because it appeared to encourage auto travel instead of mass transit. The result was a new round of reforms, which included allowing local Greenpeace groups more opportunity to choose their own projects and the formation of an advisory committee of delegates from local groups (Blühdorn, 1995; Bode, 1996b; Bergstedt, 2002; Flechner, 1999).

Greenpeace claimed that these steps reduced the tensions, but it is clear that they were not entirely eliminated. For example, a decision by the central office to cancel a returnable container campaign in which local groups had invested considerable effort led to renewed friction in the late 1990s (Flechner, 1999). Indeed, in 1999 Greenpeace appeared to move in two directions at once on the question of centralization. It replaced a three-person executive committee with a single executive secretary but also created an employee advisory assembly that could call into question—but not overturn—leadership decisions (Greenpeace Nachrichten, 1999f).

Although neither BUND nor NABU is as centralized as Greenpeace, both have also experienced internal conflicts over increasing centralization. Wolf's careful study of the national BUND organization and the chapters in Bavaria and Lower Saxony uncovered numerous tensions resulting from the national leadership's efforts to control the agenda and create more uniformity and coordination by increasing centralization. Some BUND activists, including especially those who had entered the organization during the period of confrontation, resisted these trends. They wanted more grassroots democracy, greater autonomy for local chapters, and a more confrontational agenda (Wolf, 1996; see also Bergstedt, 2002). In the late 1980s and early 1990s, complaints from disaffected members did lead to some adjustments, including establishment of an advisory council representing the interests of the *Länder* chapters to the national organization; however, these did not end the discord. Conflict surfaced again in the late 1990s, for example, over pressure on local groups by the national headquarters to hold the "Tupperware parties" called for by the organization's contract with Tupperware. The parties were one component of a campaign initiated unilaterally by the national headquarters to promote Tupperware as an environmentally responsible alternative to disposable containers (Geden, 1999).

Conflicts over increased centralization have also arisen from time to time in BUND chapters on the *Land* level. Hoplitschek (1984) described the resistance from advocates of grassroots democracy that cropped up when BUND's Bavarian chapter replaced its annual member assembly, which any member could attend, with an assembly of elected delegates from local groups, which also included heavy representation of elected officers and paid staff. Later, in Lower Saxony, a minor revolt against undue staff influence on the board of directors led to the removal of the paid executive secretary from the board (Wolf, 1996).

Resistance to national initiatives has surfaced in NABU as well. Some local groups continue to pursue primarily apolitical nature protection projects and avoid political activity, ignoring the national headquarters' strong push for more political involvement (Bergstedt, 2002; see also Chapter 9). As several of my interviewees explained, the national leadership and local groups ultimately arrived at a de facto compromise in which the national organization can pursue a more activist agenda so long as it allows local groups to pursue their own interests and is not too heavy-handed in pushing its own agenda on them; still, the conflict has never been completely resolved.

A recent example of these problems occurred when the executive secretary of one of NABU's *Land* chapters issued a press release criticizing NABU's support for wind power and citing evidence that windmills were a threat to passing birds (Hart, 2001). Resistance like this evidently had its effect. By the end of 2001, NABU was calling for a measured approach to development of wind power installations on the North and Baltic Seas using pilot projects and careful evaluation of the effects on sea birds (Naturschutz Heute, 2001b).

Lack of democracy can also open an organization to criticism from other environmentalists and the media. Indeed, public expectations that organizations presenting themselves as citizens' crusades should be democratic give special weight to such criticisms. WWF-Germany has largely escaped criticism of its lack of democracy from within—probably because individuals who want to participate in an organization where local groups and members exert influence are not attracted to it. However, a number of my interviewees from BUND, NABU, and the Greens made a point of noting the WWF's lack of democracy as one of its weaknesses, and criticism of its corporate model of decision making can also be found in the press (e.g., Die Zeit, 1991, 1995d; Der Spiegel, 1995a; Frankfurter Rundschau, 1999). Lack of democracy has also opened Greenpeace to criticism from political opponents. For example, during a dispute over North Sea oil drilling, British Petroleum criticized it as undemocratic and out of control (Bennie, 1998).

Summary and Conclusions

For most of their history, environmental organizations in Germany operated with a minimum of professionalization and centralization. Central office staffs were small, and paid employees were limited to clerical employees or a paid executive secretary. Most of the organizations' work was carried out by volunteers, and local chapters were subject to minimal control from the headquarters. An occasional leader might achieve more than average power on the basis of charisma, and a long-serving, hardworking leader or executive secretary might gain influence by virtue of extensive knowledge and contracts. Nevertheless, meager budgets, the absence of paid professional staff, the federal organization structure of most of the organizations, and difficulties in communicating with local chapters all mitigated against professionalization and the development of the oligarchy predicted by Michels' iron law. Even the Nazis and the East German socialists were unable to fully realize their ambitions to turn the organizations into effective centrally controlled tools of the state, for local groups often stubbornly pursued their historical interests in nature protection rather than state-mandated objectives.

Lack of professionalization and centralization had predictable effects on the efficiency of the organizations. During periods when they had effective leadership and a core of committed volunteers who were willing to acquire expertise and stick with a task, they could be fairly effective—especially in the period before 1970, when understaffed government agencies relied heavily on the advice of amateur experts. Yet even then, lack of paid staff and continuity often impeded their effectiveness. This pattern continued when the West German organizations reestablished themselves after the war, for they reverted to historical patterns of minimal professionalization, heavy reliance on volunteers, and decentralization of power. However, ongoing social changes eventually made it all but impossible for them to avoid professionalizing their operations if they wished to demonstrate the successes that they needed to maintain their legitimacy and attract support.

As the cycle of social movement mobilization that had dominated the 1970s and 1980s wound down, typical interest group and civil society strategies, such as public education, lobbying, land acquisition, and protection of threatened ecosystems and species, became more appealing to the public than the more confrontational strategies of the preceding decades. Moreover, the ongoing institutionalization of environmentalism was opening some of the doors to participation in Germany's semicorporatist decision-making structure to environmental organizations that were willing to work cooperatively within the system. The organizations thus increasingly found themselves invited—or at least entitled—to participate in legislative and policy implementation processes, albeit as junior partners. And in the changed social context, doing so seemed more likely to show results than continued efforts to mount protest.

Once at the table, however, the environmental organizations encountered well-funded, professional representatives from business and unions, as well as representatives of government agencies that were relying increasingly on professional experts of their own for expertise. Forced to deal with increasingly professionalized counterparts and operating in a culture that placed great emphasis on knowledge and expertise, environmental organizations found that their historic modes of operation no longer sufficed to provide the continuity, expert knowledge, and respectability they needed. Professionals were clearly needed.

Rapid membership growth beginning in the 1970s created additional pressures for professionalization and provided the financial resources to hire additional staff. Managing exploding membership rolls, effectively soliciting new members and donations, and administering larger organizations with more complicated programs all called for additional professional staff. So, too, did media relations, since attention-grabbing protest, public interest in environmental issues, and media attention were all fading. At the same time, the decline in movement mobilization and other ongoing social changes reduced the availability of volunteers willing to commit large blocks of time to volunteer work over an extended period. All of these factors combined to make it increasingly unappealing to rely on amateurs to take care of key tasks.

Two factors, however, operated to retard professionalization. First, the organizations' rapid growth during the period of confrontation had brought them many new members who had been strongly influenced by the BIs and by the counterculture, with its skepticism of professionals and bureaucratic structures and its emphasis on volunteerism and grassroots democracy. In all of the organizations except the WWF, the preferences of these members have acted as a brake on professionalization. Second, especially at BUND and NABU, opponents of professionalization could evoke the organizations' well-institutionalized traditions of volunteer effort.

Nevertheless, the pressures for professionalization ultimately proved too strong to resist, and all four of the large national environmental organizations have moved toward increasing professionalization, albeit on somewhat different trajectories. The WWF, which had relied significantly on paid staff from the beginning, enlarged its professional staff in response to its rapid growth and the expanding scope of its operations. Although founded by volunteer activists, Greenpeace Germany followed the lead of its parent international organization by moving very quickly to a highly professionalized model of operation. The professionalization of BUND and NABU began later and has not advanced as far; however, both organizations increasingly approximate the Greenpeace model, in which volunteer activists operate as adjuncts to a large professional staff of scientists, public relations specialists, fundraisers, and managers.

This pronounced trend toward professionalization brought the organizations not only real benefits, but also significant problems and resistance, and it is for this reason that I classify professionalization as the first of the strategy dilemmas facing the organizations. However committed they may be to environmental principles, professionals are prone to prefer working with other professionals over volunteers, since the latter may be only episodically available and unable to produce professional-quality results. Expanding their own roles within the organization also contributes to professionals' prestige, influence, and career advancement. For all these reasons, they have ample grounds to wish for the transfer of important and rewarding tasks from volunteer activists to themselves or to other professionals and to keep the professional staff growing. Professionals also can be expected to focus the organizations' activities on the tasks they do best: professionalized public education, lobbying, fundraising—and in the case of Greenpeace, even protest. Finally, their desire to use their professional talents and advance their careers may lead them to prefer working within the system over challenging it. All of these tendencies create feedback loops that call for ever more professionalization.

These tendencies set the stage, however, for internal conflict and for criticism from both within the organization and without. Volunteer activists, especially those who entered the fray during the environmental movement's confrontational, anti-establishment stage, often perceived professionalization as antithetical to their image of environmentalism as a citizens' movement. They complained that professionals were too career-oriented, too hungry for prestige and influence, too bureaucratic, and too willing to compromise, and even volunteers with weaker commitment to grassroots democracy sometimes resented seeing more and more of the organization's most interesting and important work being taken over by professionals. Volunteer boards also sometimes reacted negatively to staff employees who used their expertise and their continuous presence in the headquarters office to exert ever more influence over decisions and programs. The consequence has been regular outbreaks of internal conflict between professionals and volunteers and the withdrawal of involvement by some volunteers that have plagued all of the organizations except the WWF. Finally, the efforts of highly professionalized organizations such as Greenpeace to present themselves as citizens' crusades invite ridicule from their opponents and the media. Both the conflicts and these critiques threaten to undermine legitimacy and cut off needed supplies of volunteers and donations.

A closely associated dilemma involves how much centralization of influence environmental organizations should build into their structures and procedures. Historically, their lack of funds, minuscule central office staffs, and heavy reliance on volunteers in geographically dispersed local chapters meant that Michels' iron law could hardly operate to produce oligarchical structures. However, increases in the size and professionalization of central office staffs and pressures from the organizations' external environments created pressures toward centralization of influence in headquarters offices that went far beyond the dynamics described by Michels.

In this new social context, professional staff, as well as those national leaders who shared their orientation, found working within decentralized structures that emphasized grassroots influence and local autonomy to be frustrating indeed. They complained that too much time was required to reach decisions, that the organizations did not operate efficiently, and that the national headquarters could not implement campaigns uniformly throughout the organization. They also argued that lack of speed and efficiency limited the organization's ability to respond quickly and deal with the expectations of negotiating partners and the media. They suggested that concentrating power in the national office in the hands of the professional staff and elected leaders with full-time, paid positions would help to overcome these difficulties.

The WWF was already a quite centralized organization, but in response to these internal and external pressures, the other three organizations all moved toward increased centralization. Greenpeace made the transition very quickly in the early 1990s, while NABU and BUND, with their formally democratic, federal structures, lagged behind. Still, by the beginning of the new century, they too were much more centralized organizations than the BBU, the Green League, or Robin Wood.

The march toward centralization of power in staff-dominated central offices has produced some of the hoped-for advantages, but centralization has also produced disadvantages, and it remains controversial, at least within Greenpeace, BUND, and NABU. Germany's political structures are quite decentralized, so "one size fits all" campaigns planned and controlled from the central office may be ill-suited to conditions in the field. More importantly, centralization of decision-making clearly tends to undermine the involvement and commitment of activist members, who typically desire a hand in shaping organizational policy and the freedom to plan and execute their own campaigns at the regional and local levels. BUND and NABU have long histories as volunteer-dominated organizations, and all of the organizations are fond of portraying themselves as citizens' crusades, so arguments opposing centralized control enjoy considerable legitimacy there. Moreover, in the 1970s and 1980s all of the organizations were influenced by infusions of new members, many of whose worldviews had been shaped by countercultural perspectives and emphasis on grassroots democracy.

As a result, centralization of power proved to be a "hard sell," and resistance surfaced repeatedly in Greenpeace, BUND, and NABU. The critics of centralization within the organizations complained that it had made the environmental organizations disturbingly similar to their opponents in industry and government, and that centralization deprived the organizations of flexibility and the ability to innovate. These criticisms were echoed in skeptical articles in the press, damaging the organizations' credibility, legitimacy, and support.

Excluded from opportunities to develop new ideas and plan organizational programs, some potential activist members lapsed into alienated passivity. Others exercised the option to voice their discontent in bitter and disruptive internal disputes. Others chose to exit altogether by abandoning their environmental work, moving to less centralized environmental groups like BIs, or founding new organizations such as Robin Wood. Lack of democracy has not kept the environmental organizations from attracting new members, but it clearly has been a deterrent for some activists, possibly leaving the organizations even more dependent on professional staff.

Disputes over professionalization and centralization continue to surface as the organizations enter the new century, and there is every reason to expect them to continue. All four organizations have embraced levels of professionalization and centralization unprecedented by historical standards; however, these changes have brought not only advantages, but disadvantages as well, and critics continue to fight lively rear-guard actions against them. Among their key critiques has been the argument that the highly professionalized and centralized environmental organizations of the early twenty-first century seemed better suited to operate as interest groups supported by a mass of relatively passive members than as social movement organizations, a theme I explore in the next chapter.

Chapter 11

DILEMMAS OF RESOURCE ACQUISITION

The Perils of Fundraising

Present-day efforts by NABU, the WWF, BUND, and Greenpeace to influence public policy and public opinion through ambitious, wide-ranging campaigns mounted by large professional staffs require budgets of a different order of magnitude than those of the volunteer-staffed nature protection organizations of a half-century ago. Acquiring large amounts of cash has thus become an unavoidable necessity for both the national organizations (Oswald von Nell Breuning Institut, 1996; Rat von Sachverständigen für Umweltfrage, 1996) and their regional branches (Maxim and Degenhardt, 2003).

While the organizations can raise small amounts of money from sales of merchandise, there are only four possible sources of funds on the scale they require: business, government, donations from individuals, and foundations. But despite noteworthy growth since the 1990s, the foundation sector in Germany remains underdeveloped, and many foundations are government-funded (Strachwitz, 1998; Haibach, 2006). Thus, in contrast to the US, where foundation funds have been instrumental in establishing and supporting environmental organizations (Dowie, 1995; Brulle and Jenkins, 2005; Bosso, 2005), foundation gifts play only a very minor role in Germany.

Financial Support from Business and Government

The first potential source of funding for environmental organizations is German business. In principle, such support has many advantages.[1] Business donations

[1]The discussion here draws on the following discussions of the advantages of accepting financial support from business: Zillessen and Rahmel (1991); Grüßer (1995); Bosso (1995, 2005); Flasbarth

tend to be far larger than individual gifts, reducing the number of supporters who must be solicited in order to reach fundraising goals. Financial support from business also decreases the organizations' dependence on individual donations, and it increases their credibility in some quarters.

In practice, however, pursuit of business donations has proved to have many disadvantages. Some of these result from business policies and practices (see Chapter 8). The amount of support available from business generally varies with the state of the economy, producing undesirable swings in donations, and German business has generally preferred supporting sports and cultural activities to environmental projects. Businesses that do donate to environmental groups often do so more out of a desire to polish their images than to help the environment, and they increasingly expect recipient organizations to help them polish their images through sponsoring arrangements. These call for prominent display of the business' name on publications of the organization, and allowing the business to publicize its contributions. Environmental organizations that accept financial support from business may also come under direct or indirect pressure to adopt goals and modes of operation that conform to business donors' wishes. Predictably, business usually prefers to support environmental organizations that are less confrontational and seen as responsible by clients, customers, and stockholders.

Accepting donations from business can also subject environmental organizations to criticism. German environmental organizations have a long history of conflict with business, and these conflicts continue today, albeit at a somewhat reduced level. Therefore, many of the organizations' supporters are skeptical of business' goals and motivations. Moreover, some environmental activists remain sympathetic to anti-materialistic and anti-business themes from the period of confrontation. Some supporters also worry that their organizations' reputations will be tarnished if their business partners behave in environmentally irresponsible ways. One survey of BUND members, for example, showed that 20 percent preferred absolutely no cooperation with business, while others wanted it only under limited conditions (Bodenstein et al., 1998; see also Hoplitschek, 1984; Schumacher, 2003a).

Criticism from supporters, other environmentalists, and the press is most likely to arise when environmental organizations accept money from businesses whose activities are viewed as especially damaging to the environment. The critics charge that accepting financial support from such businesses legitimates their bad behavior, undermines the environmental organizations' credibility, and makes them financially dependent on business. For example, many of my interviewees cited the high representation of business leaders on the WWF's governing board (see also Bergstedt, 2002) as evidence of business influence. They contrasted this with the absence of business leaders on the boards of Greenpeace and BUND

(1995); Niermann (1996); Rat von Sachverständigen für Umweltfragen (1996); Jordan and Maloney (1997); Brand, Eder, and Poferl (1997); Bammerlin (1998); Brulle (2000); Bergstedt (2002); and Haibach (2006).

(see Chapter 9). Earlier, harsh criticism was directed at a brewery executive who served as the WWF's president even though he had opposed Germany' returnable bottles system, and at an executive secretary who had once worked in the nuclear power industry (Cornelsen, 1991; Kazcor, 1992). The WWF has also come under criticism for licensing its well-known panda logo for use in advertising ecologically questionable products, including air travel (Lufthansa), disposable diapers (Pampers), automobiles (Opel), and nonrefillable bottles (Zillessen and Rahmel, 1991; Cornelsen, 1991; Haßler, 1993; Bergstedt, 2002; Bauske, 2003).

How well-founded this criticism is remains open to debate. The WWF's list of major donors does read like a corporate Who's Who of German business, and it includes some ecologically problematic firms, including Lufthansa, BMW, and Bayer Chemicals (WWF-Deutschland, 2001b). The WWF has also licensed its logo to some questionable products; however, it did turn down the chance to license its logo to McDonald's (Heuser, 1994) and avoids cooperating with oil or chemical companies (Bauske, 2003). It also has guidelines that rule out financial support from some companies (see below). Its president also pointed out correctly to me that a relatively low percentage of its revenue comes from business donations and that the WWF does at times level sharp criticism at government or business. Evidence presented in Chapter 12 shows that there is considerable support for the latter claim; however, it also shows that the WWF is indeed more willing to compromise with business than the other three large organizations. It is hard to assess the extent to which the WWF's goals and strategies are influenced by the preferences of the businesses that support it financially. A recent article by a WWF fundraiser does acknowledge that it is harder to find corporate sponsors for technical lobbying work than for anti-poaching campaigns or campaigns to save Siberian tigers (Bauske, 2003), but it is also plausible that the WWF's goals and strategies are chosen for other reasons and that this simply makes it easier to attract more business support.

Criticism for accepting donations from business has not been limited to the WWF. NABU was criticized for accepting contributions from Lufthansa, Volkswagen, BP, and Ford and for conferring its Ecological Manager of the Year award on an individual who subsequently, as president of the Federal Association of German Industry, opposed the "ecology tax" on fuel. BUND came in for criticism for taking contributions from Ford and from a power company with nuclear power plants. And both NABU and BUND have been called to task for setting up two environmental foundations, "German Environmental Assistance" and "Euronatur," that accept funds from corporations like Daimler-Chrysler and then channel some of the money back to them in the form of grants (Bergstedt, 1998, 2002; Deutsche Umwelthilfe, 2005; Stiftung Europäisches Naturerbe [Euronatur], 2006).

Charges like these are threatening to environmental organizations because they have the potential to transfer the negative images of business sponsors to the organizations and thereby reduce support from individual volunteers or donors. The organizations' sensitivity to such criticism is demonstrated by their vigorous efforts to deflect it. In the early 1990s, a number of environmental groups formulated a set of joint standards for potential sponsors, which included commitment of the donor firm to the environment and incorporation of environmental concerns into its goals and operating procedures (Luthe, 1997). Each of the organizations also has its own guidelines for accepting donations from business, sponsorship arrangements with business, and business advertising in its publications.[2] The strongest stand has been adopted by Greenpeace Germany, which proudly proclaims that it solicits no corporate donations, refuses all advertising in its magazine, and carefully screens the corporate donations it does receive for conflicts of interest (Reiss, 1988; Wallmeyer, 1996; Greenpeace Deutschland, 2000). BUND has historically been very cautious about accepting business support, limiting the number of its sponsoring relationships, setting strict criteria, and insisting that its business partners take steps to operate in environmentally responsible ways (Zillessen and Rahmel, 1991). After a long and sometimes heated discussion, it adopted new guidelines in 1992 that ruled out contributions from the automobile, chemical, armaments, and genetic technology industries and set up a committee to monitor corporate donations (Brand, Eder, and Poferl, 1997; Felbinger, 2003). NABU's policy (Flasbarth, 1995), on the other hand, has not deterred it from accepting advertisements from such mainstays of the German consumer economy as Lufthansa (e.g., Naturschutz Heute, 2000h) and Media Markt (e.g., Naturschutz Heute, 2000i), at least when the ads contain ecological themes. When entering sponsoring arrangements or accepting donations, the WWF looks at factors such as the products a firm manufactures, whether it has an environmental management system, whether it has been the subject of persuasive critical press or Internet reports, and how its environmental performance is rated by external rating agencies (Frankfurter Allgemeine Zeitung, 1998e; WWF-Deutschland, no date; Bauske, 2003). It also recently tightened its standards (Bausle, 2003).

The WWF, NABU, and BUND offer a variety of arguments in defense of accepting donations from business and entering sponsoring arrangements. They emphasize that their business partners are environmentally sensitive (e.g., NABU, 1999, 2000b; WWF-Journal, 1999a) and insist that the businesses that provide them with funds do not attempt to influence their policies (e.g., Flasbarth, 1995). Their annual reports (e.g., WWF-Deutschland, 2001b; Bauske, 2003) make clear that business donations are a relatively small portion of their budgets, and a number of my interviewees from BUND, NABU, and the WWF were at pains to point out that they do not rely on money from business or government to support their basic operations. Finally, they argue that they establish cooperative relationships with businesses that support them, in which they help the

[2] The following sources contain general discussions of environmental organizations' policies on fundraising from business, often including examples from the four organizations emphasized here: Zillessen and Rahmel (1991); Haßler (1993); Halcour (1995); Brand, Eder, and Poferl (1997); and Felbinger (2003).

businesses to adopt environmentally sound policies (e.g., WWF-Journal, 1999a; Bauske, 2003). BUND, for example, had a long-term and widely publicized cooperation with a department store chain, which it advised about reducing packaging and removing environmentally damaging products from its inventory. The two organizations also promoted use of cloth bags as an alternative to plastic, with BUND getting a small percentage of the revenues of each sold (Zillessen and Rahmel, 1991; BUNDMagazin, 2000b). It also had a recent arrangement with the German postal service, which supported local nature preservation projects financially and received advice about using recycled paper in its envelopes (BUNDMagazin, 2004a). NABU has had a similar arrangement with a mail order house (Flasbarth, 1995; NABU, 2000b).

Nevertheless, some of the organizations' statements about donations or sponsoring arrangements can sound defensive. NABU offered an elaborate justification of its collaboration with Tupperware (Naturschutz Heute, 1998f), and the WWF is at pains to explain that potential corporate donations do not determine the organization's agenda and that licensing its logo to a corporate donor is not tantamount to endorsing all of the firm's products or activities (e.g., Zillessen and Rahmel, 1991). It made a point of explaining that a mail order house with which it had a sponsoring arrangement had removed many ecologically questionable items from its catalog, that Opel was a leader in introducing catalytic converters voluntarily (Zillessen and Rahmel, 1991), and that the stuffed animals that are marketed using its logo are made in an environmentally friendly way without child labor (WWF-Journal, 1998e, 1999a). When it instituted a cooperation with Unilever, in which the corporation was permitted to place the WWF's panda logo on frozen fish caught through sustainable fishing, it emphasized that it did not endorse all of Unilever's products (Stuttgarter Nachrichten, 1998). Even Greenpeace felt it necessary to provide an elaborate explanation of how it selects the "green" products it sells through its mail order operation (Bruns, 1996). The organizations' concern with public perception of their relationships with business was also illustrated in my interview with a fundraiser for one of them. He told me that although Shell Oil had been quite environmentally responsible in recent years, the unfavorable reputation it acquired during the controversy over its plan to sink the Brent Spar oil platform in the North Sea ruled out accepting financial support from Shell.

The organizations have also benefited considerably from government grants; however, the amount of funds available is relatively small, the competition is great (see Chapter 8), and such funds account for only small portions of the budgets of the WWF, BUND, and NABU (see Chapter 9). They have also made considerable use of government-paid employees from government job creation programs and the mandatory civilian service alternative to the draft. Reliance on such employees in the early 1990s was especially high in the East in the years immediately following reunification, as most job creation funds were redirected there (Weinbach, 1998; Bammerlin, 1998; Naturschutz Heute, 2000c; Rink, 2001). The organizations' dependence on these employees is evidenced by their dismay over plans that might have reduced the civilian alternative to the draft (DNR Deutschland Rundbrief, 2000c, 2000j).

Financial support from government has many of the same advantages and disadvantages as support from business. In general, it has provoked less criticism than support from business, but charges that accepting government funding compromises the organizations' independence or channels their efforts into activities the government is willing to fund are sometimes voiced (for overviews see Behrens, 1993; Hey and Brendle, 1994; Bergstedt, 2002; Bammerlin, 1998), as they were, for example, by critics at BUND's 1996 annual meeting (Bergstedt, 2002) or by some members of the Green League in the early 1990s (Behrens et al., 1993).

There are some grounds for this critique. Support from government for environmental projects often emphasizes specific topics as particularly grant worthy (e.g., DNR Deutschland Rundbrief, 2001c), and Dalton's (1994) study of European environmental organizations showed that organizations that received a higher percentage of their revenue from government or foundation grants were more likely to have contacts with government ministries and participate in government commissions and less likely to engage in protests. This pattern was reversed for organizations that received support mainly from membership dues or individual donations. The frequently noted timidity of the DNR is sometimes ascribed to the fact that part of its operating budget comes from government subsidies (Kazcor, 1992; Lugger, 1994; Bergstedt, 2002), and a DNR leader I interviewed conceded that this is, in fact, at least an occasional problem. Government pressure can also be more direct. BUND was threatened with loss of revenue from fines because of its opposition to a Frankfurt airport expansion, and the Green League was ordered to rewrite or withdraw brochures for a government-funded youth congress (see Chapter 8).

Greenpeace is so keen to avoid such pressures that it accepts no government funds at all (Wallmeyer, 1996), while the WWF, NABU, and BUND try to reduce the potential for undue influence and public criticism by soliciting government assistance only to support specific projects, such as land purchases or public education, not for basic operating expenses (e.g., NABU, 1999; WWF-Deutschland, 1999).

In summary, none of the organizations are dependent on business and government to fund more than a fraction of their expenses or to support their basic operations, but they vary considerably in how much income they receive from these sources. Greenpeace receives almost none, while in recent years the WWF has received as much as a third of its revenue from business and government, though it was down to a fifth in 2004 (Frankfurter Allgemeine Zeitung, 1998e; WWF-Deutschland, 1999, 2001b, 2003, 2005). In any event, limitations on the amount of money available and the drawbacks of accepting funds from business and government make it unlikely that any of the organizations will look to these sources for the majority of their support in the foreseeable future. Instead, all have turned to donations from private citizens to acquire the financial resources they require.

Mass Membership vs. Committed Activists

Today BUND, the WWF, NABU and Greenpeace all take it for granted that they will fund the bulk of their budgets though donations from a mass of individual supporters and that professional staffs paid from these funds will have major responsibility for conducting their core activities. Their growing reliance on this approach, arguably the most consequential strategy choice they have made over the last quarter-century, emerged slowly and has become one of their taken-for-granted assumptions.

Yet not all originally pursued this strategy, and it is not the only possible one. Greenpeace Germany began as a small band of committed activists without today's large professional staff. Like BIs, they raised the relatively small amounts of funds needed for their work primarily through their own small-scale, labor-intensive efforts, such as fundraising stands in public places or door-to-door solicitation. Indeed, even after two years, Greenpeace Germany had only 500 regular supporters (Meyer, 1991; Flechner, 1999). Only after Greenpeace retained a professional fundraising organization to conduct a mailout campaign did it discover that asking the general public for support bought in mailbags full of small donations and move to professionalize its fundraising operation (Reiss, 1988; Meyer, 1991; Wallmeyer, 1996). For almost half its history, WWF-Germany was an elite organization supported by a relatively small number of donors, who numbered only 5,000 in 1980, seventeen years after its founding (Haag, 1986; Dalton, 1994). The BfV and BN operated for most of their histories with very small budgets. Volunteer leaders undertook most of the organizations' work, dues were low, and new member recruitment occurred primarily through informal contacts, not professionalized fundraising. And even today, some national and regional environmental groups, including the Green League, consciously eschew mass support, believing it would undermine their commitment to grassroots democracy.

Nowadays, however, all four organizations examined here rely on mass fundraising campaigns conducted by professionals. In addition to making possible the trend toward professionalization described in Chapter 10, reliance on mass fundraising has greatly increased the organizations' revenues and number of supporters. It is, however, a strategy with both advantages and disadvantages.[3]

Attracting donors or members *en masse* offers advantages that go beyond simply generating needed cash. As pointed out by the president of NABU in an editorial titled "Success Requires Size" (Naturschutz Heute, 2000f), as well as in an article by a top WWF fundraiser (Bouman, 2003), having hundreds of thousands of supporters confers legitimacy, carries weight in negotiations with politicians

[3]The discussion of the advantages and disadvantages of mass membership draws on the following: McCarthy and Zald (1977); McAdam, McCarthy, and Zald (1988); Hey and Brendle (1994); Bosso (1994, 1995, 2005); Dalton (1994); Lugger (1994); Wallmeyer (1996); Brulle (1996, 2000); Lofland (1996); Luthe (1997); Jordan and Maloney (1997); Lake (1998); Haibach (1998b); Bammerlin (1998); Diani and Donati (1999); Shaiko (1999); Vandamme (2000); Felbinger (2003, 2005), and Fisher (2006).

and business, frees the organization from problematic dependence on funds from business and government, and enhances their credibility with the media.

There are, however, disadvantages to funding their operations by attracting as many members and donors as possible. Individual supporters are notoriously fickle in their giving habits, and there is strong competition for donations from other environmental groups and other fundraising groups of all kinds. Economic downturns or other disturbances can thus lead to unanticipated year-to-year fluctuations in donations, since when individual donations decline unexpectedly, disruptive program cuts and staff layoffs are usually not far behind. A 1993 downturn in Greenpeace's contributions, for example, led to the elimination of twenty-two jobs—a sizeable portion of the organization's staff (Meyer, 1996). The other three organizations operate with far smaller financial reserves than Greenpeace, so major downturns in donations could impact them much more quickly.

Maintaining an adequate and steady flow of donations in the social context at the beginning of the new century is an especially difficult assignment. The environmental organizations' core supporters represent a small percentage of the population, and the number of potential supporters is stable at best and may be in decline (See Chapters 7 and 8). Consequently, the organizations have turned to modern marketing techniques to recruit supporters from among the less committed. This strategy often "works," in the sense that it lengthens membership rolls and raises additional funds. However, writing a small check is a "low cost activity" (Diekmann and Preisendörfer, 1998), which requires little commitment to the cause, and the supporters attracted via mass marketing techniques are apt to be less committed than those attracted via personal networks of committed supporters. As a result, all four organizations now find themselves with a significant proportion of members who are passive and relatively uninvolved. The DNR recently estimated that overall, only about 4 percent of the members of its member groups were active (Rucht and Roose, 1999). Less than 1 percent of Greenpeace's supporters are activists who participate in demonstrations, local public education, or fundraising, and only about 10 percent of BUND and NABU supporters are active (see Chapter 9). One survey of BUND members found that about a fifth of its supporters could be classified as *purely* passive members with little environmental consciousness, while committed activists represented a small percentage (Bodenstein et al., 1998).

Because passive supporters are not usually bound to the organization by social ties or participation in decision making, it is hard to increase their commitment. Indeed, a number of the BUND and NABU leaders and staff members I interviewed lamented the decline of the old-fashioned activist member, whose work for an organization was founded on lifelong commitment to the organization and involvement in a social network with other members. One staffer noted with dismay the problems his organization had encountered in communicating with and motivating members who do not attend meetings and are not networked with other members, while the president of the same organization speculated that new models that asked people for a short-term commitment to a specific task were needed. The most important mechanism for communicating with passive members is the member magazine, but cost limitations mean that it can only be mailed a few times a year, and there is no guarantee that uninvolved members read it.

Retention of passive members is also a chronic problem. Studies of environmental organizations in the US and Britain (Bosso, 1995; Jordan and Maloney, 1997; Shaiko, 1999) show that passive members are particularly likely to drop their membership, especially if they perceive the organization as unsuccessful or as pursuing the wrong objectives, and a number of my interviewees noted that this is also a problem in Germany. By ignoring membership renewal notices, passive members force the organization to maintain a high level of effort and expenditures just to hold on to their memberships.

Recruiting masses of passive supporters can also reinforce trends toward centralization and professionalization and discourage activists. Activists generally prefer to work in organizations where members are committed to the organization's goals and willing to undertake many tasks themselves (see Chapters 2 and 10), and they often react negatively when they perceive that all the organization wants from its members is a periodic financial contribution to support the professionals. Such sentiments are especially problematic for BUND and NABU, which have long traditions of active member participation and involvement. Here, member activism is not just a way to accomplish some tasks; it is a positively valued component of the organization's identity. However, activists' resistance to mass fundraising techniques was also reported to be among the reasons for the withdrawal of some Greenpeace activists during its early years to form Robin Wood (Der Spiegel, 1998; Meyer, 1991).

Recruiting a mass of passive members may also send the message that protecting the environment requires little personal involvement or sacrifice. Grassroots activism is then supplanted by "astroturf" (fake grass) environmentalism, in which professionally controlled organizations give merely the appearance of having a mass of supporters, many of whom actually have little commitment to the cause. There is even some evidence that being bombarded by fundraising appeals that play on the emotions contributes to supporters' cynicism about charitable and environmental organizations (Luthe, 1997). Having too many passive members can also reduce an environmental organization's legitimacy, offsetting the benefits of being able to claim mass support. If the fact that many supporters do nothing more than make an occasional contribution when the spirit moves them and thumb through a group's magazine becomes widely known, the group's claims to credibility and its political influence are diminished.

Effects of Pursuit of Supporters and Donations on Organizational Strategies

The greatest disadvantage of mass fundraising is that it can deflect the organizations' priorities away from saving the environment toward raising money. Dependent on attracting a mass of supporters—many with only a shallow commitment to the cause—to finance their operations, they find themselves under pressure to act as marketers selling a product to a mass public. Numerous problems result.

The first of these is that successful fundraising mandates maintaining high visibility and a positive image. The organizations thus come under pressure to portray themselves as marching onward from success to success. One result of this is the embarrassing spectacle of each of the organizations that were in any way involved in a campaign claiming credit for its success. Organizations may choose easily achievable goals in hopes of creating the desired string of wins or to try to distinguish themselves by marking out territories not claimed by others, regardless of whether these objectives are really the most critical for the environment.

It is clear that image is very important to the four organizations. They proudly report their number of supporters and their positive reputations (e.g., WWF-Deutschland, 1999; bfa Analyse, 2000a, 2000b), and their annual reports (see citations at bottom of Figure 9.1) proudly report growth in membership or contributions and attempt to put the best face on things when they decline. When BUND's membership began to stagnate at the end of the 1990s, it switched to reporting the combined total of members and regular donors, which had continued to grow. And when NABU concluded a loose alliance with the bird watching organization in Bavaria, it was quick to include the new affiliate's members in its membership tally.

The organizations also carefully monitor their public and press images (Wallmeyer, 1996; Brand, 1999b). BUND and the WWF, for example, have contracted for national surveys to gauge public perceptions of them and their activities (e.g., Wickert Institut, 1989; EMNID Institut, 1993; WWF-Deutschland, 2000; Groth, 2003), and the WWF uses a research bureau to carefully monitor the number and nature of its appearances in print and broadcast media (WWF-Deutschland, 2000; Groth, 2003). Greenpeace has conducted increasingly elaborate studies of its public image since 1990, and it has carefully monitored its appearances and image in the print media since the mid 1980s and on television since 1993 (bfp Analyse, 2000a, 2000b). During the late 1990s, it became so concerned about media characterizations of it as a "PR-troop" that it asked its media consultants to prepare a report on this trend and how to combat it (bfp Analyse, 2000c). It also responded quickly and in detail to a Bavarian television report that it had wasted money on an unnecessary ship (Greenpeace Nachrichten, 1999e).

Organizations cultivating donations from a base that includes many passive supporters may adopt discourses of reform (Brulle, 1996, 2000) and mute their messages because public support for strong positions is less widespread than support for moderate stands and less controversial themes. They may engage in attention-grabbing theatrics or emphasize issues that resonate with the public, such as threats to nature and beloved animal species or genetically altered foods. Indeed, one former DNR leader I interviewed argued that the reliance of BUND, NABU, the WWF, and Greenpeace on financial support from the general public is actually more constraining on their actions than the DNR's frequently criticized reliance on government funds.

Greenpeace has come in for particularly strong criticism on this count.[4] Several of my interviewees, including interviewees from the other environmental organizations, characterized its trademark spectacular actions as media stunts that cater to the media's desire for striking, oversimplified stories but ignore today's more complex environmental problems. Greenpeace is also accused of choosing easily winnable battles in order to burnish its image, engaging in shoddy research, and neglecting difficult behind-the-scenes work, which is often ignored by the media. Finally, the critics claim that Greenpeace fails to cultivate its supporters' understanding of or commitment to environmental issues or to build a broad base of involved activists, preferring instead to cultivate a broad base of passive supporters.

The WWF has been faced with parallel allegations. Critics contend that it pays more attention to saving impressive species, such as tigers, elephants, and cranes, than to combating the rampant consumerism that threatens all life on the planet because the former problems play better in fundraising appeals. BUND too is sometimes accused of abandoning the militance that was characteristic of its early years in order to attract donations from the less committed (e.g., Cornelsen, 1991; Bergstedt, 2002).

The validity of such criticisms is not easy to assess. In the first place, tailoring campaigns to attract public attention is not necessarily altogether undesirable. Greenpeace, for example, argues that its spectacular campaigns focus media attention on topics that would otherwise be ignored and increases the public's environmental consciousness (Radow and Krüger, 1996; Pietschmann, 1996; Greenpeace Magazin, 1996b), and there is some evidence that spectacular actions play a key role in drawing media attention to a topic, which later extends to covering follow-up press releases and reports (bfp Analyse, 2000b).

The organizations do not make the details of how they choose their campaign themes and other activities public; however, in my interviews and their occasional published comments on the issue (Wallmeyer, 1996), leaders and fundraisers for the four organizations say that attracting media coverage and supporters is—at most—one of many factors in choosing campaigns. At the time of my interviews, fundraisers in two of the organizations participated in choosing campaigns; two did not, but one of these told me he was trying to increase his participation. The two that did participate both pointed out that they were but one voice among many.

The organizations clearly do engage in some activities that are unlikely to attract media attention or donations from potential supporters whose commitment to the environmental cause is weak or transient. All four carefully track legislation and the actions of various government agencies and take positions on technically complex issues. These include tax policy (Frankfurter Allgemeine

---

[4] See, for example, Eyerman and Jamison (1989), Kunz (1989), Brand, Eder, and Poferl (1997), Vandamme (1996, 2000), and Bergstedt (2002). The same criticisms have been echoed in the press (Die Zeit, 1991; Der Spiegel, 1991a, 1991b, 1998).

Zeitung, 1998g), European Union agricultural policy (Frankfurter Allgemeine Zeitung, 1999c), deregulation of electric utilities (Frankfurter Allgemeine Zeitung, 1999d), nature protection legislation (NABU, 2000b), bottle return laws (Naturschutz Heute, 2000k), and waste disposal (DNR Deutschland Rundbrief, 2001f). My interviewees from the organizations (see also Wallmeyer, 1996) also frequently pointed up unpopular campaigns or public stands undertaken on principle. A particularly good example was the "ecology tax" endorsed by all four, which raised the gasoline tax in the late 1990s (Die Zeit, 1997a; Frankfurter Allgemeine Zeitung, 1998c). BUND even continued to push for further increases in this tax in the face of strong public outcry (DNR Deutschland Rundbrief, 2001h) and evidence that it is widely unpopular (Bundesministerium für Umwelt, 2004).

During the 1980s, the BUND chapter in North Rhine-Westphalia spoke out against a number of popular but ecologically questionable practices, including automobile dependence, motorcycle racing, vacations in faraway lands, and golf courses (Oberkrome, 2003a). Articles that use a deep-ecology perspective to criticize the foundations of a consumer society and articles advocating potentially unpopular steps, such as avoiding consumption of vegetables from Southern Europe and Africa and limiting highly polluting air travel to absolutely necessary trips, do appear from time to time in BUND's and NABU's magazines (e.g., Naturschutz Heute, 1998a, 1998d). (However, NABU's magazine also accepted advertisements from Lufthansa touting the environmental efficiency of its planes [Naturschutz Heute, 2000h].) The magazines also sometimes publish detailed analyses of government policy likely to interest only very knowledgeable readers (e.g., Naturschutz Heute, 1998c). Greenpeace, too, has pursued campaigns that win it little press or public attention and stir up opposition (Radow and Krüger, 1996; Pietschmann, 1996; Greenpeace Magazin, 1996b). For example, it doggedly pursued a campaign to increase fuel efficiency, even though it garnered little media attention and was hardly designed to increase its popularity (bfp Analyse, 2000b).

On the other hand, interviews conducted by Brand, Eder, and Poferl (1997) in the early 1990s and my interviews a decade later indicate that the organizations' leaders and fundraisers are very conscious of the importance of a positive public image for reaching fundraising targets. One WWF leader I interviewed told me that his organization had been fortunate not to be associated in the public mind with propositions like the Greens' proposal to sharply increase gasoline taxes, which led to plummeting public opinion polls (Rüdig, 2002). And in a press interview (Die Zeit, 1995b), the executive secretary of Greenpeace Germany acknowledged that advocating hard-line proposals that require major public sacrifice can be problematic.

Greenpeace has reportedly even used a public opinion polling firm to help it select topics for its campaigns (Der Spiegel, 1991a). Its campaign against the use of chlorine in paper production was called into question within the organization because research showed it had little resonance in the media (bfp Analyse,

2000a), and it abandoned a campaign to reduce car use when it discovered how much opposition the campaign was stirring up (Rucht, 1995b). Brand, Eder and Poferl (1997) also note that even as the other organizations were criticizing Greenpeace for adapting its campaigns to public taste they were, in fact, responding to the same pressures. A good example of this is the acquisition of nature reserves. Several BUND and NABU leaders pointed out to me that buying relatively small tracts of land was really not the best use of their limited resources, but they continue to purchase and maintain nature reserves because their supporters expect this of them.

Although they do sometimes appear, neither "hard-line" articles nor self-criticism are especially prominent in the organizations' member magazines. The president of one organization explained to me that articles calling for unpopular changes in personal behavior needed to be published in appropriate "doses," and the editor of one of the magazines told me that suggestions to limit air or car travel do draw some reader criticism. Later he said reflectively that, while it would be good journalism to report not only the organization's successes but also its failures, this was not always tactically wise and that he tried to find a balance. The editor of another group's magazine said that, while he does try to educate the public, few readers are interested in the details of the organization's lobbying activity, and that he struggled to find the right mix of articles for a readership that combines both committed activists and casual supporters. Still another interviewee expressed concerns that another staff member's propensity to "point the finger [of shame]" in magazine articles at citizens whose behavior might damage the environment could be harming fundraising efforts. Finally, one of the fundraisers I interviewed said that his organization's magazine was aimed too much at committed activists and needed to modify its content to be more appealing to the more passive members recruited through mass mailings or door-to-door solicitations. BUND deals with this issue by publishing two magazines, *BUNDMagazin* for general members and *BUNDSchau*—now published only online—for activists.

The organizations also exercise discretion in choosing activities to profile in their fundraising letters. Several of the fundraisers I interviewed noted that not all of their organization's activities are equally appealing to potential donors and that they attempt to profile the most attractive projects or components of projects in their mailings (see also Haibach, 2006). One went on to explain that the best themes are easy to understand, have emotional appeal, and make the need for money obvious. Topics like agricultural reform are important, he said, but they must be funded from general revenue, not emphasized in special fundraising appeals. The president of another organization emphasized that effective fundraising requires highlighting the organization's accomplishments, not just endless pounding on the problems. Another president, although careful to deny that his organization ever developed projects specifically to appeal to donors, readily conceded that it featured projects with donor appeal in solicitations for members and donors.

Issues of self-presentation also arise in connection with the organizations' Internet sites. In the late 1990s, non-profit organizations in Germany began developing web sites as a method to raise funds, communicate with members and the public, and market mail-order merchandise, and their use has increased rapidly (Böker, 2001; Haibach, 2006). By 1998, Greenpeace, the WWF, and NABU already had web sites, as did BUND's large regional branch in Bavaria; however, all were relatively crude, and only the Greenpeace site had extensive content. Sabrina Broselow Moser and I conducted a content analysis of these four Web sites, together with Robin Wood's in 1998 (Markham, 1999; Markham and Broselow, 2001). It suggested that the organizations were including themes similar to what they place in their member magazines. We found that the web pages described a natural world threatened by environmentally unsound trends and practices—a theme mentioned in over a third of all paragraphs; however, fewer than 2 percent of paragraphs predicted ecological catastrophe. The organizations often portrayed themselves as active and effective advocates of reforms that would meliorate environmental problems by limiting unsound practices, but they infrequently pointed the finger of blame at business or government. Only 5 percent of paragraphs cited business actions as harming the environment, and only 1 percent directly accused business of dishonorable motives. Just 2 percent of paragraphs mentioned government actions that harmed the environment. Nor did the Internet pages often moralize or prescribe extreme actions. Only 3 percent of paragraphs mentioned a moral duty to help the environment, under 1 percent suggested a radical restructuring of society, and only 1 percent mentioned replacing automobiles with less ecologically damaging transportation. Nonconfrontational solutions to environmental problems were mentioned more often than confrontational ones by all of the organizations except Greenpeace. The message from the Internet pages thus appeared well-tailored to appeal to the typical supporter: a well-educated, urban, relatively prosperous, upper middle-class citizen who is somewhat environmentally conscious (Markham, 1999; Markham and Broselow, 2001).

Growing Reliance on Mass Membership

Reliance on contributions from hundreds of thousands of donors to support large staffs of professionals and a wide palette of goals and activities clearly has significant drawbacks for the organizations. Nevertheless, all four clearly believe that the benefits outweigh the costs, for all now energetically seek to maximize the number of members or donors. Indeed, the leader of one of them told me that his goal was to reach a million members!

Several factors have contributed to convergence on this strategy. First, as described in Chapter 7, the diminished relative priority of environmental issues in the public mind and the fading of the era of countercultural protest have made it more difficult to use the most obvious alternative strategy, mass protest. Second, for reasons described in Chapter 10, the organizations have come to rely on professionals to carry out their wide-ranging activities and programs, and this approach is not cheap. Third, reliance on financial support from business and government has many disadvantages, and the funds available are quite limited. In the absence of a well-developed private foundation sector, the large, national environmental organizations have turned to the only available alternative source of funding, soliciting financial support from a mass of private citizens.

This approach, which evolved gradually over several decades, has now become thoroughly institutionalized. It is viewed as the normal way to do things. It diffuses from one organization to the other through the conferences, workshops, and publications of the German Fundraising Association (Urselmann, 1998; Haibach, 2006), through the movement of fundraising professionals from one nonprofit organization to another (several of the fundraisers I interviewed had previously raised funds for other organizations), through the efforts of consulting firms (Die Zeit, 1996a), and through books and articles that describe fundraising techniques (e.g., Urselmann, 1998; Scheuch, 1999; Rosegger, Schneider, and Hönig, 2000; Haibach, 2006). Several of my interviewees also pointed out that the very effectiveness of the professionalized fundraising and membership recruitment tactics used by other environmental groups and by other nonprofits left them little choice but to adopt similar strategies or be left behind.

Fundraising Strategies

Today all four organizations make heavy use of sophisticated fundraising techniques. Greenpeace and the WWF professionalized their fundraising departments during the 1980s, and BUND and NABU began hiring professional fundraisers shortly thereafter (Cornelsen, 1991; Der Spiegel, 1998; Bergstedt, 1998). All four now have large, sophisticated, and well-resourced fundraising operations. By the early 1990s, Greenpeace was reported to have a computerized file of 1.5 million addresses and to have captured 1.6 percent of the entire German donation market (Der Spiegel, 1991a), and in 2000, its fundraising director won the Social Marketing Association's Fundraiser of the Year Award (Greenpeace Nachrichten, 2000e). The WWF employed fourteen staff members in its fundraising department in the late 1990s (WWF-Deutschland, 2000), and a 2003 article lists six staff members in direct marketing and five in corporate fundraising (Bouman, 2003).

Taking their cue from the US, where fundraising techniques were developed and refined in the 1960s and 1970s (Mitchell, 1989; Bosso, 1995; Haibach, 2006), all four organizations make heavy use of direct mail to likely supporters, often using leased or purchased lists of persons who have given to similar organizations before or have high disposable incomes (Böker, 2001; Greenpeace Nachrichten, 2002d; Bouman, 2003; Haibach, 2006). Greenpeace in particular has refined its methods for finding new donors to add to its mailing list to a high art. It lists its phone number in every large city phone book, persuades newspaper publishers to run its small ads free of cost in the otherwise dead space at the bottom of columns, purchases magazine advertising, runs short spots before films, and places information stands in public places (Meyer, 1991; Greenpeace Nachrichten, 2002d).

The organizations' basic mail solicitation strategy would be familiar to anyone who receives mailings from an environmental organization in the US (Shaiko, 1999). Letters to donors, sent several times a year, highlight a specific environmental problem with wide appeal, emphasize the organization's crucial contribution to solving it, and end with an appeal for funds. They may also offer the recipient the opportunity to sign a petition or participate in some other small way (Meyer, 1991; Wallmeyer, 1996; Haibach, 2006). These campaigns are carefully planned. According to my interviews, all four organizations use outside agencies to prepare or edit copy for the mailings (see also Reiss, 1988), and they carefully monitor the cost of each mailing and the total number and amount of donations that it brings in (e.g., WWF-Deutschland, 2000).

A new development in fundraising is use of the Internet. The Internet sites that Sabrina Broselow Moser and I analyzed in the late 1990s were relatively crude, but all already included fundraising appeals. By 1999, a year after our research, the WWF's site was recording 1,500 hits a day and had raised €25,000, as well as successfully mobilized readers to participate in petition drives (WWF-Deutschland, 2000). Since then, all four organizations have invested in elaborate, professionally designed sites (Naturschutz Heute, 2002d; Felbinger, 2005) that solicit donations online. The fundraising potential of the Internet sites is limited by the fact that members or the general public must take the initiative to visit the site, but organizations seek to persuade them to do so by featuring their Web addresses prominently in their printed materials. As early as 2000, the WWF's site recorded a half-million visits (WWF-Deutschland, 2001b). Two national surveys, both conducted in 2002 (European Opinion Research Group, 2002; Gruneberg and Kuckartz, 2003), showed that 6 to 7 percent of the population regularly sought information about environmental issues on the Internet, and the percentage has no doubt increased since then. The potential of the technique is illustrated by the impressive total of €400,000 that WWF raised via the Internet in 2004 (WWF-Deutschland, 2005). NABU even maintains statistics on which of the pages on its Internet site receive the most hits, which allows it to track visitor interest in various themes (Naturschutz Heute, 2005e).

The organizations skillfully use differentiated appeals to various market segments, including potential new members, lapsed members, persons interested in particular environmental issues, and specific age groups (Felbinger, 2003, 2005). The latter include the WWF's "Young Pandas," a similar NABU group for children, and Greenpeace's "Greenteams" for youth. NABU and BUND offer family memberships, BUND has reduced membership fees for students and the unemployed, and Greenpeace has "Team 50 Plus" for older adults (Greenpeace Deutschland, 1995; Greenpeace Nachrichten, 1998; WWF-Deutschland, 2000; Michelsen et al., 2001; Felbinger, 2005). More recently the organizations have taken note of the growing number of bequests to charity and targeted supporters who might include them in their wills, offering them brochures or telephone consultations (WWF-Journal, 1999c; BUNDMagazin, 1999b; Naturschutz Heute, 2003e; Felbinger, 2005). BUND even designated a special fundraiser for this purpose (BUNDMagazin, 2005k). Revenues from this source are subject to sharp swings from year to year (Felbinger, 2005), but Greenpeace received almost €1.7 million in such bequests in 2003 (Greenpeace Nachrichten, 2004c), and the WWF received €2.7 million in 2004 (WWF-Deutschland, 2005). It also appears that the various organizations tailor their appeals to their constituencies. Greenpeace's Internet site stresses the organization's ability to pressure the mighty to change, while BUND stresses its expertise (Felbinger, 2005) and the WWF touts celebrity endorsements from tennis star Steffi Graf and others (Bouman, 2003).

The WWF and NABU, as well as some BUND chapters, offer supporters the chance to become patrons of particular projects in return for a larger-than-average donation (Frankfurter Allgemeine Zeitung, 1998e; WWF-Deutschland, 2002; Naturschutz Heute, 2002c; NABU, 2005a; Felbinger, 2005). The WWF, for example, raised €1.6 million in 2004 from patrons who obligated themselves to give at least €30 a month, and from 365 persons who gave at least €1,000 per year (WWF-Deutschland, 2005). The organizations also give major or longtime donors special perquisites, such as special reports on WWF projects, a Tiffany-sponsored WWF panda pin, membership in the "NABU-Club," guided tours of a Greenpeace ship or NABU nature reserve, or special group outings (BUNDMagazin, 2000g; Naturschutz Heute, 2000e; WWF-Deutschland, 2001b, 2002; Haibach, 2006). Their appeals to potential supporters encourage them to authorize annual or monthly deductions from their GIRO account—the equivalent of a US checking account (Markham, 1999)—which research shows increases retention of support over the long term (Jordan and Maloney, 1997; Urselmann, 1998; Felbinger, 2005). Although regular contributions to Greenpeace and the WWF do not confer voting rights, this distinction is often obscured in the organization's communications with supporters and potential supporters (Felbinger, 2005).

German law forbids "cold calls," so phone solicitation of donations is a less prominent feature of environmental fundraising in Germany than in the US. Nevertheless, the organizations do use phone banks to contact previous donors or those who have returned postcards indicating an interest in their work, and fundraising guidebooks include sections on how to get people to provide phone numbers in ways that allow them to be called legally (Urselmann, 1998; Böker, 2001; Haibach, 2006).

Unsolicited e-mail solicitation of strangers is illegal in Germany (Böker, 2001). The organizations could use e-mail to contact their members; however, as of 2005, it appeared that none of them were doing so. I belong to all four groups, but none have solicited my e-mail address, and I have never received an e-mail fundraising appeal. The WWF and NABU have informational listserves, but the messages do not contain appeals for donations.

Like some of their US counterparts (Fisher, 2006), BUND, NABU, and the WWF also make extensive use of outside agencies to recruit new supporters.

These agencies employ teams of young people to staff information stands in public places and canvass door-to-door. The stands were very visible in Berlin while I resided there in 1999–2000. They were uniformly staffed by attractive, lively, and articulate young people. As of 2000, all three organizations were using the Wesser Job-Haus agency in Stuttgart, whose Internet site at that time (Wesser, 2001) solicited students for this work, advertising an earning potential of about €2,000 per month. The WWF also used the Dialog Direct agency (WWF-Deutschland, 2000). Often the agencies receive most of the membership dues that the new members pay for the first two years (Haibach, 2006). Greenpeace does not solicit door to door, but it has set up its own subsidiary firm with salaried employees to distribute information and solicit donations at information stands (Greenpeace Nachrichten, 2002d, 2003b).

Environmental organizations can also attract supporters by offering them additional membership benefits that have little to do with advancing the goals of the organization (Shaiko, 1999; Felbinger, 2005). BUND, NABU, and the WWF all offer opportunities to participate in hikes or nature excursions, including trips to foreign lands (e.g., Bouman, 2003). In 1997, BUND rolled out a plan to offer members a wide variety of services arranged through BUND, including travel insurance, health insurance oriented to natural healing, and accident insurance oriented to pedestrians and cyclists (BUND, 1998). In 1999, it added a service bureau to assist members with building, financing, remodeling, and furnishing ecologically sound housing (BUND, 2000), and in 2005, it began offering a discount rail travel card in cooperation with the German railway system (BUNDMagazin, 2005a) and discounts on a cell phone service (BUNDMagazin, 2005g). It also operates a travel service (Bund Naturschutz Service, 2007). NABU provides similar services, including the discount rail card (Naturschutz Heute, 2005g), its own credit card (Naturschutz Heute, 2005c), ecology tours through an affiliate (Naturschutz Heute, 2005d), and discounts on commercial services such as mobile phones (e.g., Naturschutz Heute, 2000g). The WWF even offers a retirement annuity (WWF-Magazin, 2004). BUND, NABU, and Greenpeace offer members "green electrical power" (BUND, no date, 2001; Frankfurter Allgemeine Zeitung, 1999i; Greenpeace Nachrichten, 2001b; Naturschutz Heute, 2000j), and Greenpeace advertises the opportunity to take study trips on one of its ships (Greenpeace Nachrichten, 1999a). All four organizations also sell merchandise and books, some with the organization's logo and distribute stickers with their logo. Several offer certificates to patrons of their projects, and the WWF offers free gifts keyed to the level of the contribution (Felbinger, 2005). Finally, donations to environmental organizations are tax deductible, and the organizations often mention this in their fundraising appeals; however, there is evidence that the majority of taxpayers do not claim the deduction, suggesting that it is not a key motivator for most (Böker, 2001; Felbinger, 2005).

It is hard to imagine that these benefits are the most important factor in attracting members or donations, and survey results indicate that other motives are more important (Felbinger, 2005). Nevertheless, they may well play a role for some supporters, and the president of one of the organizations cited offering services like environmentally sound power as a significant part of his organization's membership recruitment and retention strategy. The merchandising efforts also generate some revenue, though they are not a major factor in the organizations' budgets.

These sophisticated, professionalized marketing strategies have proved highly effective in attracting members and donors, making them the mainstay of the environmental organizations' fundraising efforts. Indeed, their skillful use of these techniques can help to explain how they have managed to overcome the free rider dilemma and remain successful, even during a period when environmental issues are no longer uppermost in the consciousness of the German public (Jordan and Maloney, 1997; Haibach, 2006). Use of outside agencies to recruit new members at stands and door to door has proved particularly effective. Several interviewees from NABU, BUND, and the WWF cited this technique's effectiveness as the major reason for their organization's continued growth (see also Bergstedt, 1998; Blühdorn, 2002).

Despite their effectiveness, these techniques do have drawbacks. Both direct mail and door-to-door solicitation are expensive, and rates of return on mailings to nonmembers can be as low as 1 to 2 percent (Dalton, 1994; Bosso, 1995; Jordan and Maloney, 1997; Shaiko, 1999). Nor do the costs end when new members are recruited, as new recruits must soon be persuaded to renew their membership. All of the fundraisers I interviewed said that nonrenewal rates are particularly high for members or donors recruited through mass marketing approaches, and the president of one organization noted that the result is recurring costs to replace members or donors who fail to renew (see also Bosso, 1994; Jordan and Maloney, 1997; Shaiko, 1999).

As Greenpeace discovered when it pioneered professional fundraising, large expenditures to recruit members or solicit donations also invite criticism from the organization's own supporters and outside critics, who contend that too much of the money donated to help the environment is actually diverted to the pursuit of further donations. Similar criticism has been directed at the other organizations (Meyer, 1991; Hey and Brendle, 1994; Urselmann, 1998; Bouman, 2003), and there is evidence of concern about this issue in public opinion polls (Felbinger, 2005). The percentages of the four organizations' budgets spent to solicit new members are, in fact, not especially high. BUND, for example, reports that direct expenditures on member recruitment and fundraising comprise about 5 percent of its budget (BUND, 2005), while Greenpeace claims to spend only about 6 percent (Greenpeace Nachrichten, 2002d; Greenpeace Deutschland, 2005a).

Nevertheless, fundraising costs can be a sensitive matter (Haibach, 1998b, 2006). Some environmental organizations affiliate with organizations that certify their conformance to standards of good practice in fundraising (Luthe, 1997; Urselmann, 1998; Felbinger, 2005; Haibach, 2006), and annual reports (e.g., NABU, 2005a; WWF-Deutschland, 2005) sometimes obscure the amount spent on fundraising and member recruitment by grouping it with other expenditures under headings like "membership service." They are at pains in their annual reports and Internet sites to emphasize their efforts to minimize administrative costs and fundraising expenses, and all have their statements professionally audited. Greenpeace also offers to mail a detailed financial statement to anyone who requests it (Felbinger, 2005). Their public statements about their fundraising expenses can also sound defensive (e.g., Die Zeit, 1996a; Bouman, 2003), and my interview questions about this topic were the only ones that provoked defensive reactions. One interviewee chose this occasion to inquire further about the purpose of my study, and another, visibly discomfited, said that the amount spent could not be revealed because it might embarrass the organization (ironically, the information was available in the organization's annual report). The reasons for such concern are obvious. Spending hundreds of thousands of euros to attract donations and members does not square well with efforts to project environmentalism as a citizens' crusade.

Solicitation of members and donors by paid agencies, a technique used by all of the organizations except Greenpeace, clearly provokes mixed emotions within the organizations. Several of my interviewees worried that supporters recruited this way are not only more likely to drop out, but also apt to be passive supporters (see also Bodenstein et al., 1998), and one expressed the concern that the new techniques might soon reach their limits (see also Blühdorn, 2002). Several NABU and BUND interviewees said that they wished that they could forgo the agencies and reemphasize motivating existing members to actively recruit new ones. One characterized use of the agencies as a regrettable necessity resulting from need for funds and changing social conditions, and the president of another organization expressed similar sentiments. He had initially opposed use of the agencies and admitted that using them had provoked some criticism, but he also pointed out that the agencies had proven very effective. The president of a *Land* chapter offered similar comments. He said that he personally disliked door-to-door solicitation but noted that one of the other *Länder* was using it with great success. Other interviewees noted that use of agencies has become standard practice, and one president went on to explain that his organization had turned to using an agency only after efforts to staff stands with local members had proved unsuccessful and the organization realized that modern fundraising had to be done by professionals.

When use of the agencies was introduced, some BUND and NABU activists criticized their use as a threat to the organizations' credibility and as incongruent with the historical image of the organizations as comprised of local groups of committed activists. They also complained that the practice was likely to undermine reliance on social networking to attract and retain members. There were heated discussions at *Länder* annual meetings, and several voted not to use the technique (Frankfurter Rundschau, 1996; Haibach, 1998a; Bammerlin, 1998). Door-to-door solicitation also provoked press criticism, earning the agencies the sobriquet "*Drückerkolonne*" (literally, pressure squads), a term normally applied to high pressure door-to-door salespeople (Bergstedt, 2002).

Most interviewees from the national offices of BUND and NABU said that by 2000 resistance from within had died down somewhat as the technique proved its worth and became institutionalized. Nevertheless, the president of one organization noted that some local groups still refuse to allow it. In 2001, the leader of one *Land* chapter told me that criticism had recently caused them to change agencies and limit the aggressiveness of their efforts, and presidents of two others told me that their chapters had eliminated door-to-door appeals and limited the agencies to solicitation from stands. Perhaps in an effort to increase the number of new members attracted through social networks rather than by professional solicitors, NABU, BUND and the WWF have all initiated campaigns to encourage existing members to recruit new ones—somewhat ironically offering free gifts to members whose recommendations resulted in enlisting new members (WWF-Journal, 2002a; Naturschutz Heute, 2002b; BUNDMagazin, 2005b).

The additional benefits the organizations offer supporters also come in for criticism from some activists, who see them as diverting energy and attention away from the organization's core goals—or at times even as contradicting them. When BUND's Bavarian chapter attempted to extend its hiking program to include trips abroad in the late 1970s, it encountered so much criticism that it was forced to scale back the program. The critics argued that long-distance tourism in highly polluting jet planes was inconsistent with an ecologically responsible lifestyle (Hoplitschek, 1984). Much later NABU encountered similar criticism when it ran an advertisement in its member magazine offering member discounts with a telephone service provider. The ad coincidentally but prominently displayed a phone manufactured by a firm that is active in the nuclear power industry, resulting in enough criticism to elicit a subsequent apologetic article in the member magazine (Naturschutz Heute, 1998b).

Summary and Conclusions

Ongoing changes in the environmental organizations' environment over the past-quarter century—including their decreasing capability to mobilize their supporters for active protest and recruit committed volunteers, the increasing complexity of environmental problems and professionalization of their opponents in business and agriculture, and the partial lowering of the barriers to their participation in Germany's semicorporate decision-making circles—have combined to make reliance on large, professionalized staffs an increasingly attractive option. As described in the previous chapter, all four groups have thus turned to professionals to develop their campaigns, lobby government, work with the media, and educate the public. While this strategy allows them to continue their efforts on a scale that might not have been possible otherwise, it requires budgets that are, by historical standards, quite large. This has made the funding they need to sustain their operations, pay for their far-flung campaigns, and pay their professional staffs a crucial resource on which they are highly dependent, for failure to obtain the required cash would require major adjustments in their structures and operations and threaten their very existence. Therefore, maintaining their revenue streams is a top priority for all four and they must consider their other objectives in this light.

Sales of merchandise provide relatively little revenue, and the foundation sector is underdeveloped in Germany, so there are only three potential sources of revenues of the magnitude the organizations need: business, government, and individual donors. All four of the organizations rely, at least to some extent, on the first two, although Greenpeace purposively holds its dependence on government and business to a minimum. Greenpeace's policy is the most extreme, but the other organizations have also discovered that relying on business or government funds has many drawbacks. The amount of money available is limited and fluctuates with economic cycles, and accepting business and government support channels the organizations' efforts into whatever projects government and business are willing to fund. Relying on business or government grants also opens the organizations to criticism that they have sold out. Such criticisms are especially telling for two reasons. First, they are made in the context of a society inclined toward corporatism, where receipt of financial support from government is associated with participation in a consensus-oriented political system and pressures on the interest groups to compromise their objectives and keep their memberships under control. Second, except in the case of the WWF, receipt of business and government funding runs counter to the organizations' own traditions and their identities as independent citizens' crusades.

Criticisms of undue reliance on funding from government or business threatens not only the organizations' legitimacy, but also the continued support of their most committed supporters, many of whom became active during the period of confrontation and still maintain their penchant for social movement mobilization and their skepticism of the German establishment. These committed supporters provide all four organizations with much needed financial support, and for all but the WWF they are a major source of the volunteer labor needed to supplement the efforts of the organizations' professional staffs.

Were truly massive infusions of business or government funds available, the WWF—and perhaps NABU and BUND as well—might consider making these funds their major source of support; however, this is anything but a realistic prospect. Consequently, the organizations have to carefully assess the possible impact of accepting business and government support on their primary revenue source, donations from individuals. Greenpeace responds to this dilemma by all but refusing donations from government and business. The WWF, BUND, and NABU solicit and accept some money from these sources, but they set conditions on the amount and nature of the support they accept and take other measures to deflect criticism of the support that they do accept.

Unable to fund their rapidly professionalizing operations through donations from business, government, or foundations, the organizations have, of necessity, turned to individual supporters. The strategies the nature protection associations had traditionally used to attract supporters—low dues and recruitment through interpersonal networks—were barely adequate to support them as low-budget organizations relying on volunteer labor. Supporting a large staff of paid professionals this way proved to be simply out of the question. The problem of increasing the flow of funds from individual donors was mitigated in the 1970s and early 1980s by the waves of new supporters delivered to the organizations' doorsteps by the new social movement mobilization—which included the environmental and anti–nuclear power movements—and the flowering of countercultural perspectives and postmaterialist values. The movement, with its mass support, apocalyptic framings of environmental problems, and dramatic protests, brought the organizations not only committed activists, but also free media publicity and favorable press treatment. A positive press and favorable public image increased their visibility and legitimacy and, thereby, their membership rolls and donations.

As the 1980s shaded into the 1990s, the cycle of new social movement mobilization began to run its course. The amount of free publicity available through media diminished greatly, new donors and volunteers became steadily harder to find, and the number of donors showed signs of stagnation or even—in the case of Greenpeace—serious decline. Clearly, innovative new solutions for the free rider problem were required. To keep the membership rolls growing and the donations flowing, the organizations turned increasingly to sophisticated direct mail techniques borrowed from the US and to the use of paid agencies to recruit supporters. These techniques allowed them to continue to grow and increase their budgets, even in the relatively unfavorable climate that prevailed during the years immediately around the turn of the twenty-first century; however, these solutions brought with them problems and unanticipated consequences for the organizations' goals and operating strategies.

Acquiring the volume of donations the organizations needed to support themselves required expanding their support base beyond the limited pool of environmental activists to incorporate many passive supporters. These supporters were not part of the tight social networks that had characterized the nature protection organizations, and they had little in common with the social movement activists of the preceding decades. They were often disinclined to work actively for the cause, and the organizations hardly expected them to do so.

New methods and new problem framings have been required to recruit mass support in this changed climate. Careful management of the organizations' visibility and positive image in the media and public mind is now the order of the day. Supporters are recruited through door-to-door solicitation and professionally designed mail appeals that do not demand that they become extensively involved. The organizations design their appeals to portray themselves as effective agents of change dealing successfully with serious problems, but confrontational approaches or apocalyptic themes are hardly to be found. This approach allows supporters to feel good about themselves and soothe their troubled consciences by the low-cost activity of authorizing a small bank transfer. Use of differentiated appeals to various market segments and selected incentives, such as reduced-price insurance, consumer tips, and ecotourism, are all part of a package designed to turn citizens from free riders into donors.

Their reliance on this approach has moved the organizations further away from the grassroots models of action typical of the early stages of social movements and toward functioning as professionalized social movement organizations—or even as "mail order" interest groups, organizations to which the term social movement hardly applies. Although originally more diverse, all four appear to be converging around a three-pronged model of operation. It features, first, a professionalized staff, large by historical standards, which plans and carries out much of the organization's work, especially at the national level. Second, except in the case of the WWF, the efforts of the paid staff are complemented by the work of volunteer activists. These activists constitute a relatively small percentage of the organizations' supporters, but they carry out a significant fraction of its work, especially at the local level. A chance to work for a cause they believe in, the social rewards of participation, and—except in the case of Greenpeace—opportunities to exert influence motivate volunteers to contribute far more energy to the cause than do the mass of passive supporters. Finally, all of the organizations have a large mass of relatively passive supporters, whose major contributions are financial support and occasional participation in a petition drive or boycott. Greenpeace is perhaps the most open about its reliance on this strategy. Its supporters lack even the right to vote, and donors are recruited with pleas and slogans that emphasize financial support, not personal activism: "Help us so that we can go forward" (Kunz, 1989) or "The check as [your] ballot" (Wallmeyer, 1996). That is, the organizations have moved toward a model in which professionals plan and execute campaigns that are then "marketed" to their members and the public, who provide the financial support.

While this model allows the organizations to survive and accomplish worthwhile work, it brings with it difficulties. In the first place, the trend toward recruitment of masses of passive supporters and the tactics used to accomplish it can, for old activists and those still committed to social movement mobilization models, come to symbolize all that is wrong with the organizations. The critics complain that environmental organizations have become institutionalized fundraising machines rather than citizens' crusades and that they soften their messages and use theatrics, interest-catching campaigns, and self-promotion to keep the donations from their uninvolved supporters flowing—a classic case of means/ends displacement. The critics also charge that the organizations spend too much money on fundraising and rely on questionable fundraising tactics, and that they implicitly teach the public that environmental action can be left to the professionals and requires little more than writing a check. They invoke images from a better, albeit somewhat mythologized, time, when most work was carried out by committed volunteers and members were recruited on the basis of commitment to the cause. Except in the case of the WWF, the criticism has come not only from outside critics, but from the groups' own activists, particularly those who long for the days when volunteers and grassroots activists played a greater role in governance and day-to-day operations.

That fundraising needs dominate every decision is less clear than the critics suggest. In fact, the organizations do sometimes take principled stands that are unlikely to win them mass support, though it would be hard to argue that they do so routinely or without some trepidation. Instead, they appear to be engaged in a delicate balancing act in which they try simultaneously to appeal to both their activist and more passive supporters and to benefit the environment. Nevertheless, to the extent that the critics can make their charges stick, the organizations stand to lose credibility, influence, and some of their more activist supporters. Thus far, the debatability of the charges and assiduous image management by the organizations have allowed them to avoid this fate, although Greenpeace has at times been seriously damaged by the charges. On the other hand, a move back to goals and strategies of the "good old days" has the potential to deprive them of many of the less involved supporters they have recruited in recent years, a dilemma indeed.

In any event, the organizational consequences of heavy reliance on professionalized fundraising for the internal dynamics of the organizations have proved hard to avoid. The growing proportion of passive members clearly contributes to the tendencies toward the professionalization and centralization described in the previous chapter, and the professionalization of fundraising and new member recruitment adds yet another contingent of professionals to the staff and supports the overall trend toward professionalization. These developments suggest the emergence of a series of feedback loops, in which sophisticated fundraising, a growing percentage of passive members, and professionalization all reinforce one another.

There is also little doubt that the trends toward professionalization and centralization discourage activist participation. Here again there appear to be mutually reinforcing feedback loops, this time involving a lowered level of volunteer activism, reliance on sophisticated fundraising techniques, and professionalization. These developments may also be linked to the adoption of a less confrontational approach to environmental problems, and to competition with one another for supporters, the primary topic of the next chapter.

Chapter 12

Dilemmas of Goals and Strategies

Confrontation, Cooperation, and Competition

The third set of dilemmas environmental organizations face centers around choosing goals and the strategies for reaching them. Goal setting is problematic because abstract goals, such as saving the environment or protecting nature, have to be translated into operative goals (Perrow, 1961), such as increasing public knowledge of global warming or blocking a particular legislative initiative. In view of the multiplicity of conditions that might be defined as environmental problems, the uncertainty of scientific knowledge about the causes and effects of these conditions, and uncertainty about what initiatives might have the most effect on politics, industry, and citizen behavior, choosing among goals and strategies is no small task. Yet this is not all. The organizations cannot and do not make choices about goals and strategies in a social vacuum. Constituencies inside and outside the organization often differ about the priorities they assign to various goals and their assessments of the efficacy and acceptability of various strategies. Not surprisingly then, environmental organizations are often characterized by long and sometimes passionate debates about goals and strategies.

Of special interest in this chapter are issues that arise repeatedly in these debates. The literatures about interest groups, social movements, organizations, and civil society reviewed in Chapter 2, as well as literature about the environmental organizations, suggest that three such issues surface repeatedly: cooperation vs. confrontation with business and government, cooperation vs. competition with other environmental organizations, and focusing on a small number of environmental problems versus many. Like the dilemmas of structure and the dilemmas of resource acquisition discussed in the preceding chapters, these dilemmas appear and reappear because the horns of each dilemma have both advantages and disadvantages and because constituencies that are important for the organizations' survival and success frequently disagree about how to assess them.

Cooperation vs. Confrontation with Business and Government

Environmental organizations must choose a place on a continuum of goals and strategies ranging from confrontation to cooperation with government and business.[1] These choices are consequential because government and business are, simultaneously, potential allies and sources of resources for environmental organizations and their frequent opponents. Business' and government's priorities are rarely identical to those of environmentalists; however, they are also not invariably diametrically opposed.

As illustrated in Chapters 3 to 7, both cooperative and confrontational approaches are well represented in the history of German environmentalism, and the organizations have sometimes shifted between these approaches or employed elements of both simultaneously. For most of its long history, the BfV's leadership had many overlaps with government and business, and its strategies were generally nonconfrontational. During the 1980s, it moved gradually to a more confrontational stance, a transition marked by serious internal conflict. BUND's predecessor organization in Bavaria had a long tradition of cooperation with the authorities, but it became more confrontational during the 1970s. BUND itself was founded as a compromise between the confrontational BBU and conservative nature protection groups, and like NABU, it has always displayed elements of both strategies. Greenpeace emerged directly out of the confrontational atmosphere of the 1970s, but it has recently moved toward a somewhat less confrontational approach.

Confrontational strategies–including noisy or violent demonstrations, site occupations, boycotts, and Greenpeace-style spectacular actions—capture media coverage, dramatize problems, bring environmental scandals to light, and focus public attention on them. When they receive favorable press coverage and capture public sympathy, they create pressure on government and business to make concessions, and the threat to order represented by demonstrations may induce opponents to compromise. Public sympathy can yield new supporters and donations, and government or business concessions can burnish an organizations' image for effectiveness. By sharpening differences and building solidarity, con-

[1] A great deal has been written about the advantages and disadvantages of confrontational strategies by environmental organizations, both in Germany and elsewhere. This discussion draws primarily on the following sources: Hoplitschek (1984); Dölle and Lüginbühl (1985); Mayer-Tasch (1985); Weinzierl (1991); Dalton (1994); Dowie (1995); Schaeffer (1995); Weigand (1995); Koopmans (1995); Halcour (1995); Wolf (1996); Oswald von Nell Breuning Institut (1996); Brand, Eder, and Poferl (1997); Jordan and Maloney (1997); Bammerlin (1998); Lahusen (1998); Bodenstein et al. (1998); Diani and Donati (1999); Brulle (2000); Nölting (2002); Bergstedt (2002); and Dryzek et al. (2003).

frontations can also be motivational for activists. Finally, environmental groups that function as "outsiders" are more likely to introduce new framings of issues and new tactics.

On the other hand, levels of confrontation that are motivational for activists may be viewed as excessive by less committed supporters and the general public. Moreover, radical goals and heavy use of confrontational tactics may drive away moderate supporters, who worry that being identified as an environmentalist will reduce rather than enhance their prestige in the community. Confrontation that is extreme or escalates into violence is particularly problematic because its effects on public opinion are so hard to predict. They depend on the nature of the protest, whether and how it is repressed, and how it is reported and framed in the media. During the period of confrontation over nuclear power in the 1970s and early 1980s, for example, environmental and anti-nuclear activists discovered that eruptions of violence—especially if the government and media could reasonably attribute the events to the involvement of the far left or label it as senseless—reduced their legitimacy and public support (Mayer-Tasch, 1985; Joppke, 1993). Greenpeace became so concerned that it refused to take part in anti-nuclear protests (Streich, 1986; Kunz, 1989), and at the height of protests in Bavaria, BUND's Bavarian chapter issued press releases disavowing connections to the far left and asserting loyalty to the German constitution. It also withdrew from actions that appeared headed for violence (Hoplitschek, 1984). A more recent incident illustrates the problems well: a protest against a nuclear waste shipment in which a sixteen-year-old girl had her arm chained inside a pipe embedded in concrete poured around a railroad track provoked not the hoped-for support, but sharp press and government criticism (Berliner Zeitung, 2001d). Finally, as Greenpeace has discovered (Die Zeit, 1991; Vandamme, 2000), confrontational tactics that once captured media attention tend to lose their novelty over time. This requires ever more dramatic—and from the standpoint of public opinion more problematic—actions to attract media coverage.

There are good reasons for concern about the effects of confrontational tactics on public opinion. Even during the 1980s, the period when environmental consciousness protest was at its height, a series of Eurobarometer surveys suggested that the public had reservations about confrontation. Three-quarters to nine-tenths of respondents approved or strongly approved of nature protection organizations, while only about half approved of more radical ecology groups (Dalton, 1994). Highly confrontational strategies can also evoke criticism or ridicule. The violence-tinged rhetoric of leftist-influenced BIs during the period of confrontation provoked questions about their fitness to participate in a democracy (Thaysen, 1984), and a 1986 Bavarian government publication went so far as to compare Robin Wood to a terrorist organization (Scholz, 1989). A decade later, Germany's environmental minister characterized Greenpeace's purposive violations of law as a threat to democracy (Die Zeit, 1996d), and the president of a *Land* chapter of one of the organizations told me that even today some citizens remain suspicious of environmental organizations because of their 1980s demands for radical action to save the German forests, which later turned out to be based on exaggerated predictions about the forests' imminent demise. The negative effects of radical goals and confrontational strategies on public opinion became more pronounced as environmentalism became institutionalized and as government and business became more willing to compromise—and more adept at greenwashing. In this emerging social climate, environmental organizations that maintained a bunker mentality encountered strong criticism (e.g., Die Zeit, 1991; Frankfurter Rundschau, 1991).

Protests have also sometimes precipitated legal actions, or threats of such action, by opponents seeking compensation for damage to their property or reputations. Greenpeace, the most confrontational of the organizations, has been a frequent target of lawsuits. One of the best-known occurred in the 1988, when Bayer chemicals sued Greenpeace for over $200,000 as compensation for damage caused when its activists clandestinely installed a faucet on a discharge pipe to allow Greenpeace and the general public to take water samples to test for pollutants (Reiss, 1988; Eitner, 1996). Later, the chemical giant Hoechst sued Greenpeace for placing its CEO's face on a widely circulated poster with the legend "Everyone talks about the climate; we ruin it" (Frankfurter Allgemeine Zeitung, 1999e). More recently, a dairy firm was successful in obtaining a temporary injunction to stop Greenpeace from satirizing its admitted use of genetically altered crops as cattle feed (Greenpeace Nachrichten, 2004a). Greenpeace's legal precautions and a judiciary generally sympathetic to nonviolent protest on issues that evoke widespread public concern have thus far generally kept the organization out of harm's way (Reiss, 1988; Rucht, 1995b; Günther, 1996; Bammerlin, 1998). Nevertheless, the legal expenses involved are significant, and Greenpeace has set up a legal defense fund to help finance its legal expenses and now requires that staff attorneys clear all planned demonstrations and media campaigns in advance (Greenpeace Nachrichten, 2003a). Robin Wood has also settled a number of legal complaints with relatively small payments (Scholz, 1989).

Government can also act forcefully against organizations it views as a threat. The anti–nuclear power demonstrations of the late 1970s and 1980s were countered by massive police deployments and violence, and recent confrontations over nuclear waste shipments have again been greeted by massive police responses (e.g., Berliner Zeitung, 2001a). There have also been government threats to deprive environmental groups of their tax-exempt status or force them to pay costs resulting from protests. In 2005, Greenpeace was threatened with revocation of its certification as a public service organization, which would have prevented donors from deducting donations to it from their income tax (Greenpeace Nachrichten, 2005a, 2005e). The government also prosecuted Greenpeace in 1998 for welding a steel box to a railroad track to prevent movement of a nuclear waste shipment (Frankfurter Allgemeine Zeitung, 1998a), and the minister of the interior attempted to get Robin Wood to pay the costs of removing protestors who had blocked the tracks with concrete or chained themselves to railroad tracks (Berliner Zeitung, 2001b; DNR Deutschland Rundbrief, 2002a).

Governments have also at times used more subtle pressure against the organizations. Greenpeace had to struggle for years to obtain approval for conscientious objectors to do their alternative service as Greenpeace volunteers (Reiss, 1988), and more recently, it complained that conservative communities in southern Germany routinely refuse to approve permits for its fundraising booths, even as they freely issue them to charitable organizations (Greenpeace Nachrichten, 2002d). The Frankfurt-area commuter rail system once even refused to accept BUND's ads urging the public to reduce energy waste and air pollution by minimizing air travel (Die Zeit, 1995i). Business and government may also simply decline to cooperate or negotiate with environmental groups that use extreme statements or heavy protests. By defining them as outside the mainstream and refusing to work with them, they undermine the organizations' claims to be credible, legitimate representatives of the public interest. Dalton's (1994) study of European environmental organizations, for example, found that more confrontational organizations had fewer positive relationships with business associations and government economic ministries. The DNR's rather conservative president, on the other hand, was welcomed for regular meetings with the CDU's Chancellor Kohl for many years (Röscheisen, 2006). Even symbolic issues can matter in how politicians respond to environmental organizations. One CDU politician I interviewed complained that environmentalists often undermined their credibility with their dress and grooming, which he viewed as a relic of another era and as inappropriate for conducting serious business. On the other hand, he praised the WWF, the least confrontational of the national organizations, as more credible and more open to constructive compromise.

Positive working relationships with government are too big an advantage for the organizations to sacrifice casually. As one longtime BUND leader pointed out to me, refusal to negotiate or cooperate with government can lead the public to see the organizations as marginal groups that only know how to protest and accomplish little or nothing. Organizations that become too confrontational also risk losing their reputations as sources of reliable information or constructive partners in dialog. This deprives them of opportunities to be represented in government commissions and be consulted informally, of access to information from government sources, and of government grants. BUND and NABU also risk forfeiting the legal status that entitles them to be consulted about local projects that might impact nature. Cooperation with business too can have important benefits that the organizations hesitate to give up. Through such collaborations, the organizations accomplish more than just protecting the environment. They also increase their legitimacy as constructive players in the system.

Nevertheless, the results of collaboration with government and business are rarely unequivocally positive. When they forgo confrontation and operate within the system, environmental organizations can easily find themselves in negotiating relationships where they are outmatched by their opponents' financial resources, staffing, access to information, and scientific credibility. Moreover, by participating in negotiations, environmental organizations implicitly commit themselves to the premise that environmental issues are technical issues to be resolved by negotiation over technical details. Not only is this at times counterfactual, it can also be demotivational for activists. Environmental organizations that cooperate or negotiate with business and government also run the risk of being coopted by politicians or businesses whose main interest is not constructive results but improving their images.

Only environmental organizations that are willing to adopt moderate positions, compromise, and play within the existing rules are apt to be welcome in negotiations with business and government. In this light, it is hardly surprising that several of the more conservative politicians I interviewed cited the WWF's greater willingness to compromise as their reason for preferring working with it instead of with BUND. Negotiations also often lead to ambiguous results and to marginal, incremental change. Such outcomes are unsatisfying for activists and hard to explain to the public. In recent years, critics of compromise and negotiation have frequently pointed to industry's successful use of economic stagnation as a lever to obtain rollbacks of environmental regulations as evidence that negotiation with business and government is counterproductive. Even some of the leaders of the environmental organizations whom Brand and his colleagues interviewed in the mid 1990s cautiously echoed this viewpoint (Brand, Eder, and Poferl, 1997).

In extreme cases, the opportunities for collaboration or negotiation offered to environmental organizations may be only pro forma, or the concessions required to obtain a seat at the table can be so great as to make participation impossible. There are numerous examples of this pattern. When major anti-nuclear protests began in Germany, one of the government's first steps was to try to establish a "dialog" on nuclear energy; however, the BIs soon rejected it as merely an effort to coopt them (Rucht, 1980). During the 1980s, critics in BUND's Bavarian chapter complained that participation in many government committees was pointless because most of the decisions had been made in advance and the organization's participation undermined the efforts of more confrontational groups (Hoplitschek, 1984). This same pattern recurred recently, when several environmental organizations refused to take part in a mediation process about adding a new runway at the Frankfurt airport. They claimed that the rules of the game had been rigged so that they would have no influence but would have to endorse the final decision (BUNDMagazin, 1998e). Still another example occurred in 2000, when the DNR and other major environmental organizations pulled out of a government-sponsored study group on energy because they could wring only the smallest concessions out of the energy company representatives who dominated the group (Röscheisen, 2006). DNR representatives also pulled out of an oversight committee for genetically modified organisms when they found themselves blocked at every turn and concluded that their presence only served as an alibi for business as usual (Röscheisen, 2006).

Devoting time and energy to preparing position papers and participating in negotiations and hearings can also deflect the organizations' efforts away from other important activities. A particularly striking example is the provision of German law that allows local environmental groups to submit position papers on projects that might injure nature (see Chapter 8). Local chapters of BUND and NABU both have this right, and they exercise it extensively and have worked hard to expand it. Nevertheless, the procedure can become a trap when the groups spend so much time preparing position papers and attending hearings–often with little to show for their efforts—that they have little time left for other activities (Grüger and Usersfeld, 1986; Weiger, 1987a, 1987b; Cornelsen, 1991).

Another price of negotiation or other cooperation with business or government is the resulting criticism from environmental activists, especially those whose opinions were shaped during the confrontational era of the 1970s and early 1980s. For example, a local elected official from the "Fundi" wing of the Greens with a long history of movement activism told me that BUND and NABU had lost their effectiveness because they were no longer willing to go to the streets and spent too much time trying to work within the system. The president of NABU told me that such criticism was exaggerated because NABU is actually only minimally dependent on business financially; however, he also noted that a small minority of members—and a somewhat larger proportion of activist members—criticize virtually every NABU cooperation with business and are equally critical when NABU praises government for doing the right thing. The president of one of NABU's *Land* chapters echoed these views. He said that a minority of vocal members whose views were shaped by many years of confrontation with government and business continually push for more confrontational stands. The organizations ignore these supporters at their peril, since they make up a significant portion of their leaders and strongest supporters.

Unreformed activists have a long litany of specific complaints. The DNR, in particular, has been criticized as too willing to compromise on account of its close ties to government and business. In one particularly striking example, environmental activists attacked the DNR's sponsorship of the 1992 German Environmental Day, which they saw as dominated by business and government, by protesting in front of the information booths of environmentally questionable industries and distributing an alternative newsletter (Bergstedt, 1998, 2002). BUND and NABU have run into occasional criticism for having government officials in leadership positions and the movement of some of their former leaders into elected or appointed government positions (Bergstedt, 1998; Maxim and Degenhardt, 2003).

More recently, criticism has been leveled at their eager embrace of the "Agenda 21" program. This program, an outgrowth of the 1992 United Nations–sponsored Rio environmental conference, was aimed at developing sustainable communities in Germany, often through grants for demonstration projects. Critics charged that government and business interests were so deeply involved in the development of the program that it degenerated into pointless discussions and reports and that the environmental organizations had compromised their principles in pursuit of grant money and the prestige of being involved (Bergstedt, 1998, 2002; Geden, 1999).

Internal divisions over compromise with business and government have been especially prominent within BUND. In 1992, it found itself split between supporters and opponents of its participation in the German Environmental Day (Bergstedt, 2002). Some BUND members also criticized BUND's decision to negotiate with Unilever, whose products it had previously criticized as ecologically unsound (Die Zeit, 1995c), while its cooperative relationship with a department store chain led to charges that it was indirectly endorsing consumerism (Die Zeit, 1995e; Bodenstein et al., 1998; Bergstedt, 2002). More recently, a decision by BUND's Thuringia chapter, following the advice of the national board, to abandon a legal effort to block construction of a reservoir in return for a large payment into a foundation for environmental protection precipitated a storm of criticism. The critique, which came not only from within BUND but also from outside, led to the development of new guidelines to preclude similar cases (Spielmann, 1997; Weiger, 1997; Bergstedt, 2002). The WWF and NABU have also received smaller but similar settlements (Der Spiegel, 1997b).

Criticism of the WWF (e.g., Bergstedt, 2002) has often focused on the large number of business leaders on its board, but its efforts to cooperate with business have also come under criticism. Its decision to participate in the 2000 World's Fair in Hanover, for example, drew especially strong criticism because, after long internal debates, the other organizations had concluded that the fair was simply too business-oriented for them to participate (Die Zeit, 1997d; Bergstedt, 2002). Greenpeace publicly condemned the exposition (Greenpeace Nachrichten, 2000d), and NABU also published a critical article (Naturschutz Heute, 2000n). In interviews, several leaders of BUND and NABU also criticized the WWF's decision to participate. The WWF defended its decision to participate vigorously, but the tone of its statements (e.g., WWF-Journal, 2000a) was defensive.

The WWF is not the only organization to be criticized by the others. In a tense meeting in 2004, BUND leaders criticized NABU leaders for cooperating with Volkswagen (BUNDschau, 2004f). And in a recent interview, Thilo Bode, formerly executive secretary of Greenpeace Germany and Greenpeace International, sharply attacked BUND and the other environmental organizations for cozying up to government and not being confrontational enough (taz, 2004). Ironically, however, Bode had been executive secretary of Greenpeace International when it came under strong criticism for entering negotiations with the same oil companies it had earlier strongly criticized over North Sea drilling (Bennie, 1998).

To protect themselves from criticism, the organizations have developed detailed guidelines for their cooperations with business and acceptance of donations from business (see Chapter 11). The sensitivity of the issue is illustrated by the two years of study and meetings that were required for BUND to arrive at its policy. The organizations also offer detailed explanations to justify their accepting funds from business and cooperating with business (see Chapter 11). Part of the reason for the organizations' defensiveness is concern that activists who perceive the organizations as too tame might redirect their energies or support to more radical groups (Hermand, 1991; Rootes, 1999b). This pattern appeared in the 1970s, when movement activists initially bypassed the nonconfrontational nature protection organizations and flocked instead to BIs and the BBU. In the late 1980s and 1990s, disaffected members of NABU who wanted a more activist agenda withdrew from the organization (Möller, 1993; Bergstedt, 1998), and recent waves of protest against shipments of nuclear waste and globalization hint that the pattern of the 1970s could repeat itself (Kolb, 1997; Hunold, 2001a, 2001b; Rootes, 1999a, 2002).

Coping with the Cooperation vs. Confrontation Dilemma

Because cooperation and confrontation each have important advantages and disadvantages, the organizations find that neither exclusively confrontational nor accommodationist strategies can be pursued without incurring significant costs. In this situation, decisions about goals and strategies hinge partly on the preferences of leaders and on organizational traditions and ideology. However, they are also constrained by the preferences of the organizations' various constituencies—activists, passive supporters, the media, and negotiating partners in government and business.

By the end of the twentieth century, support for confrontational environmentalism was on the wane, and the available evidence (see Chapter 8) suggests that most supporters of environmental organizations today prefer a mix of cooperation and confrontation. Only a minority long for the strongly confrontational approach of the 1970s and early 1980s. This means, as interviews with several organizational leaders explicitly confirmed, that the organizations can no longer count on their ability to call out masses of people for demonstrations. One national leader cited the "pathetic" turnout at a recent rally in Berlin as clear evidence that a purely confrontational strategy is no longer viable, and the president of a *Land* chapter cited several recent actions, some as mild as petition drives, that his chapter had abandoned due to fear that participation would be too low.

It is therefore not surprising that all four organizations now deal with the confrontation vs. compromise dilemma by attempting to steer a middle course.[2] While they can still be uncompromising on selected issues, such as nuclear power or genetically manipulated food, all but the WWF are now less prone to confrontation than they were during the polarized period of the late 1970s and early 1980s. Choosing a middle course provides them with the potential to obtain some of the results their supporters want, meanwhile retaining the threat of protest actions as an incentive for opponents to compromise and avoiding becoming captives of business, the state, or the rules of the political game. All now partici-

[2]Virtually every commentary on the recent history of the environmental organizations notes the decline in use of confrontational strategies. See, for example, Rucht (1989), Cornelsen (1991), Hey and Brendle (1994), Blühdorn (1995), Oswald von Nell Breuning Institut (1996), Rat von Sachverständigen für Umweltfragen (1996), Brand, Eder, and Poferl (1997), Pehle (1997), Bammerlin (1998), Brand (1993, 1999a, in press), Bergstedt (2002), Dryzek et al. (2003), and Rogall (2003).

pate in hearings, parliamentary commissions, and planning processes at all levels of government. They have also moved from blanket condemnations of consumerism and business to targeting specific firms or industries, and they work with environmentally conscious businesses and producers of environmentally sound technologies on projects such as introducing cleaner production technologies and reducing packaging.

BUND, for example, cooperated with a brewery to develop recyclable and stackable beer crates (BUNDMagazin, 2005d), with a computer manufacturer to develop recyclable computer equipment (Frankfurter Allgemeine Zeitung, 1998b), and with a cosmetics and toiletries firm to replace excessive packaging with returnable containers (Brand, Eder, and Poferl, 1997; Bodenstein et al., 1998). NABU mounted a joint effort with local bottlers to promote returnable bottle use (NABU, 2000b) and worked with the German rail system to redesign overhead power lines to reduce bird electrocutions (Cornelsen, 1991). The WWF has cooperated with firms that produce renewable energy (WWF-Magazin, 2003a). It has also worked with Berlin's power company to place solar panels atop Berlin's train stations (WWF-Deutschland, 1999) and with a manufacturer of solar-powered motor scooters to make them available for loan in the Freiburg train station (DNR Deutschland Rundbrief, 2000f). Greenpeace worked with the furniture manufacturer Ikea on guidelines for avoiding use of tropical woods in its product line (Die Zeit, 1999c), developed a demonstration model of a fuel-efficient car (Die Zeit, 1995g; Frankfurter Allgemeine Zeitung, 1997c), and cooperated with an East German firm to produce a refrigerator free of the coolants that damage the ozone layer (Die Zeit, 1996b).

There have also been several longer-running cooperations with business. The Hertie department store chain (now Karstadt) had a longstanding cooperation with BUND, which advised it about reducing packaging and replacing environmentally damaging products on the shelves with less damaging ones (Zillessen and Rahmel, 1991; Die Zeit, 1995e; BUNDMagazin, 2000b), and NABU had a similar relationship with the Otto mail order house (Zillessen and Rahmel, 1991; Flasbarth, 1995; NABU, 1999).

Finally, there have been cooperations with trade associations, such as a declaration of common principles between BUND and the chemical industry (Rat von Sachverständigen für Umwelfragen, 1996) and BUND's alliance with a manufacturers' trade group to advocate legislation favoring power generation systems that use waste heat to heat buildings (DNR Deutschland Rundbrief, 2001e). The Association of Young Entrepreneurs also joined BUND in promoting an energy tax (Rat von Sachverständigen für Umweltfragen, 1996; Pehle, 1997).

The organizations vary, however, in how much confrontation they prefer. Almost all of my interviewees ranked Greenpeace and BUND as the more confrontational organizations and the WWF as most prone to cooperation. An analysis of newspaper reports of protest actions between 1988 and 1997 (Rucht and Roose, 2001b, 2001c) supports this conclusion. It showed that Greenpeace and BUND were far more likely to be recorded as the organizers of environmental protests than NABU or the WWF. This result parallels Dalton's (1994) earlier finding that European environmental organizations founded during the period of confrontation remain more confrontational than older organizations with roots in nature protection.

Greenpeace originated with spectacular protests, and it remains the most confrontational of the organizations. For example, it has mounted a long series of demonstrations against atomic power (e.g., Greenpeace Nachrichten, 1999b), and in 2001 it dumped a freight container of genetically altered corn in front of the health ministry to protest against introduction of genetically altered foods in Germany (DNR Deutschland Rundbrief, 2001k). In 2002, it started a campaign sharply critical of Esso, the European marketing brand of Exxon/Mobil, in which it wrote the corporation's name with dollar signs substituted for the s's and accused the corporation of gross irresponsibility and greed in ignoring global warming (Greenpeace Nachrichten, 2002a).

Nevertheless, by the mid 1990s Greenpeace leaders were already saying that more than just spectacular protest was necessary to sustain public support and that demonstrations and confrontations were productive only if combined with concrete proposals for a less environmentally destructive economy and lifestyle and willingness to cooperate with industry and government (Bode, 1996a, 1996b; DNR Deutschland Rundbrief, 2000h). This was not just talk. Greenpeace has worked hard to develop concrete policy suggestions and a dialog with business and government. It has contracted with ecological research institutes to produce studies of fuel efficiency and transportation policy (Die Zeit, 1996c) and energy taxes (Greenpeace Deutschland, 1995; Lurisch, 1995) and has even engaged in behind-the-scenes negotiations with oil companies (Bennie, 1998).

BUND today also pursues a compromise course in dealing with the confrontation versus cooperation dilemma. This represents some moderation of its stance a quarter-century ago, although even in more confrontational times it had retained its emphasis on practical nature protection at the local level and sought to preserve a good working relationship with the authorities and maintain its reputation for expertise (see Chapter 5). A 1989 speech by BUND's longtime president, Weinzierl, provides an apt example of BUND's mixed approach in recent years. He first cited BUND's joint declaration of principles with the Federation of German Industry as an example of progress, but then denounced German society's lack of concern about environmental protection (Weinzierl et al., 1990).

BUND's effort to balance cooperation and confrontation, its turn away from challenging the basic assumptions of the economic and political systems, and its tendency to embrace precepts of ecological modernization are also on display in the centerpiece of its program at the end of the twentieth century, the book-length report *Germany: Ready for the Future* (BUND/Misereor, 1996; see also BUND, 1995). BUND's vice-president, Angelika Zahrnt, an economist who is now the organization's president, played a key role in producing the report, which was written in cooperation with the Catholic development organization Misereor and the Wuppertal Institute for Climate, Environment, and Energy. A follow-up report appeared in 2002 (BUND/Misereor, 2002). These reports set forth a vision of a German future built around sustainability, ecological modernization, and reductions of wasteful consumption, not a radical system critique (Rat von Sachverständigen für Umweltfragen, 1996; Wolf, 1996; Bergstedt, 2002). BUND thus emphasizes seeking out partner firms in business that wish to develop ecologically sound consumer products and are committed to ecological modernization and sustainability (Felbinger, 2003).

In line with this approach, Zahrnt argued in a 1995 interview that alliances with firms that are truly willing to change can help to push forward positive change in society; however, she also warned that BUND must be careful not to let itself be used as an "ecological fig leaf" (BUND, 1995). Several years later, as incoming president, she succinctly summarized the organization's approach as follows: "Confrontation when necessary, critical advice where appropriate, cooperation when possible" (BUNDMagazin, 1999c). BUND does indeed continue to use confrontational strategies at times (e.g., Brinkmann and Lüdke-Höher, 1994; BUNDMagazin, 1998c), and in a recent interview Zahrnt once again echoed old themes about the need for simplified lifestyles and reductions in consumption (Zahrnt, 2002).

NABU, historically a nonconfrontational nature protection organization, evolved into a more politicized organization during the 1980s, although it never became as confrontational as BUND (see Chapter 5). As several of my interviews pointed out, its approach today parallels that of BUND: occasional confrontation, willingness to cooperate with business and government in some situations, and advocacy of ecological modernization and sustainable development without radical system critique. NABU engages in numerous cooperative and sponsorship arrangements with business (see above and Chapter 11), but it can also be confrontational. In 1999, for example, SPD Chancellor Schröder became so angered by harsh criticism from NABU that he abruptly canceled his planned speech at the organization's centennial celebration (Frankfurter Allgemeine Zeitung, 1999b). This action contrasted strongly with another speech a year later that was full of effusive praise for the retiring president of the DNR, who was known for his cautious, nonconfrontational approach (DNR Deutschland Rundbrief, 2000k). More recently, NABU harshly criticized the economics minister for placing economic growth ahead of environmental protection (DNR Deutschland Rundbrief, 2003b). A number of NABU leaders I interviewed also noted that the national leadership still prefers a more activist stance than many local groups, but they reported that conflict over this issue has died down as both sides learned to compromise.

The WWF, with its close ties to business, is the organization most to willing to cooperate with business and government and the least likely to use confrontational tactics. As its president told me in an interview, the organization's nonconfrontational style is intentional. It rests on the assumption that more can be accomplished this way, a position he also repeatedly advocated in the organization's publications (e.g., WWF-Deutschland, 2001b; WWF-Journal 2000a, 2000e; Kohl, 2003). The same position was advocated by the chair of WWF-Germany's board of directors on the occasion of WWF-Germany's fortieth anniversary. He argued that progress in "small steps" is the most effective strategy (WWF-Journal, 2002e). In other public statements (e.g., Frankfurter Allgemeine Zeitung, 1997a; WWF-Journal, 1999a), WWF leaders stress taking a "realistic" approach. One top WWF leader I interviewed, for example, justified WWF's decision to buck the unanimous stand by the other environmental organizations and participate in the Hanover World's Fair by citing the need to be part of the public dialog (see also WWF-Deutschland, 2001b). The WWF also stood alone in its position that farmers should be compensated when environmental regulations limit the use of their land (Frankfurter Allgemeine Zeitung, 1998d). It regularly places prominent business leaders and politicians on its board (Weigand, 1995; Bergstedt, 2002), and it annually names an Ecological Manager of the Year in an effort to encourage ecologically sound behavior (WWF-Journal, 1999b, 2000b; Bauske, 2003).

Not surprisingly, the WWF's criticism of government and business tends to be restrained (Brand, 1999a), but it does at times adopt a more confrontational posture. It engaged in a sharp dispute with German logging interests over whether to support the Forest Stewardship Council seal for sustainable forestry or a milder version favored by the industry (see also Frankfurter Allgemeine Zeitung, 1998h), and it organized a petition drive to protest the Kohl government for backsliding on Germany's commitment to reduce greenhouse emissions (WWF-Journal, 1998b). In 1999, it sank a huge replica of a 353 million mark banknote at the site of a proposed dam and locks at the mouth of the Ems River to symbolize the wastefulness of the project (WWF-Deutschland, 1999). These steps toward increased political activity paralleled developments at WWF-International, which seems to have become somewhat more open to protest strategies in recent years (Diani and Donati, 1999).

Despite its appeal as a principle, steering a middle course cannot be easy. Not much information about the internal deliberations of the organizations on this topic is available; however, Wolf has chronicled the struggles of leaders of BUND's chapter in Lower Saxony and the national organization to strike a balance between a level of criticism of government policy that would satisfy their more militant members without costing them the support of their more traditional members, damaging their ties with government, or reducing the public's respect for their expertise. The chapter in Lower Saxony even came near to a split over this issue in the late 1980s. Wolf also describes the struggles of BUND's Bavarian chapter to find a way to participate in protest and alliances with more radical environmental groups without being taken over by extremists (Wolf, 1996).

Greens in Government: Cooperation vs. Confrontation?

The tension between cooperation and confrontation was heightened between 1998 and 2005, when the German Green Party was part of Germany's governing coalition. Germany's multiparty parliamentary system, in which entry into the legislature requires only 5 percent of the national vote and almost all governments are comprised of coalitions of parties, makes trying to influence policy by forming a small political party a viable and attractive option for environmentalists (see Chapter 8), and this has been the strategy of the Green Party. The Greens emerged at the end of the 1970s and entered the Bundestag in 1983. They agreed to join the Social Democrats as the junior partner in the governing coalition in 1998. As the junior partner in coalition with a much larger party, however, they were able to implement only a small part of their agenda. Consequently, the party's "realo" leadership had to contend with steady criticism from the "fundi" wing, as well as from many environmental activists (see Chapter 7).

As the president of Greenpeace International (Frankfurter Rundschau, 2000) and BUND's press spokesperson (Die Zeit, 1998) pointed out in separate interviews, the Greens' participation in the government posed many new issues for organizations accustomed to approaching the government as outsiders. There were many incentives and opportunities for cooperation with the Greens, but overly close cooperation involved considerable risk for the organizations, primarily because many of their supporters were "green" but not "Greens." There is, of course, considerable overlap between Green Party voters and environmental organization members (Bundesministerium für Umwelt, 1996, 1998), and the leaders of the environmental organizations rated the performance of the Greens as better than the other parties (Dalton, 1994). Nevertheless, all four organizations have many supporters and leaders who are members or activists in other parties. They also have a long history of nonpartisanship, which arose out of fears that being perceived as an ally of any one party might drive away supporters who belong to the others and make enemies of the other parties (Kazcor, 1992; Steenbock, 1996a; Bode, 1996b; Bergstedt, 2002).

Their dilemma centered around two issues. The first was how to deal with elections. Unlike their US counterparts, German environmental organizations do not make campaign contributions, as campaigns are almost entirely publicly financed, and they endorse neither specific candidates nor parties (BUNDMagazin, 2002b; Naturschutz Heute, 2002e). Several of my interviewees explained that party discipline in Germany is so tight that it would make no sense to endorse individual candidates and that endorsing the Greens would alienate large numbers of the organizations' members and tie them to a small party with an uncertain future. Indeed, when a top staff member of BUND's Bavarian chapter decided early on to run as the Green candidate for mayor, the organization instituted a policy forbidding this (Hoplitschek, 1984). And Greenpeace once pointedly distanced itself from one of its local groups that had carried a banner criticizing the Christian Democrats in one of its protests (Reiss, 1988).

Rather than influencing elections directly, the environmental organizations typically attempt to influence the actions of whatever government is in power through nonpartisan lobbying, petitions, public education, and protests (Lahusen, 1998). They do, however, sometimes explicitly compare the party platforms of the various parties. BUND, for example, published a very detailed analysis of the environmental platforms for the 2002 national election, which implied support for a Green/SPD coalition but stopped short of a direct endorsement (BUNDMagazin, 2002c), and a series of articles published right before the 2005 election adopted the same approach (BUNDMagazin, 2005h, 2005i, 2005j). NABU's 2002 assessment, though also tilted toward the Green/SPD government, was somewhat more balanced (Naturschutz Heute, 2002e), and its analysis of the 2005 election noted some weaknesses in the performance of the SPD/Green government but displayed a clear tilt toward its policies (Naturschutz Heute, 2005h). Greenpeace rarely expresses itself on the topic of German political parties. A 2002 article by its executive secretary did contain mild praise of the SPD/Green government's accomplishments, especially in comparison to the proposals of the other parties, but it also criticized the government's shortcomings (Greenpeace Nachrichten, 2002c). Shortly before the 2005 election, Greenpeace published a brochure reporting responses of party leaders to a series of questions about environmental issues addressed to each of them (Greenpeace Deutschland, 2005c). And in 2005, the DNR, BUND, NABU, the WWF, and Greenpeace issued a joint statement with their demands for the 2005 national elections (DNR Deutschland Rundbrief, 2005a).

The second issue was when and how strongly to criticize the SPD/Green government. This ticklish question was not completely new. Conflicts and misunderstanding between environmentalists in government and environmentalists outside the government surfaced almost immediately after Green Party members and environmentalists affiliated with other parties began to assume positions in the *Länder* governments during the 1980s (Renn, 1985; Müller-Rommel, 1993; Koopmans, 1995; Zahrnt, 1993). Jo Leinen's 1985 move from the post of leader of the BBU to the position of environmental minister in the Saarland, for example, brought him under fire from environmentalists, and a Greenpeace cofounder and top leader who assumed the same post in Lower Saxony five years later had a similar experience (Rucht and Roose, 2001c).

After 1998, this pattern repeated itself at the national level. Not long after the Greens entered the national government, the party's leader, Joschka Fischer, was complaining publicly that the environmental organizations needed to be realistic about what the Greens could achieve, not hypercritical (WWF-Journal, 1999f) and calling for more willingness to compromise (DNR Deutschland Rundbrief, 2000e). All three of the Green Party leaders I interviewed a year later complained that BUND, the regime's sharpest critic, was too consistently critical to be a welcome partner (see also Dryzek et al., 2003). One high-level party staffer showed visible anger, arguing that most of BUND's membership and the German public supported the government's proposal for a phased elimination of nuclear power, while BUND's out-of-touch leadership continued to push for an immediate shutdown of all nuclear plants (see also Frankfurter Allgemeine Zeitung, 1998f, 1999g; BUNDMagazin, 1999a, 2000e). Pointed criticism of BUND was voiced with equal vigor by the SPD's environmental expert, while a CDU leader I interviewed claimed that BUND's—and to a lesser extent NABU's—leaderships were out of touch with their own members and too close to the SPD/Green government.

The Greens' complaints about BUND's stance were not invented out of whole cloth. As early as 1999, BUND's president began to criticize the new government for timidity and slow progress not only on nuclear power, but also on protecting the public from radiation, energy tax reform, identifying new national parks, and other issues (BUND, 2000). In 2001, she authored a sharply critical article characterizing a draft of a new party platform meant to guide the Greens for the next two decades as timid and lacking ecological vision (BUNDMagazin, 2001a), and in 2003, she argued that the SPD/Green government was allowing the traditional domination of government by a closed alliance between industry, unions, and business (BUNDMagazin, 2003a). Another very critical article focusing on the government's transportation and climate policies appeared in 2004 (BUNDMagazin, 2004d), and one sharply criticizing its one-sided focus on economic growth at all costs came out in 2005 (BUNDMagazin, 2005f). One top BUND leader told me that despite the Greens' resentment of criticisms like this, the organization could give up its basic positions and become a cheering section for the Greens.

BUND's critique of the SPD/Green government was the strongest, but it did not stand alone. Greenpeace's newly appointed executive secretary harshly criticized the government in a 1999 interview (Greenpeace Nachrichten, 1999c), and in 2004, the four major environmental organizations issued a joint press release criticizing its failure to live up to its promises (NABU, 2004b). The strongest criticism was aimed at the compromise on ending nuclear power production, an area where the Greens had markedly softened their earlier positions (Hunold, 2001a; Hoffman, 2002). Like BUND (BUNDMagazin, 2001b), NABU, Greenpeace, and the DNR supported protests against the resumption of nuclear waste shipments (Hunold, 2001b).

Nevertheless, the organizations did sometimes offer qualified praise of the government's achievements, as BUND and NABU did in assessments of progress during the first years of the new century (DNR Deutschland Rundbrief, 2003a, 2004a, 2004c). They also continued to maintain contacts with opposition leaders, such as the CDU's Angela Merkel, who subsequently became the new chancellor (DNR Deutschland Rundbrief, 2003d).

Cooperation vs. Competition with other Environmental Organizations

NABU, BUND, the WWF, and Greenpeace compete with one another for financial support, press attention, and public esteem. Therefore, they must take one another into account as they choose their goals and strategies. Their modes of relating to one another could be analyzed along various dimensions; however, the theoretical literature reviewed in Chapter 2, existing studies of the four organizations, and interviews with their leaders all suggest that the most important of these is whether they choose cooperative or competitive relationships with one another. Once again, both approaches have benefits and drawbacks.[3]

If the organizations choose to cooperate, they can speak with one voice and bring to bear their combined resources. This is especially important in the German context, where other sectors, such as industry and labor, are represented by large national associations. These national associations can be formidable opponents, and the environmental organizations can exert more leverage against such opponents and be a more convincing presence in the media and in government decisions if they work together. Moreover, government, other interest groups, and the media often prefer to work with interest groups that speak with a single voice, and are frustrated by disunified interest groups. It was partly this factor that led government ministries concerned with the environment to assist with the establishment of the DNR and BBU and to provide them with financial support (see Chapters 4 and 5 respectively); however, in view of the BBU's virtual disappearance and the continued weakness of the DNR, government ministries continue to complain that they are confronted with excessive and conflicting input from too many different environmental organizations (Rat von Sachverständigen für Umweltfragen, 1996; Pehle, 1997, 1998).

Long-term cooperation could also make possible a division of labor in which organizations developed their own niches and took the lead on specific issues, relying on the support of the others to strengthen their hand. Such a strategy could also reduce duplication of effort and allow more efficient use of resources. Organizations with relatively narrow goals could also specialize their fundraising appeals, potentially attracting more supporters to the environmentalist cause. One interviewee suggested, for example, that NABU would be more effective if it could focus on its strengths in nature protection, where the interests of many of its members lie, instead of trying to duplicate areas already covered by BUND. Finally, a large number of specialized organizations might well prove more adaptable and flexible in dealing with changing circumstances than would a few omnibus organizations.

There are, however, numerous obstacles to cooperation, many of which center around the organizations' need for independent identities. Concerns about potential loss of identity and traditions make mergers especially hard to contemplate, but even small-scale cooperations, such as joint press releases, can reduce the visibility or obscure the identities of the individual organizations involved. And because of the organizations' high dependence on individual donations, retaining visibility and a distinctive image are matters of vital importance. Interviewees from the organizations repeatedly pointed out that a clear organizational

---

[3]This section draws on the discussions of the cooperation versus competition dilemma in the following: Böckman and Effers (1985); Cornelsen (1991); Flasbarth (1993); Rucht (1993a, 1993b); Hey and Brendle (1994); Heuser (1994); Bosso (1995, 2005); Oswald von Nell Breuning Institut (1996); Rat von Sachverständigen für Umweltfragen (1996), Pehle (1997); Brand, Eder, and Poferl (1997); Bammerlin (1998); Diani and Donati (1999); Boström (2001); and Felbinger (2003).

identity, such as Greenpeace's "David versus Goliath" image or BUND's carefully nurtured reputation for scientific expertise, and media visibility are key to their success. As one interviewee succinctly put it, "All the environmental organizations want to be opinion leaders and be visible in the media. They need media attention to be effective in fundraising and vice versa." The organizations carefully monitor their public images (see Chapter 11), and declines in membership or failure to keep up with competitors, such as BUND's slower growth than NABU during the 1990s, occasion concern and self-criticism (Wolf, 1996).

The organizations' needs for visibility and a clear identity have set off numerous conflicts among them. BUND, during its early years in North Rhine-Westphalia, was the target of sharp criticism from the BfV for diverting member support from the older organization (Maxim and Degenhardt, 2003), and a decade later a series of heated exchanges occurred when NABU adopted its present name, which BUND leaders thought was too similar to their own (Wolf, 1996, Bammerlin, 1998). Greenpeace's success in converting its clear profile and high level of media visibility into fundraising success also generates resentment. One former BUND leader stated rather bluntly that the other organizations view Greenpeace as an organization of publicity hounds, whose highly visible spectacular actions are tailored to fundraising rather than results.

Several interviewees also suggested that concerns about visibility are a major reason why the four organizations frequently resist having the DNR take the lead or coordinate campaigns. Indeed, on occasion, the organizations even choose to coordinate their activities among themselves rather than work through the DNR (see Chapter 8). Two interviewees who were intimately familiar with the DNR's efforts to assume leadership confirmed that BUND and NABU prefer that it act merely as a service bureau so that they can retain their own identities, while a DNR leader complained to me that several of the DNR's member organizations were so intent on media visibility and claiming credit for every accomplishment that they found it difficult to work with the DNR—and sometimes even with one another. Two of the four large environmental organizations solve the problem by simply ignoring the DNR. Greenpeace was a member for only a short time in the 1980s, and the WWF withdrew in 2000.

Cooperation among the organizations can also be rendered difficult by differences in their understandings of environmental problems, their concrete objectives, their histories and traditions—including past conflicts between them—and their styles of operation or communication. Differences like these have always made cooperation between the general environmental organizations and the fishermen's and hunters' associations difficult; however, these same kinds of differences can also help to explain the pattern of collaborations among the environmental organizations themselves. Rucht and Roose (2001a, 2001b, 2001c), for example, found that Greenpeace was far less often cited as a partner in joint campaigns among environmental organizations than BUND or NABU. (The WWF was also lower-ranked on this count than the other two, but this may have occurred because of its lower overall level of involvement in campaigns.) A top

BUND leader told me that Greenpeace's emphasis on quick action and staging spectacular events gets in the way of collaborating with NABU and BUND, partly because their democratic decision-making process operates more slowly (see also Oswald von Nell Breuning Institut, 1996; Felbinger, 2003).

When the organizations choose to compete rather than cooperate, they not only fail to reap the benefits of cooperation, they encounter other costs as well. If they disagree about substantive matters, their opponents can exploit their differences. For example, Germany's former Environmental Minister and current chancellor once countered Greenpeace's criticism that the government's climate policy was inadequate by pointing out that the same policy had been praised by the WWF (Der Spiegel, 1997a). Such problems arise repeatedly in the procedure that entitles organizations with environmental interest to submit statements about government plans that might damage nature (see Chapter 8). BUND and NABU sometimes submit separate and partially contradictory statements, and the positions of the environmental organizations are not infrequently contradicted by those of hunters, fishermen, or hikers (Cornelsen, 1991; Weiger, 1987b; Rat von Sachverständigen für Umweltfragen, 1996). Repeated disagreements between environmental organizations—or sometimes between factions within the same organization—over, for instance, pushing for windmills because they save energy versus opposing them because they disfigure nature and harm birds have produced similar embarrassment (Rucht and Roose, 1999; BUNDMagazin 2000b, 2003c; NABU, 2005a). When reported in the press, the unseemly appearance of disunity can tarnish the reputations of all involved. Competition can also lead to bickering among organizations, as when the other organizations criticize Greenpeace for drawing support from networks of local activists the other organizations helped to develop but then declining to cooperate with them (Klein, 1996).

Coping with the Cooperation versus Confrontation Dilemma

All four organizations have adapted to the cooperation versus competition dilemma by pursuing a strategy that mixes occasional cooperation with continuing competition for visibility, supporters, and donations. This strategy was readily acknowledged by their leaders in interviews (see also Flasbarth, 1992; Hey and Brendle, 1994). One fundraiser, for example, summarized the limits of cooperation succinctly by saying, "We don't cooperate on fundraising." And when asked whether there was competition among the organizations for members and donors, another responded with a spontaneous "I'll say!" The president of one of the organizations acknowledged that there is considerable competition for donations and public recognition and added that good member service was another area where the competition was played out.

The organizations' high dependence on contributions from the public and consequent need to be visible in the media means that cooperation is apt to remain limited; however, one of the organization's presidents described the present situation as a "sporting" competition among organizations that also value one another's efforts, and a number of interviewees indicated that the balance in recent years has shifted somewhat toward cooperation (see also Oswald von Nell Breuning Institut, 1996; Rat von Sachverständigen für Umweltfragen, 1996; Brand, 1999a). The president of one of the organizations was at pains to tell me that when he or another group's president obtained important information in conversations with government officials, it was shared with the others. There are occasional meetings of the leaders, but regular institutionalized exchanges of information do not take place.

BUND, NABU, and the WWF have also engaged in a series of cooperative efforts on the national level; however, these tend to be ad hoc and temporary. Commonly planned campaigns remain exceptions (Hey and Brendle, 1994; Lahusen, 1998; Brand, Eder, and Poferl, 1997; Brand, 1999a). The organizations have issued joint press releases and statements, including a joint statement by NABU and BUND on agricultural policy (DNR Deutschland Rundbrief, 2001b) and a joint statement by all four organizations plus the Green League on saving public lands from the GDR (DNR Deutschland Rundbrief, 2002d). They have jointly sponsored workshops and conferences, such as a joint environmental day conference and festival cosponsored by BUND and NABU (BUNDMagazin, 2001c) and a joint conference on water standards staged by BUND, NABU, and the Green League (BUNDschau, 2004d). BUND and the WWF staged a joint protest urging the Russians to sign the Kyoto Protocol (BUNDschau, 2003b), and the leaders of BUND and NABU made a joint bicycle ride around Germany before the 2002 elections to increase the visibility of environmental issues (BUNDMagazin, 1998d).

There have also been various cooperative projects among the organizations. The WWF and NABU share the operation of a crane information center (von Treuenfels, 2003), and the organizations also worked together on the effort to establish a new national park in Hesse (WWF-Journal, 2000d). They have also cooperated in public relations efforts and lawsuits to fight construction plans the organizations believed would damage the Ems River (Flasbarth, 1993; WWF-Deutschland, 1999; WWF-Journal, 2000c) and in a joint effort to develop a plan for protecting the Elbe. The last example, however, also illustrates the obstacles to collaboration. The Green League, originally one of the partners, withdrew because it believed that the three other organizations had negotiated too much away (Die Zeit, 1996f, 1999d; Frankfurter Allgemeine Zeitung, 1999f).

There is also good potential for cooperation among the organizations, particularly NABU and BUND, at the *Land* and local levels; however, even here, there can be competition for press attention and supporters (Kluth and Kraetzchmer, 1985; Bammerlin, 1998). During the 1970s, the BUND and NABU chapters in Lower Saxony experimented with mutual representation on one another's boards and regular joint consultations; however, this experiment also demonstrated the limits of cooperation. Competition for members and donations undermined proposals to establish reciprocal membership status or to forgo founding new local groups in the same local areas (Wolf, 1996). The most common form of local and regional collaboration over the years has been coordination by BUND and NABU of the position statements on projects that might harm nature that they are allowed to submit under German law, but the organizations have also worked together to influence land use planning (e.g., BUND, 2000). According to interview data, BUND and NABU in Baden-Württemberg also share some board members, conduct seminars together, and have held joint press conferences.

In general, Greenpeace has been the least willing of the four organizations to cooperate with the others (Klein, 1996; Rat von Sachverständigen für Umweltfragen, 1996; Bammerlin, 1998). It does occasionally join the other organizations in position statements and press conferences, such as a joint statement calling for immediate closure of nuclear power plants because of the danger of terrorist attack (DNR Deutschland Rundbrief, 2001l). It also joined BUND, NABU, the WWF, and several international development organizations to promote a North Rhine-Westphalia lottery to raise money for their projects (BUNDMagazin, 2003e, Naturschutz Heute, 2003d), and it recently cooperated with BUND, a research institute, and the Greens' foundation to cosponsor a "McPlanet" conference on environmental threats resulting from globalization (BUNDMagazin, 2003c). Accusations that it is unwilling to cooperate are clearly a sensitive issue for Greenpeace. The information packet it provided me in response to my written interview questions included a two-page statement saying that Greenpeace's activities complement the activities of local BIs and that its local groups frequently cooperate with them. The statement argues that the environmental movement as a whole is most successful when a diversity of organizations each contributes what it is best suited to contribute.

A weaker form of cooperation involves informal but shared understandings about the division of labor among the organizations. Several of my interviewees—echoing those interviewed earlier by Brand, Eder, and Poferl (1997)—noted that the WWF, NABU, and BUND were generally willing to leave the use of spectacular actions to Greenpeace, while they placed more emphasis on lobbying and other activities (see also Lahusen, 1998; Felbinger, 2003) and that there was a general understanding that BUND would take more militant political stands than NABU or the WWF (see also Weinbach, 1998). One WWF leader, for example, stated directly that his organization was willing to let other organizations take the lead on lobbying. BUND's frequent willingness to allow NABU and the WWF to initiate action on nature protection issues also fits this pattern. Indeed, during the late 1970s, the BUND and NABU chapters in Lower Saxony even attempted to formalize just this understanding (Wolf, 1996).

Cooperation vs. Competition in European Affairs

In recent years, the European Union's rapidly growing influence on German environmental policy and its implementation has created a new version of the cooperation versus competition dilemma. The German environmental organizations must now take into account not only their relations with German peer organizations as they seek to influence events in Germany, but also possible collaboration with environmental organizations in other European countries in efforts to influence the EU.

The advantages and disadvantages of cooperation at the EU level parallel those at the national level. Working cooperatively through the EU-wide networks of national environmental organizations described in Chapter 8 can reduce duplication of efforts and allow sharing of information. Moreover, the united voice of environmental organizations in various countries resonates more loudly than potentially fragmented or contradictory input from individual national organizations. On the other hand, the European networks require financial support from their members, and participation can also involve time-consuming efforts to reach compromises on the priorities of various issues (Roose, 2002). Moreover, it is especially hard for the organizations to claim credit for accomplishments at the European level, so cooperative work does little to establish their identity (Röscheisen, 2006).

The German organizations have recognized the potential importance of the European Union for some time. BUND created a working group for European affairs in 1989. By the early 1990s, NABU had a volunteer coordinator for international affairs and an executive secretary who was active in European affairs, and Greenpeace had a paid coordinator for European affairs. Nevertheless, the amount of attention and energy the organizations devoted to European affairs was comparatively small (Hey and Brendle, 1994). Recent years have seen continued indications of interest and activity in EU politics. BUND conducted a special fundraising campaign in 1997 to finance efforts to influence the new EU treaty being framed in Amsterdam. NABU reemphasized its EU work in a 2001 reorganization, and it has cooperated with Bird Life International on petition drives (Naturschutz Heute, 2000l). Nevertheless, the environmental organizations remain focused on their work at the national level. They rely heavily on the DNR for information about EU politics, and there is relatively little effort to influence the EU by lobbying EU agencies, the EU parliament, or the German government directly (Rucht, 1997, 2001; Roose, 2002; Röscheisen, 2006). The German organizations almost never mobilize for protest at the European level. In a study based on newspaper data from 1970 to 1994, only about 5 percent of all protests in Germany focused on other European countries or European affairs, and less than 1 percent focused on the EU per se. Antinuclear protests, counted separately, were almost never directed at the EU (Rucht, 2001).

Insofar as they do work at the European level, the German organizations have rather consistently opted for a cooperative approach. None of them have set up an independent office in Brussels. Instead, all have affiliated with Europe-wide networks of environmental organizations (see Chapters 8 and 9). BUND, NABU, and the WWF are members of the EEB, although relatively few Germans attend its meetings (Röscheisen, 2006), and BUND and NABU are members of the DNR, which attempts to coordinate the EU activities of the German environmental and nature protection organizations. NABU is affiliated with the European office of Bird Life International and BUND with the European Friends of the Earth. Greenpeace and the WWF are also linked to the European offices of their international parent organizations. This relatively high level of cooperation stands in sharp contrast to the organizations' episodic and sometimes reluctant cooperation at the national, regional, and local levels.

A variety of possible reasons for this pattern have been cited.[4] The first is simply that the organizations have been unable or unwilling to make the heavy investments in staffing, travel, office space, information gathering, and expertise required to work effectively within the complex structures of the EU. They affiliate with umbrella groups in Brussels simply because this approach offers them a cheaper way to keep abreast of European developments and exert at least some influence.

Second, the organizations find it difficult to shift their focus to EU work because national orientations and cultures are deeply embedded in the minds of their staff and supporters—as well as in procedures, skills, and capabilities developed over decades of work in Germany. Greenpeace, for example, discovered that protest tactics honed in Germany are not very effective at the EU level because of the absence of a truly European public opinion and the difficulties involved in mobilizing protests against an amorphous entity far from home. Although the European networks and the DNR provide them with ample information about EU politics, the organizations lack the time and expertise to digest it, and they lack experience drafting EU regulations and working with the EU bureaucracy. Finally, pressing issues at the national level often seem more important and make more immediate claims on limited time and financial resources than European issues.

Third, although the Environmental Directorate of the European Commissions has been eager to receive input from European level environmental groups and actually assisted with the establishment of the EEB, it has been much less open to input from individual national organizations. The organizations could, of course, seek to lobby the European Parliament on their own, but non-German members of Parliament would probably not be especially receptive to appeals from German organizations.

Fourth, in sharp contrast to the situation at the national level, cooperative work at the EU level does not deprive the organizations of opportunities to polish their public images. When the organizations work alone within Germany, achieve successes, and publicize them, their image with the German public improves—and with it prospects of obtaining the resources they need. This is generally not the case at the EU level. EU politics are covered less thoroughly than

---

[4]The discussion of the organizations' propensity to work cooperatively through networks at the EU level draws on Hey (1994), Hey and Brendle (1994), Rat von Sachverständigen für Umweltfragen (1996), Rucht (1997, 2001), Long (1998), Roose (2002), Rootes (2004a), and Röscheisen (2006).

German politics in the press, and most Germans are not well informed about them. Consequently, any successes that a German organization did manage to score in Brussels by going it alone would be unlikely to result in increased support and donations at home. There is no well-defined European public to which they can report their successes, and Europeans in other countries are unlikely to donate to German organizations. Under these circumstances, the organizations have little to gain from independent work at the EU level, and the money and time they save by joining EU-wide networks can be invested instead in national-level activities, which have more payoff.

Limited vs. Broad Agenda

The last of the seven dilemmas the environmental organizations face is whether to limit themselves to a single objective or a small number of objectives, or whether to adopt a broader focus and address numerous environmental problems. Once again, there are advantages and disadvantages to each choice.[5]

Most environmental organizations, even the largest, must operate within the constraints of limited budgets and staffing, so limiting the number of issues they take on may be the only way to ensure that they can actually have an impact. Specialized organizations can develop deeper expertise about the issues they emphasize and tactics especially well suited to these issues. One interviewee who knew all four organizations well told me that the WWF's effectiveness in lobbying is enhanced because it focuses its efforts on only a few issues, and a prominent Green Party politician commented that NABU's efforts to cover almost the whole range of nature protection and environmental issues had turned it into a "sundries store," implying that its broad diversity of goals kept it from doing a good job of anything. Environmental organizations with limited objectives may also enjoy a clearer profile with the public. Finally, it can be argued that a set of specialized organizations can complement rather than compete with one another and can therefore attract more supporters in total.

Nevertheless, pursuit of only a limited set of goals also has drawbacks. Many environmental problems are interrelated, and an organization that focuses only on a single problem or a small set of problems may be unable to address them realistically without extensive coordination with other organizations that address related problems. Organizations with only a few goals also have more difficulties attracting the ear of the government or other interest groups, especially in a system that prefers to deal with a single voice from each major interest group.

General-purpose organizations, on the other hand, may find it easier to address key issues that have little appeal to potential supporters or are difficult to explain. As described in Chapter 11, they can "market" their more attractive

[5]For discussions of this issue see Dalton (1994), Heuser (1994), Rat von Sachverständigen für Umweltfragen (1996), Wolf (1996), Pehle (1997), and Brand (in press).

programs to potential supporters and use the support they obtain to support less popular but important ones. By stating their goals broadly and somewhat ambiguously, they can also appeal to a broader range of citizens, which gives them the potential to attract more supporters. On the other hand, pursuing too many topics simultaneously can make the organization's image a fuzzy one, precipitate conflicts among goals, and spread resources among so many issues that the organization is not effective in any area.

This research purposively focuses on environmental organizations with multiple objectives, primarily because they are by far the largest and most influential. Organizations that pursue a single environmental goal or a small number of goals have remained much smaller and less prominent than the organizations included here. This suggests that, at least in terms of members and influence, the pursuit of multiple goals has proven to be the more effective strategy.

Still, the beneficial effects of adding ever more goals are far less clear when one examines only the four largest organizations. The profiles in Chapter 9 show considerable variation in the number of goals the organizations pursue. Greenpeace, in particular, has always tried to limit itself to a relatively small number of themes. Yet, at least in terms of the number of supporters, budget, and visibility, it is by far the most successful of the four. Despite new emphases on topics like climate change and preserving ocean ecosystems, the WWF remains solidly anchored in nature protection. Although ranking last in number of supporters, it now has the second-largest budget. BUND and NABU, on the other hand, address virtually every environmental concern imaginable, although NABU's longtime emphasis on nature protection means that its public image is more specialized. BUND, the organization with the widest range of goals, now ranks last in revenues and has experienced problems with stagnating support.

The relative success of the organizations cannot, of course, be traced to any one factor, but there is some evidence that the embrace of too many goals can be a problem. Both BUND and NABU have experienced internal conflict over the priority of nature protection and environmental protection. BUND's president told me that some supporters join mainly to do political work, while others want to engage in local nature protection projects, resulting in a kind of organizational schizophrenia and disinterest in national projects by some local groups (see also BUNDMagazin, 2000b). A number of interviewees from both inside and outside NABU made essentially the same point, citing the greater priority of nature protection goals at the local level and of environmental issues and political participation at the national level.

It is clear that organizations limited to one or a few goals are the "niche players" (Bosso, 2005) of German environmentalism. While they may be able to secure their continued existence and be effective in some respects, none have amassed a large enough pool of supporters or a large enough budget to wield great national influence or be highly visible in the national media. On the other hand, comparisons among the four large organizations indicate that taking on every possible environmental issue is of dubious benefit.

Summary and Conclusions

By the 1990s, the German environmental organizations found themselves in a social context quite different from the one that had obtained during the 1970s and 1980s. The cycle of social movement mobilization centered around new social movements was rapidly winding down, countercultural ideologies were losing their visibility and appeal, and competition from the BIs had diminished. Although the environment remained a concern for many citizens, it was no longer at the forefront of public consciousness, especially in comparison to economic issues. Environmentalism was now partially institutionalized in the practices of business and government, and by the end of the decade, the Green Party was part of the governing coalition. For environmentalists willing to moderate their rhetoric and strategies, there were many new opportunities for input into the system and for realization of some of their goals. Clinging to confrontational social movement strategies, on the other hand, invited marginalization, ridicule, and decline. The organizations could no longer easily call masses of supporters to the streets, and confrontation could offend potential negotiating partners in business and government or the masses of new supporters they had recruited.

Largely in response to these altered circumstances, the four organizations moved to professionalize their operations, which in turn greatly increased their need for financial resources. To meet the spiraling demand for funds, they turned to professional fundraising directed at a broad base of donors. Many of the new supporters attracted by their increasingly sophisticated fundraising operations, however, identified far less strongly with the environmental cause than the activists of previous decades. The organizations were able to retain many activists from this preceding period of social movement mobilization, but attracting new activists became steadily more difficult.

Taken together, these changes provided strong incentives for a shift away from confrontation and toward work within the system. That is, the changes nudged the organizations away from functioning as social movement organizations for a confrontational social movement and provided incentives to function as more conventional interest groups. As interest groups working within an established system with pronounced neocorporatist tendencies, the organizations had to be willing to compromise in order to achieve a portion of their goals and gain a modicum of legitimacy. Moreover, they had to convince their supporters that this was the best course of action. By touting their accomplishments and their standing as accepted interest groups that could deliver results, the organizations were able to strengthen their standing with their supporters, especially the relatively passive donors who increasingly populated their ranks. They now centered their activities on carefully organized and planned "campaigns" (Osti, 2007). These were coordinated efforts involving public education, media campaigns, lobbying, research reports, and occasionally even demonstrations. Reliance on these new strategies has in turn reinforced trends toward professionalization, as professionals are needed to execute them well (see Chapter 10).

Cooperative participation in a system based on negotiation and compromise has also called for new framings of environmental problems. The reforms that the organizations propound today are significant, and their proposals are frequently unpalatable for business and government. Still, present-day discourses focused on sustainable development and ecological modernization are far removed from the radical ideologies of "deep ecology" and system transformation through grassroots political mobilization that were prominent a quarter-century ago. As a result of these changes in strategies, goals, and ideologies, the mainstream environmental organizations at the beginning of the new century could be described as social movement organizations only in the very broadest and least restrictive senses of the term. They functioned instead primarily as well-organized public interest groups with a significant support base, but rather limited influence and access.

Yet the pendulum did not swing all the way to the other extreme. Goals statements with radical undertones and protests remain a part of the organizations' action repertoires, even if they are deployed less often. Confrontation did not completely disappear because it retains very real benefits. First, it can be motivational for committed activists, especially those who cut their teeth during more confrontational times. Second, the organizations remain only partially incorporated into the inner circles of decision making. This encourages them to use the full range of interest group tactics, including occasional demonstrations and strong rhetoric. Skillfully used, confrontation can still dramatize issues, attract media and public attention, and remind opponents that the environmental organizations can and will fight for their core concerns.

Confrontation retains a place in action repertoires for a third reason as well. For Greenpeace, BUND, and NABU, at least, it has become an institutionalized component of organizational cultures and identities. For two decades, confrontations with the German establishment served to sharpen and symbolize organizational and movement identities and values, and many core supporters, especially middle-aged activists, have continued to expect their organizations to be willing to fight for their goals. These deeply rooted habits of thought are not easily expunged merely because it has become convenient to do so. Confrontational strategies have also become institutionalized in organizational knowledge, practice, and procedures, and organizations with sunk costs in pursuing specific goals using specific strategies cannot turn on a dime. Greenpeace, born and grown to maturity during the period of confrontation, faces special problems in this regard, which may explain its continued reliance on spectacular actions and confrontational rhetoric. BUND shares in this tradition of confrontation to a significant extent; however, like NABU and the WWF, its roots extend back to earlier periods, and it never fully abandoned the goals and strategies of a less confrontational era.

The newfound emphasis on cooperation has, in fact, paid off for the organizations in significant ways. It has legitimated them with business and government, improved their access to information, increased their credibility with the public, and even allowed them to obtain limited financial support from government and business. Using their increased influence, they have also been able to achieve some real gains for the environment. These gains have increased their credibility and allowed them to demonstrate their effectiveness to their supporters, enhancing their fundraising efforts.

Yet cooperation has disadvantages as well. Even professionalized environmental organizations often lack the expertise and resources to match government and business in negotiations and lobbying, and the time and resources required to participate may require them to sacrifice other important efforts. Negotiating partners from government and the economy may be more interested in polishing their images or coopting environmental groups than in real change, and incremental progress and long, technical negotiations make poor copy for press releases and fundraising letters. At times, the compromises required as the price of a seat at the table may be so great that the organizations can accomplish few of their goals and their supporters become discouraged. Disenchanted activists can become critics, accusing the organizations of selling out. Worse still, they may abandon the organizations in disgust or transfer their allegiance to more confrontational groups.

The organizations have thus found themselves walking a tightrope between too much confrontation and too little. The dilemma became particularly acute during the period when the Greens were part of the governing coalition, for efforts to mix cooperation with occasional strong criticism resulted in conflictual relations with the Greens and a difficult situation for all. But whether the Greens are in the government or not, the organizations must remain confrontational enough to keep the pressure on their negotiating partners and hold on to the donations and volunteer labor their most activist supporters provide. At the same time, however, they must display enough willingness to compromise to be acceptable to negotiating partners from other interest groups and to gain the accomplishments, prestige, and legitimacy they need to attract funding from these sources and their less confrontational supporters.

The four organizations' solutions to this dilemma are not identical, for they are conditioned by their unique traditions, alliances, and the opinions of their supporters. Greenpeace embraces confrontation as part of its heritage but emphasizes its willingness to work with sincere partners. The WWF stresses the merits of cooperation, but deploys pointed rhetoric or demonstrations on occasion. Yet, in practice, all seek to find a balance, and all have been relatively successful at finding compromises that let them attract support and continue to pursue their objectives.

The organizations also find themselves walking the tightrope as they attempt to regulate their relationships with one another. Examined as open systems, the organizations have similar goals, require the same kinds of resources—money, legitimacy, and influence—and must acquire these resources from a limited set of suppliers. They also turn out many of the same "products," including lobbying, public education, practical nature protection efforts, and acquisition and maintenance of nature preserves. Their activities are observed and evaluated by the same reporters, political leaders, and business leaders. These evaluations, which are inevitably comparative, influence the organizations' chances of survival and success.

Consequently, while there is much to be said for close cooperation—avoiding duplication, speaking with a united voice, and the opportunity to institute a division of labor—stubborn realities, centered around issues of organizational visibility and identity, limit how much cooperation is actually achieved. The organizations' dependence on donations from the general public and their desire for a large base of supporters make maintaining a clear public profile a matter of vital importance for each. However desirable cooperation with other organizations might be on other grounds, joint projects with the other organizations or the DNR inevitably threaten to submerge the identities of the individual organizations involved. Although threats to visibility are the main obstacle, difference in specific goals, organizational cultures, and styles of operation, as well as members' identification with "their" organization, also stand in the way of close cooperation. The results are what one would expect. Although examples of cooperation are easy to find, they occur in limited numbers and are entered into with considerable caution.

The picture with regard to the cooperation versus competition dilemma is quite different in EU governance, where many consequential decisions are made and a new structure of Europe-wide lobby groups is emerging. There, cooperation is the rule, not the exception. The high costs of operating independently and the unresponsiveness of the EU bureaucracy to national-level organizations help to explain this outcome, but there is another noteworthy factor. European-level work contributes far less to maintaining a distinct organizational profile than the organizations' much more visible work at the national level. Influencing the distant and often almost impenetrable world of EU politics is important, but it does not make good copy for fundraising letters or press releases. In this case, cooperation becomes an asset, not a liability.

The last of the seven dilemmas concerns whether the organizations should focus on many different environmental problems or only a few. All four of the organizations included here have concluded that the net benefits of the former exceed those of the latter and have adopted relatively broad missions. This has obviously served them well in terms of attracting members and donations, since they are larger and more successful than any of Germany's single-purpose environmental organizations. On the other hand, the four largest organizations do vary in the breadth of their goals, and BUND, the organization with the widest array of goals, has been the least successful in recent years, while Greenpeace with its somewhat more limited mission attracts the greatest support. It is possible that more could be accomplished for the environment by a set of organizations with relatively narrow goals, a clear division of labor, and close coordination and cooperation, but the obstacles to cooperation make it unlikely that such a system will evolve in the foreseeable future, for niche players rarely achieve as much visibility as organizations with goals that span the full range.

The organizations' resolutions of the three dilemmas of goals and strategy thus clearly influence one another. Moreover, how they deal with these dilemmas influences and is influenced by how they resolve the dilemmas of organizational structure and resource acquisition discussed in Chapters 10 and 11. The final chapter takes up these questions in more detail in the context of a broader examination of the organizations' histories and the application of organization theory to analyzing their strategic choices.

Chapter 13

CONCLUDING OBSERVATIONS

This final chapter undertakes four tasks. First, I look back at the history of German environmental organizations to identify patterns and formulate generalizations useful for understanding the German experience and, potentially, environmental organizations in other Western nations as well. Second, I develop a model, grounded in organization theory, which identifies the dilemmas the four most important environmental organizations faced at the beginning of the twenty-first century and helps to explain their strategies for coping with them. Third, I examine the implications of this research for improving theory about environmental organizations. Finally, I reflect briefly on the possible futures of German organizations over the next few decades.

Lessons from History

A historical study in Germany is especially revealing because twentieth-century Germany has been characterized by a wider array of social and political contexts than any other Western nation. Rapid industrialization and urbanization, depression and runaway inflation, world wars, Nazi dictatorship, state socialism, postwar devastation and reconstruction, the "economic miracle," major confrontations over nuclear power and other environmental issues, noteworthy successes in institutionalization of environmentalism, and reunification and economic stagnation have provided a complex and changing tapestry of external circumstances and challenges. I attempt here not to retell the entire story but to develop comparisons and generalizations that can help to make sense of the historical material.

The Persistence and Mutability of Environmental Concern

Perhaps the most striking feature of environmentalism in Germany is its sheer staying power. Concerns about environmental problems and efforts to combat them have survived all of the diverse and rapidly changing conditions mentioned above. While at times significantly weakened by the challenges, environmentalism has never disappeared from the scene. Clearly the concerns that lie at its root—nature protection, human health and well-being, resource conservation, and preservation of wildlife and natural beauty—have the capability to attract and hold attention in both good times and bad.

Environmental themes have not merely survived; they have displayed impressive mutability. Early in the century, anti-modernist crusaders assimilated environmental themes into their struggles for homeland protection, linking nature protection to preservation of traditional architecture and ways of life. During the same era, romantic wanderers extolled nature's restorative effects on bodies and souls degraded by urban life. Working-class intellectuals and hikers linked hiking, nature protection, and nature study to their critique of capitalism and to their effort to build a working-class culture. Nazi ideologists elaborated existing links among anti-urbanism, protection of rural life, and nature protection into an elaborate ideology of "blood and soil," which traced the strength of the German *Volk* to its roots in its native forest and soil.

The end of World War II left the environmental cause ideologically adrift; however, by the beginning of the 1970s, new framings that united concerns about population growth, resource exhaustion, runaway technology, consumerism, centralization of power, and nature protection were being propounded by both right- and left-wing ideologists. The result of their efforts was a view that united these formerly separate streams of thought under the rubric of environmental protection, a framing that continues to dominate the scene today. They also forged a strong but less lasting alliance between environmentalism and opposition to nuclear power.

In the GDR, the government succeeded in forcing the assimilation of nature protection into the dominant ideology of state socialism. However, this accomplishment proved short-lived. Only three decades later, concerns about urban pollution and the destruction of nature emerged with renewed vigor—this time in alliance with the calls for free speech and economic reforms that toppled the Honecker government.

Right-Wing Ecology

Clearly then, the concerns that we today understand as environmental can be melded with a variety of other issues and ideologies. Furthermore, they are far from the exclusive property of the left. No society illustrates this better than Germany, where environmentalism has repeatedly been linked to powerful ideologies and movements of the right. Long before the Nazis, the socially reactionary homeland protection movement had included nature protection among its key goals. Nazi ideology linked nature protection to protecting the archetypical German landscapes from which the German peasantry allegedly drew its strength, and the National Socialist government passed progressive nature protection legislation and supported the efforts of nature and homeland protection organizations with legitimacy and state subsidies.

The World War II defeat dealt right-wing environmentalism a blow from which it has never fully recovered, but it did not disappear. Some right-leaning leaders remained prominent in the postwar environmental organizations, and, although BUND and the Greens tilt decidedly to the left today, both had prominent right wings during their early years. Right-wing environmental parties, think tanks, journals, and organizations remain active at the margins of German politics, and the long history of right-wing ecology makes it unwise to conclude that they will never again rise to prominence. In short, the German case clearly demonstrates that overgeneralizations from current political alignments in Germany and the US, which identify environmentalism almost exclusively with "progressive" social movements, are misleading.

Environmentalism, Environmental Organizations, and Social Movements

German environmentalism has obviously been frequently associated with powerful social movements, but the historical record makes equally clear that it is quite misleading to equate environmentalism with the "environmental movement" or to view all environmental organizations as SMOs. Environmental organizations have continued to operate in Germany during periods when environmental movement mobilization was minimal or nonexistent. They continued to function, for example, under state supervision during Nazi rule and in the German Democratic Republic, even though only during the last years of the GDR—and at no time during the Nazi period—would it be reasonable to speak of environmentalism as a social movement. Environmental organizations also continued their work in West Germany during the politically quiescent years after World War II. During this period, environmental organizations are best understood as weak interest groups operating at the margins of the semicorporatist political system, or as civil society organizations dedicated to preserving nature and open space. Speaking of an environmental "movement" during these years stretches most definitions of social movements well beyond their limits. Environmentalism at the beginning of the twenty-first century may be entering another period of quiescence, when environmental organizations again function as interest groups and organizations of civil society.

The rise and fall of environmental movement mobilization over German history has been more closely related to changes in the political opportunity structure, the rise of new framings, and the activities of successful movement entrepreneurs than to the extent of environmental degradation per se. National Socialism and state socialism, for example, created political opportunity structures that made movement mobilization virtually impossible—at least until the final years of the GDR. On the other hand, new framings of environmental issues, such as homeland protection and the countercultural ideologies of the 1970s, stimulated movement activity, while successful movement entrepreneurs like the BH's Rudorff and the BBU's Wüstenhagen played key roles in launching new movements and new SMOs.

A particularly interesting aspect of German history is the propensity of German environmentalism to ally itself with social movements that had broader agendas than environmentalism per se. Before World War II, environmentalism was intertwined with anti-modernist homeland protection and with the workers' movement. During the wave of new social movement mobilization during the 1970s and 1980s, it became so closely linked to the anti–nuclear power movement that the two could hardly be separated.

Whether allied with social movements or not, large national organizations have been an important structural feature of German environmentalism throughout the twentieth century, and the rise and fall of social movements can only partly account for their growth or decline. Some of them, the BH, the Friends of Nature, BUND, the BBU, and Greenpeace, were originally established as social movement organizations. Others, though, have purposefully avoided functioning as SMOs for the movements of their eras. Examples include the BfV and the BN before World War II, the WWF during the 1970s and 1980s, and the GNU during the final years of the GDR. The objectives of these organizations would be much better characterized by theories of civil society or interest groups. They typically pursued programs that combined public education, acquisition of nature reserves, and working cooperatively with the authorities, not participation in mass protest or the embrace of radical ideologies.

Interestingly, almost all the organizations that avoided social movement participation had nature protection as their focus. The broad appeal of nature protection across class lines, including its appeal to elites, may help to explain this (Markham and van Koppen, 2007). That is, these organizations may have occupied a niche in the organizational landscape that allowed them to draw support from nature lovers who were not committed to the movements of the day.

On the other hand, the decision to eschew participation in powerful social movements that included environmental goals among their objectives was often a painful one. Participation in a rising movement can help environmental organizations forge new alliances and attract new supporters and funds, potentially increasing their influence and effectiveness. Failure to align themselves with rising movements can isolate them and make it harder to acquire needed resources—the fate of organizations such as the German Forest Protection Association and the Isar Valley Association during the 1970s and 1980s, or the GNU during the GDR's rising environmental movement in the late 1980s.

Active movement participation, on the other hand, can lead to problems of its own, especially when environmental organizations are confronted with movements that combine environmental and nonenvironmental themes. Joining with such movements can lead to goal displacement for the environmental organizations, since they are rarely strong enough to control the movement. Forging alliances with highly mobilized movements, which frequently have radical or confrontational agendas, can also threaten environmental organizations with the loss of their legitimacy, moderate supporters, and influence with government and business. Finally, there is always the risk that the movement will fail or go into decline, leaving the organization high and dry.

All of these dilemmas faced the environmental organizations during the rise of National Socialism, and their concerns about them probably account for most groups' decision to hold the Nazis at arm's length until 1933. The same problems appeared in West Germany during the 1970s and 1980s, when the rather staid nature protection organizations had to decide whether to join a burgeoning social movement with objectives that went far beyond their old agendas. The BfV and the BN, which took the plunge and transformed themselves into SMOs, encountered considerable internal resistance from members who opposed this decision, and the BN forfeited its privileged access to government. Both organizations benefited in the long run, however, via growth in membership and resources. The WWF, on the other hand, elected to steer clear of the movement. It was spared the internal conflict that affected the BfV and BN, but it also benefited less from the movement.

For environmental organizations with long lifespans, affiliation with a social movement often turns out to be transitory event, as social movements often fade with time. Indeed, some organizations have functioned, in different eras, as SMOs for more than one movement. However, the history recounted in this book makes clear that environmental organizations that have functioned as SMOs do not necessarily disappear or decline into mere institutionalized remnants once the movement fades. The League for Homeland Protection and Friends of Nature survived the early twentieth century social movements that spawned them, and BUND and NABU give every indication of outliving the recent wave of highly mobilized social movement activity. Social movement organizations with less financial support and less formalization, such as the BBU, Robin Wood, and the Green League, have had less success in surviving the movements with which they were affiliated.

Success and Failure in Adapting to Change

Adapting to changing conditions involves more than merely coping with the rise and fall of social movements, and the history recounted in this book provides numerous instances of successful and unsuccessful adaptations to change—suggesting that both the population ecology and resource dependence models contain elements of insight.

Failures to adapt successfully to changing circumstances would no doubt figure even more prominently in the historical narrative if more information were available about smaller, shorter-lived organizations. However, the material presented in this book provides many good examples of unsuccessful strategies. These include the Friends of Nature's inability to adapt to the rise of National Socialism, the problems many nature protection organizations experienced in dealing with the West German environmental movement of the 1970s and 1980s, the GNU's failure to survive the GDR, and the difficulties that the BBU, Robin Wood, and Green League experienced coping with the institutionalization of environmentalism and the decline of the environmental movement at the end of the twentieth century. As population ecology models predict, these organizations, whose goals, strategies, and structures proved ill-suited to changing conditions, suffered declining influence and stagnating support. The Friends of Nature and the GNU failed to survive at all (although the former was ultimately resurrected), and this may well be the fate of the BBU.

Information about the organizations' internal deliberations with regard to the challenges they faced is often unavailable or incomplete, but there is considerable support for the hypotheses from population ecology that sunk costs and resistance to altering deeply institutionalized identities and practices played a key role in failures to adapt. The socialist leanings of the Friends of Nature, for example, were too deeply embedded in its ideology and network of relationships to be disguised by last-minute course adjustments after 1933. And after the war, most of the organizations, weakened and ideologically adrift, were unable to develop a positive adaptation to changed circumstances. Instead, they reverted to deeply institutionalized prewar patterns in an effort to stabilize their situation. In vastly changed circumstances, however, these familiar patterns proved adequate to allow only their bare survival. Still later, as the GDR collapsed around them, leaders of the GNU found that they had little experience or knowledge that would be relevant to working in a radically changed world. Moreover, some leaders and supporters actively resisted major changes in goals and strategies. And in a related example, since reunification none of the organizations have adapted well to the special challenges of winning support and influence in eastern Germany. Indeed, some of the very strategies and structures that had made them successful in the Federal Republic proved to be impediments in the East.

The historical record also contains examples of another pattern of change suggested by population ecology: declining organizations being shouldered aside by new ones better suited to changed environments. The best examples of this situation are Greenpeace and BUND. As new organizations with goals and strategies well adapted to new conditions, both grew very rapidly during the period of confrontation, partly by attracting supporters from older organizations less well adapted to the new social context. Recent years have brought another example of this pattern: the emergence of the WWF as a major force in German environmentalism. The WWF remained a relatively marginal player until polarization began to recede during the 1980s and 1990s, but it then began to grow more rapidly than the other organizations.

There are, however, also cases of successful adaptation to changing circumstances of the sort predicted by resource dependence and contingency theories. In these cases, organizations successfully modified their goals, internal structures, and strategies or made successful efforts to influence their environments. For example, all of the organizations except the Friends of Nature accommodated relatively quickly to the changes required by the Nazis, and several launched partially successful efforts to preserve a degree of independence by playing competing parts of the Nazi bureaucracy off against one another. Their receipt of substantial financial support from government and their government-legitimated monopoly over their spheres of action no doubt eased the transition for them. During the 1950s, environmental groups in the GDR also accommodated themselves to changes required by the new government and profited from its support; however, resistance was greater in this case, perhaps because the changes required were more radical. In both cases, it appears that some local chapters struggled to retain their old goals and strategies, paying only lip service to the requirements that they had to meet in order to secure legitimacy and funds.

Other cases of successful adaptation involved the BN and the BfV during the period of confrontation. In the 1970s, the BN transformed itself from a conservative organization with close ties to the Bavarian government into an activist organization with a broadened agenda much more suited to the emerging environmental movement. And during the 1980s, NABU changed itself, albeit painfully and with considerable internal conflict, from a rather staid nature protection association with a strong local orientation to a more activist organization fighting the pollution problems of an urban society and lobbying for political change. Recent years have provided additional examples of successful adaptation to changing circumstances, which I discuss in detail in the following section.

In some historical instances, the survival strategies of organizations with multiple goals involved deemphasizing their environmental goals. Both the BH and the Friends of Nature chose to reduce the prominence of environmental objectives in the postwar years, when environmental issues were not high on the national agenda. As environmental issues became more prominent, both organizations attempted to reemphasize the environment, but without obtaining much benefit, probably because other organizations had occupied their former niches. Organizations with exclusively environmental goals, on the other hand, have little to gain from abandoning their "bread and butter" issue and have not done so.

While the number of cases is rather small, the analysis above suggests that successful adaptations are most likely when sufficient slack resources (Cyert and March, 1963) are available to sustain an organization through a period of major change. Determining the extent to which the successful adaptations arise out of strategic planning versus ad hoc experimentation requires detailed information about the internal deliberations of the organizations that is often unavailable; however, the BN and BfV examples above imply that skillful leaders who lead organizations through periods of change and adaptation may be just as important as the interest group or social movement entrepreneurs who established them in the first place.

The Importance of History: Environmental Organizations as Institutions

Although successful adaptations to changing circumstances do occur, the historical analysis makes clear that environmental organizations are not generally free to instantly adopt whatever new objectives, strategies, and framings of environmental problems might be most efficient for acquiring resources and accomplishing their goals. In part this occurs because of sunk costs invested in existing skills and equipment. But there is more. Organizations have histories, traditions, and established roles in the constellation of other organizations; consequently, radical transformations of their goals, structures, and strategies frequently encounter stiff resistance. Government agencies and competitors—inconvenienced or threatened by change—may criticize them for stepping outside of their established roles. And more importantly, employees, volunteers, and donors may see radical change as a step away from goals and activities that have great symbolic and emotional significance to them.

Examples of such resistance abound. Early in its history, the BH was split between traditionalists, who had close ties to the organization's founding father and wanted to maintain a strong anti-progress stance, and compromisers, who believed that, on practical grounds, more could be accomplished by working for improving the design of planned facilities than by opposing them outright. Later, in the postwar period, most of the national organizations, seeking a way to sustain themselves, immediately reverted to historical goals and patterns of action in hopes that tried and true patterns would still resonate with supporters. Still later, resistance to changes at the BN, BfV, and DNR during the period of confrontation was frequently anchored in claims that new proposals flew in the face of traditional organizational roles and would undermine support from long-term supporters, who valued their traditional nature protection emphasis and non-confrontational approach. Even today, BUND and NABU continue to purchase land for nature reserves, even though it is not cost-efficient to do so, because their supporters expect this of them.

Recurring Issues

The century-long history of the German environmental organizations reveals not only transient issues that have come and gone but also a set of problems that have posed recurrent problems. Most are rooted in theoretical issues discussed in Chapter 2.

Acquiring Funding

The first of these issues, difficulties in obtaining adequate funding, is rooted in the exigencies of working for the environment through formalized, national-level organizations. Even when most of the work is conducted by volunteers, working at the national level requires more money than does purely local organizing, and working effectively often requires still more. Government, wealthy donors, and business have, in general, declined to support national-level environmental organizations very generously. Their disinterest in supporting more confrontational social movement organizations is easily understandable, as such organizations often threaten established economic and political interests; however, they have been only slightly more willing to open their pocketbooks to organizations that work for nature protection and commit themselves to working within the system. The reasons for this pattern have not been explored very thoroughly in past research, but Chapter 11 provides a first approach to the question by looking at the likely reasons at the turn of the twenty-first century.

Various solutions to the funding problem have been tried over the last century, none of them fully satisfactory. Through most of their histories, the BH, BfV, and BN relied on modest state subsidies and dues from members, who were mainly recruited from social networks within the upper middle class. Recruiting within these networks helped to overcome the free rider problem by offering social and prestige rewards for membership. Still, the resulting revenue stream was small, both because of the limited size of this support base and because of heavy reliance on overworked volunteers to recruit new members. The BfV was able to recruit more members because major donations from the wealthy family of its president allowed it to charge only nominal dues. This approach overcame the free rider problem by making membership virtually free, but it enrolled many inactive members and produced no significant gain in revenues.

Reliance on these revenue sources was compatible with the organizations' strategy of working within the system and emphasizing relatively noncontroversial nature protection projects. It worked against taking strong stands on behalf of the environment—especially during the National Socialist era, when more generous subsidies were accompanied by tight state control. This same pattern later characterized the GDR's state-sponsored GNU.

Federated organizational structures apparently exacerbate the problems national headquarters offices experience with obtaining adequate funding. The BH's financial difficulties, for example, were heightened early in the twentieth century when it adopted a federated structure and became financially dependent on its local chapters, which jealously guarded their limited funds against claims by the central headquarters. This same problem was to plague BBU seven decades later, and it has also been a continuing problem for the DNR.

During its first quarter-century, the WWF pioneered a different approach to fundraising, focusing its fundraising efforts on a select circle of wealthy donors. Although it proved possible to raise a respectable amount with this approach, the method did not produce a flood of revenue, and it tied the organization closely to the preferences of this constituency. More recently, the WWF has joined the other large organizations in implementing another innovation, mass fundraising campaigns designed and operated by professionals and aimed at the general public. In comparison to previous approaches, these campaigns, described in Chapter 11, have been very successful, but they have also produced many new problems.

Getting the Work Done: Reliance on Volunteers

For most of their histories, the national environmental organizations relied on volunteers to accomplish almost all of their work. Officers and board members served without compensation, and underpaid (or even unpaid) executive secretaries, assisted by tiny clerical staffs, struggled to keep the books and maintain the flow of communications to members. The efforts of a core of dedicated volunteer activists made it possible for these organizations to accomplish more than one might expect, and their task was made easier by the fact that neither government nor business associations yet possessed large staffs of professionals. Nevertheless, the time and energy of volunteers, no matter how committed they may be, is limited, and the available evidence suggests that environmental organizations that were completely dependent on volunteer labor—romanticized visions of the past to the contrary—did not usually function especially reliably or efficiently. The problems with keeping the organizations operating without a large paid staff were serious before World War II, but things became still worse after the war. Other priorities competed for their volunteers' time, and government officials who had once constituted a large proportion of volunteer activists did not return in large numbers.

Cultural and economic changes and social movement mobilization in West Germany beginning in the late 1960s enlarged the pool of people willing to do volunteer work for the environment. Although many of them preferred to work within BIs, the national environmental organizations also benefited from their energy. More recently, however, developments described in Chapter 7 have again made attracting volunteers difficult.

The roots of the organizations' problems with attracting volunteers lie chiefly in the free rider problem. Environmental organizations ask their volunteers to take time and energy away from leisure time and other role obligations to work for outcomes that, if achieved, will be equally available to nonvolunteers. Moreover, potential volunteers may suspect that their efforts are unlikely to measurably increase the chance of attaining these outcomes.

Over the course of the century environmental organizations have explored virtually every available avenue in their efforts to attract volunteers. Many volunteer activists have been members of close-knit social networks centered around environmental work—whether comprised of reactionary advocates of homeland protection, new social movement activists, or simply nature lovers—and many new volunteers have always come from these networks. Throughout their histories, environmental organizations have tried to reinforce these social rewards by offering volunteers opportunities to work together on projects and socialize at outings, lectures, and meetings.

Like volunteer-based organizations everywhere, environmental organizations also try to elevate the prestige of their most active volunteers via fulsome praise in their publications and ceremonies. Especially before World War II, activists were typically drawn from relatively privileged social strata, and the volunteers' work for the organizations probably enhanced their status in their communities. Indeed, one of the reasons the environmental organizations faced so many difficulties recruiting volunteers after the war may have been the diminution of these rewards due to the organizations' association with the Nazis and their seeming irrelevance in an era dominated by other concerns. In every period, leadership roles and challenging assignments have usually been available to hard-working activists—providing another motivation for them to ignore the free rider problem. However, as pointed out in the next section, the recent professionalization of the organizations has reduced the availability of these rewards. The few existing studies of the importance of social rewards, prestige, and chances to exert influence as motivators for activists indicate that these rewards are of limited importance, but more research is needed. There are essentially no data for earlier periods.

Until quite recently environmental organizations generally offered their supporters few material incentives, often just a newsletter and the opportunity to participate in occasional outings. In recent years, these offerings have expanded to include discounts on various goods and services and advice about ecologically sound lifestyles. Benefits like these are equally available to activists and the most casual of supporters, but there are also sometimes special benefits for top volunteers and donors. Some supporters no doubt value these benefits, but, as public interest organizations, the organizations cannot offer their members the kinds of selective financial incentives available to supporters of trade or professional associations.

The rhetoric of environmental organizations of every type in every period has been aimed to attract volunteers. References to the "collective bads" that will ensue if no action is taken, claims that the organizations contribute meaningfully to solving problems, and pleas to "do your part" appear repeatedly in their publications and in their leaders' speeches. In periods when social movements are highly mobilized, such as the homeland protection movement and socialist movements before the war and the confrontations of the1970s and 1980s, framings typically become more apocalyptic and rhetoric more florid, but efforts to convince supporters that the situation is grave are ever-present.

Centralization of Power

Conflicts over centralization of power have also been a recurring issue for environmental organizations. Not enough information about their internal politics is available to determine whether recurrent efforts to centralize power stem from their leaders' desire for self-aggrandizement, as predicted by Michels, from a more Weberian emphasis on efficiency, or from other causes. What is clear, however, is that such efforts have been made repeatedly—and that they have been almost as often resisted by local groups and activists who preferred grassroots activism and decentralized structures.

The BH's national office and the Friends of Nature's international headquarters, for example, encountered strong resistance to their efforts to centralize control during their early years. After 1933, centralized structures were imposed on the organizations from above, but the organizations resisted through argumentation and passive resistance. Similar resistance occurred during the early years of the GDR, and the determined avoidance of centralized power by the GDR's oppositional environmental movement and the Green League illustrates the hostility that centralized control had evoked. Resistance to increasing centralization and professionalization over the last quarter-century has also provoked a series of conflicts at NABU and BUND. Much more serious conflict erupted repeatedly at Greenpeace, the most centralized of the organizations that uses volunteers. An early outcome of these disputes was the founding of a splinter organization, the purposefully decentralized Robin Wood.

Efforts to coordinate the work of environmental organizations via strong umbrella organizations have also stirred resistance, even though umbrella organizations can enhance coordination and provide a more effective voice in corporatist decision-making and media relations. For over half a century, the DNR's member organizations have remained unwilling to cede it much influence, and resistance to centralization of influence from within the BBU evoked passionate, long-running conflict.

Cooperation vs. Confrontation

Disputes over whether to cooperate with government, business, unions, and other interest groups or whether to confront established power from the outside have been a recurrent pattern in German environmental organizations. Those that choose the former course are best described by theories of interest groups and civil society, while the latter approach means functioning as SMOs. Each strategy has important advantages and disadvantages, described in detail in Chapter 12. Consequently, a given strategy's emergence as dominant for particular organizations in specific periods was often conditioned both by external factors, such as the nature of the political opportunity structure, and internal factors, including the strength of commitments to existing strategies and the political acumen and internal influence of the factions advocating each approach.

Consequently, environmental organizations have varied greatly in their reliance on confrontational strategies. In the early twentieth century, the BfV and BN sometimes used relatively pointed rhetoric to mobilize support, but neither developed a social movement ideology or allied itself with other social movements. Instead, they focused on practical nature protection, public education, and lobbying within the system. With its strong connections to reactionary thought and movements, the BH could have functioned as a confrontational SMO; however, it always included a faction more inclined toward compromise, and its early experiences with confronting the German power structure contributed to the ascendancy of this faction. The Friends of Nature, by contrast, continued to roundly condemn capitalism's destruction of nature and the inadequacies of government policy and eventually paid the ultimate price for its adherence to principle.

During the Nazi era and through most of the history of the GDR, confronting the system from the outside was hardly a viable option; however, during the final years of the GDR, a regime weakened by economic stagnation and loss of support from its citizens and the USSR became more vulnerable, and a growing wave of environmental action and protest ensued. The government-sponsored GNU thus found itself split between its city ecology groups, which wanted to participate in the movement, and a conservative leadership that feared loss of resources, legitimacy, and financing.

In postwar West Germany, two factors made confrontation an unattractive strategy for the environmental organizations. First, the nation accorded overwhelming priority to reconstruction, economic growth, and political stability. Second, environmentalists were unable to devise a new framing of environmental issues that could attract mass support or to link them to themes, such as opposition to nuclear weapons, around which mobilization was in fact occurring. Not until a quarter-century later did the rise of new framings and a new wave of movement mobilization against an unyielding state make confrontational strategies a viable option. Even then, however, BUND and NABU's decision to adopt a more confrontational posture precipitated strong internal conflict. Recent years have seen a decline in the incentives for confrontation, and most of the organizations have moved away from it. I describe these developments in the next major section.

Specialized vs. Limited Goals

Environmental organizations have also had to decide how broad a palette of environmental goals they wish to pursue. Pursuing diverse goals has many advantages, described in detail in Chapter 12, but so too does emphasizing only a few.

The history of the large, national organizations shows a fairly clear trend away from pursuing specialized environmental goals toward a wider range. During the early decades of the twentieth century, the BfV, BH, and Friends of Nature all limited their goals to selected aspects of nature protection—although the BH and Friends of Nature also pursued goals unrelated to environmental protection. The BN was a general nature protection organization, but it was regionally specialized. National-level organizations dedicated to fighting pollution or conserving resources did not emerge until much later. In postwar West Germany, a few new organizations with specialized goals, such as the German Forest Protection Association, were founded. However, the founding of the DNR, an umbrella organization with very broad nature protection goals, and of WWF-Germany, a national-level organization with a steadily expanding range of nature protection goals, signaled some shift toward broadening the palette of concerns.

It was the rise of the environmental movement, however, that ushered in national organizations with a far broader range of concerns. The BBU emphasized combating the problems of an urban environmental society, but some of its member BIs also pursued nature protection. The BN broadened its objectives greatly during the 1970s, and BUND and Greenpeace worked from the beginning on both nature protection and environmental protection. They were soon joined by NABU, which widened its concerns first to protecting nature in general and then to both nature and environmental protection. Although they continued to differ somewhat in the range of their concerns, all four of the largest national environmental organizations thus entered the twenty-first century with a rather wide range of goals.

The reasons for this trend are anything but self-evident, and the topic has been little explored in the literature. It is possible, of course, to argue that less has changed in Germany than first meets the eye, for the BH and Friends of Nature already had broad agendas in the early part of the century; these agendas simply linked nature protection to issues we do not label as environmental. This leaves unanswered, however, the questions of why advocates of environmental protection were so slow to organize large national organizations and why a decisive shift toward organizations with broadened goals occurred in almost every Western society with the rise of the environmental movement (van Koppen and Markham, 2007).

Dilemmas Facing Environmental Organizations at the Beginning of a New Century

One of the major goals of this book is to demonstrate the utility of organization theory for understanding environmental organizations. In principle, it would be possible to do this by using it to analyze in detail the goals, strategies, and structures of each of the large national organizations in each time period, but this would require far more space than is available, and the analysis would sometimes be hindered by the absence of key information. In this section, I undertake instead to illustrate the utility of organization theory through a comprehensive analysis of the four largest organizations as they entered the twenty-first century. The analysis, drawn from Chapters 7 to 12, is organized around the open systems model described in Chapter 2, but it incorporates other contributions from organization theory as well and adds to them some insights from social movement and interest group theories.

According to the model presented in Figure 2.1, environmental organizations must obtain money, legitimacy, prestige, and influence from their environments in order to persuade volunteers, paid staff, and paid suppliers to provide the labor, services, and information they need in order to continue their work. They obtain money, legitimacy, prestige, and influence from individuals and other organizations in their environment that observe and evaluate what they do. Among the most important of these are politicians, political parties, government agencies, the media, business, labor, the general public, and their members and supporters. Were all the evaluations positive, the organizations would presumably find it easy obtain the resources needed to continue their work, but in point of fact, these actors frequently disagree about what environmental organizations should be doing and how they should be doing it. Moreover, ongoing social change means that these assessments are subject to change.

External Conditions

Consequently, analysis of the dilemmas the environmental organizations face must begin with the context in which they operate. Chapter 7 identifies a number of key changes that have influenced the availability of various resources and the likelihood that various strategies will pay off in terms of money, legitimacy, prestige, and influence from various sectors of the environment, while Chapter 8 describes the configuration of key external conditions at the beginning of the new century. Some of the most important of these external conditions are shown in Figure 13.1.

The partial institutionalization of environmentalism into major German institutions has opened new doors for environmental organizations willing to work

Figure 13.1. Key Variables in the Functioning of Environmental Organizations

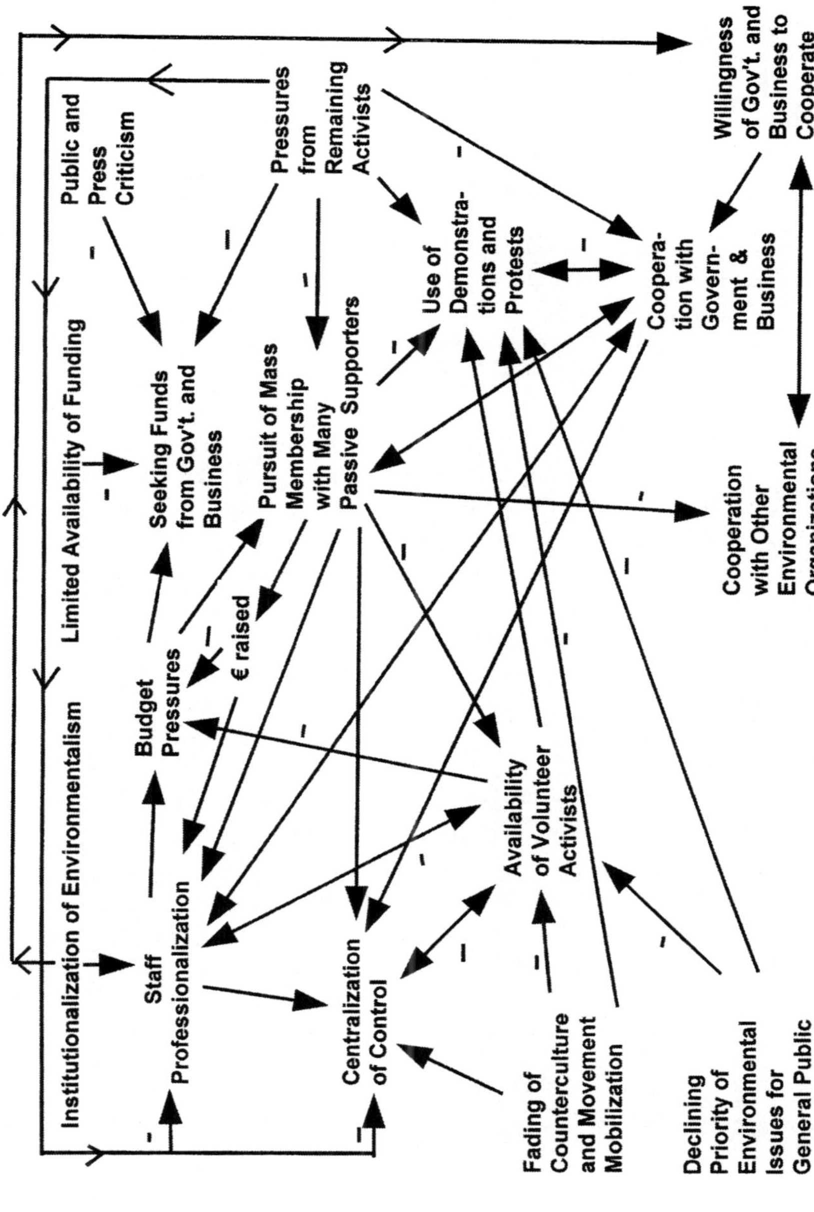

within the system. They are now invited to participate in political decision-making (albeit often as junior partners) and to cooperate with businesses and government. Doing so has the potential to increase their legitimacy, prestige, and access to funds from business, government, and citizens who are not dedicated activists. Ongoing institutionalization, including the ecological modernization of many government and business practices, also tends to undermine claims that environmental apocalypse is on the way, that business and government are unwilling to reform, and that only confrontational strategies work. Organizations that rely primarily on confrontational strategies and extreme rhetoric are thus increasingly disadvantaged in their quest for money, legitimacy, prestige, and influence.

Institutionalization has not, of course, solved all of Germany's environmental problems. It remains a densely settled land with an economy tilted more toward industry than most of its peers', a high level of consumption, an industrialized agricultural sector, and growing automobile dependence. It suffers from continuing problems with loss of open space, declining biodiversity, and air, water, and soil pollution. There remains, in other words, much for environmentalists to do and many avenues they might use to evoke public concern, if only the right framings and strategies can be found.

Institutionalization of environmentalism in government policy, including the rise of the Green Party, led to meaningful new legislation and to reforms that gave environmental organizations easier access to government. Nevertheless, environmental organizations continue to be treated as junior partners in neocorporatist decision-making—that is, they are required to pledge allegiance to principles of moderation and compromise in return for a seat at the table and remain generally unable to push through ambitious programs of change. Their rights to participate are often limited and available only after the major decisions have been reached, and the legal system provides them with only limited recourse.

The German political system, with its complex division of powers among the branches of government, the federal government and the *Länder*, and the multiple political parties, provides numerous points of access, but working successfully within the system requires sustained effort and expertise. More recently, the increasing influence of EU environmental politics and policy has added a new level of complexity to environmental policy making. The characteristic German emphasis on expertise also contributes to pressures for professionalization.

The trend toward decreasing coverage of environmental issues in the media and the diminishing number of independent, environmentally oriented media have contributed to the reduced priority the German public accords environmental issues and made it more difficult to mobilize environmental movements. Working with the media under these circumstances requires more sustained effort and expertise than it once did. Spectacular actions are still more likely to attract media attention than reasoned argumentation, but in a changed environment, the media coverage of confrontation is probably less likely to be sympathetic.

The decline of polarization and protest and the fading of the counterculture beginning in the 1980s has made social movement mobilization around environmental issues more difficult and successful protests harder to organize. It has also made it harder to attract new activists, as ideological fervor and ties to social movement networks no longer provide strong motivation. Partly because new activists have become harder to recruit, activists from earlier generations continue to be a significant factor in the organizations' internal politics. Although many have softened their views over the years, commitment to grassroots democracy, skepticism of government and industry, and propensities toward a confrontational approach persist among members of this cohort. Finally, the fading of the counterculture and reductions in polarization have also reduced, although not eliminated, criticism of the environmental organizations from outside.

Initiatives taken during the last quarter of the twentieth century transformed Germany's main environmental problems from readily understandable, visible ones to less easily understood problems, such as loss of biodiversity at home and in distant lands, a degraded ozone layer, and climate change. Moreover, Germany was beset at the turn of the twenty-first century by new and more immediate problems resulting from reunification, a stagnant economy, and high unemployment. These have diverted public attention away from environmental issues and given weight to portrayals of environmentalists as out-of-touch extremists whose proposals undermine growth.

Beginning in the 1990s, all of these changes combined to reduce the comparative salience of environmental issues in the public mind. While concerns about the environment did not disappear, public opinion polls suggested that the public believed that they were improving and that other issues were more important. These changes made fundraising more difficult, volunteers more difficult to attract, and mass protests harder to organize.

According to open systems models, the large, national environmental organizations must also take other environmental groups into account as they choose their goals and strategies, since they are potential competitors for donations, legitimacy, reputation, and influence. As documented in Chapter 8, Germany has a wide range of environmental groups—single-issue groups, regional and local groups, BIs with that emphasize grassroots organizing, fringe groups on the left and right, and umbrella organizations such as the DNR. The environmental organizations must take into account the Green Party, which is also a potential competitor for volunteers and public support.

Cooperation with these other groups offers many potential advantages in terms of efficiency and increased overall visibility for environmentalism. On the other hand, there are also drawbacks, including the risk of reducing the individual organization's visibility or reputation for effectiveness with the media and broader public.

Strategic Dilemmas and Efforts to Resolve Them

This constellation of external conditions, in combination with the organizations' histories, has produced the set of seven dilemmas—discussed in detail in Chapters 10 to 12—with which NABU, the WWF, BUND, and Greenpeace must somehow come to terms. While they are not the only important decisions the organizations make, they are especially significant for four reasons. First, because there are important advantages and disadvantages to both horns of each dilemma, decisions that involve them frequently generate internal disputes. Second, the dilemmas are never resolved once and for all but instead crop up repeatedly in various guises. Third, the decisions the environmental organizations make about them define the organizations' very nature and affect their ability to acquire key resources. Finally, decisions about how to deal with any one of the dilemmas often affect decisions about the others.

Because the four organizations have similar general objectives and operate in the same environment, it is not surprising that they manifest a limited range of variation in how they resolve the seven dilemmas—especially in comparison to the full range among all types of environmental groups past and present. Their adaptations are, however, not identical, as they differ somewhat in their specific goals, constituencies, and histories.

Professionalization vs. Volunteers

In comparison to environmental organizations before the 1980s, to the BBU during the period of polarization, or to present-day groups such as Robin Wood and the Green League, all four of the large national organizations are highly professionalized, especially at the national level. Their increasing professionalization was a response to changes described in the previous section. The institutionalization of environmentalism created a complex set of laws and regulations, a set of administrative agencies and business research departments staffed by professionals, and masses of technical data not easily mastered by amateurs. Moreover, government and business have become more willing to work with environmental organizations if they display significant expertise. As the media became less inclined to cover environmental issues, a more professional approach to media relations was needed, and soliciting new members and donations from a less interested public also required expertise and sustained attention. These tasks might conceivably have been performed by skilled volunteers, but the declining priority of environmental issues in the public mind, the decline of social movement mobilization, the perception that environmental conditions are improving, and economic stagnation and uncertainty have combined to reduce the number of volunteers willing to sacrifice their time month after month to environmental activism.

Professionalization came somewhat sooner to the WWF and Greenpeace than to BUND and NABU, and BUND and NABU remain somewhat less professionalized, perhaps because their long traditions of volunteer activism could be invoked by opponents of professionalization. Activists in BUND, NABU, and Greenpeace have resisted ongoing professionalization because it deprives them of interesting tasks and influence, but they have been unable to halt its forward march. Indeed, by undermining the motivation of volunteer activists, increased professionalization may, ironically, have created the need for still more professionalization.

Centralization vs. Democracy

The WWF lacks significant local groups, makes little use of volunteers, and has always had a centralized structure. All the other groups, however, have evolved toward increased centralization. This can be attributed in part to government, business, and media preferences for working with organizations that can make decisions quickly and speak with one voice, and to the decline of the countercultural emphasis on grassroots democracy. The primary causes, however, appear to be internal. Michel's iron law may well be involved here, but the major internal factor has been professionalization. Professionals generally find work within centralized structures with clear lines of authority better suited to their work style than grassroots democracy, and they may prefer gathering the reins of power into their own hands in order to secure their positions.

Centralization is not, however, without drawbacks. Centralized structures can make adapting to regional or local conditions more difficult; the press and critics often carp about alleged citizens' crusades that are actually hierarchically administered organizations; local leaders frequently dislike concentration of power in the headquarters office; and activists often wish to have more say in their assignments and in determining goals and strategies. Formally democratic and federalized organizational structures, as well as traditions of volunteer activism that are key aspects of the organizations' identity, have restrained the march of centralization at BUND and NABU, but they have not halted it. Instead there has been continuing tension over tendencies toward centralization, punctuated by occasional outbreaks of active conflict. Greenpeace, a highly centralized organization that nevertheless depends heavily on committed volunteers to accomplish many tasks at the local level, has experienced even more conflict over this issue.

Financial Support from Business and Government

The organizations' evolution toward professionalization greatly increased their need for funds. Seeking financial support from government is a tradition as old as the environmental organizations, so it is hardly surprising that their quest for funds sometimes leads them to the doors of government agencies and to business. Financial support from these sources could not only help to fill their coffers and diversify their sources of income but also increase their legitimacy and prestige. Yet accepting such support, especially from businesses that engage in environmentally destructive behavior, can subject them to criticism from the press, the general public, and their own committed activists, who worry that accepting financial support from these sources will divert the organizations from their mission or subject them to undue influence. BUND, NABU, and Greenpeace can hardly ignore such criticisms, as their activists continue to supply them with needed labor and expertise.

Greenpeace, which accepts essentially no support from business or government, has avoided criticism on this count. The WWF has no activist volunteers and is thus able to avoid internal criticism, which may account for its somewhat heavier reliance on such donations. Nevertheless, it does come in for criticism from the press and other environmental organizations. In any case, none of the four organizations raises more than a small portion of its budget from business and government, in part due to the sheer unavailability of funding from these sources.

Mass Membership vs. Committed Activists

Faced with heavy demand for funds and significant constraints on obtaining them from business and government, the organizations have turned to donations from individuals. They can raise significant amounts from the relatively small proportion of the population that is actively interested in environmental issues; however, although many such persons are economically comfortable, few are wealthy. Moreover, economic stagnation has constrained the total volume of giving, and there is much competition from other environmental organizations and from other types of groups.

Faced with this situation, the organizations began to cast their net as widely as possible and seek donations from a broad public, even at the cost of recruiting many passive supporters without a strong commitment to the environmental cause. Using sophisticated mass mailings and door-to-door solicitation, they have managed to cope with the free rider problem well enough to attract an impressive number of supporters, which adds to their political influence and considerably increases their revenues. Adoption of these techniques represented a major organizational innovation, and it quickly diffused from organization to organization. The new fundraising strategy also created new positions for professionals in fundraising, press relations, and member relations. The result has been a feedback loop in which professionalization and fundraising from a mass public reinforce one another.

Significant constraints on this spiral come from practical limitations on the amount of funds the new techniques can raise and from criticism of the process from activists inside and outside the organizations. The critics point out that many of the supporters won with the new techniques have relatively low commitment to the environmental cause, that the public is being taught that they need only send money and professionals will take care of the rest, and that grassroots involvement and democracy are undermined in favor of professionalization and centralization. Predictably, the internal disputes have been most pronounced within BUND and NABU, which have long traditions of volunteer involvement.

Overall, however, this organizational innovation has moved the organizations toward a common pattern of operations. A paid, professional staff accomplishes most of the organization's work at the national level, a core of unpaid activists (missing in the WWF) take care of most tasks at the local level, and a much larger number of passive supporters provide financial resources and evidence of broad support.

Cooperation vs. Confrontation with Government and Business

The social changes described above created many incentives for the organizations to move away from the confrontational approach that had characterized the 1970s and early 1980s. Incentives to work within the system were further increased by professionalization, since the skills professionals bring to the table are more oriented toward working within the system than confronting it. The move toward seeking support from a broad public, including many persons whose commitment to the environmental cause is shallow, created additional incentives for a less confrontational approach because such supporters are more difficult to call out onto the streets. Insofar as the organizations receive financing from government and business, this too has exerted pressure against confrontation.

At the same time, there are also good reasons not to completely eliminate confrontation from organizational repertoires. Even relatively nonconfrontational interest groups can use protests and demonstrations to underline their demands and provide evidence of public support. Moreover, efforts to work within a system in which they are at a power disadvantage do not always pay off with results the organizations can use to increase their reputation for effectiveness and flow of donations. In these situations, too, other tactics are needed. Finally, many volunteer activists and some donors expect the organizations to fight hard for change, not just function as yet another interest group. For them, confrontation has symbolic significance and is motivational.

Caught between these contradictory forces, all four organizations have opted to steer a middle course. None wholeheartedly embraces either endless confrontation or close cooperation with business and government at any price. Because it makes no use of volunteers and has the most conservative donor constituency, the WWF is predictably somewhat more inclined than the others toward cooperation, but it does sometimes take firm positions and stage demonstrations. Greenpeace, with a tradition of spectacular action that is part of its identity for volunteers and donors, continues to cultivate a confrontational image. Nevertheless, in recent years, it has increased its emphasis on suggesting solutions, lobbying, and negotiations. By adopting this middle way, the organizations hope to persuade all of their constituencies to accord them legitimacy, influence, and donations—while avoiding offending any of them.

By increasing their emphasis on working within the system, the organizations have reinforced tendencies toward professionalization and centralization, since expertise and designated spokespersons with authority are needed to function within the system. The compromise orientation is also compatible with a fundraising strategy that seeks to attract donations from both activists and more causal supporters.

Cooperation vs. Competition with Other Organizations

The organizations' external environments and their decisions about how to resolve the other dilemmas also set up difficult choices for them regarding cooperation among themselves and with other environmental organizations. Cooperation offers significant advantages, described in Chapter 12, and the businesses and government agencies that work with environmental organizations often prefer that they work together. The organizations' move toward mass fundraising from the general public, however, has shifted the balance against extensive cooperation. Successful fundraising requires that the organizations and their achievements be publicly visible, and they find it hard to get credit for accomplishments achieved via work through an umbrella organization like the DNR, since even if the media report their victories the contributions of individual organizations are apt to be obscured.

When they engage in cooperative efforts, the organizations frequently exaggerate their own contributions in communications with members and donors, but this is obviously far from a perfect solution. Because fundraising has become such a crucial matter for them, the organizations have adopted a strategy of only occasional, limited cooperation. They are particularly resistant to efforts by the DNR to act as the spokesperson for the entire environmental movement. Greenpeace and the WWF do not belong to the DNR, and BUND and NABU were instrumental in forcing it to retreat to the role of service bureau. On the other hand, when the four organizations work at the EU level, where there is little to gain in terms of visibility or fundraising advantage, they willingly cooperate through the DNR, the EEB and other networks.

Limited vs. Broad Goals

The last of the dilemmas involves whether the organizations should focus their attention on a small number of goals or a broader spectrum. Although Greenpeace and the WWF define their goals somewhat more narrowly than NABU and BUND, all four are general environmental organizations. The organizations' size and success indicate that this formula was well suited to conditions around the turn of the twenty-first century. This represents a change from earlier eras, when the largest national environmental organizations all focused solely on nature protection, albeit often a relatively wide range of nature protection goals. Possible explanations for this change include the proven appeal of framings that unite protection of nature with protection of human health and well-being, as well as the desire to appeal to as broad a public as possible by offering something for everyone; however, the reasons for the current ascendance of organizations with very broad goals deserves further exploration.

Concluding Remarks

The analysis above shows that the choices the organizations make about each of the dilemmas result not only from exogenous variables but are also a consequence of how the other dilemmas have been resolved. The result, summarized in Figure 13.1 above, is a complex series of feedback loops that have created a degree of equilibrium in how the organizations operate, at least for the moment. Looking at things this way helps to explain the convergence of the four organizations on similar, though not identical, goals, strategies, and structures. It also suggests that present modes of operation are unlikely to be easily changed until major shifts in external conditions make it hard for the organizations to sustain themselves with their current goals, strategies, and structures.

Implications for Improving Theory

The analysis in the preceding section and the historical comparisons from the first part of the chapter have important implications for building useful and predictive theory about environmental organizations.

In the first place, these analyses lend support to the argument, first advanced in Chapters 1 and 2, that theories designed to explain social movements, systems of interest group politics, and civil society are ill suited to studying environmental organizations *as organizations.* There are two major reasons for this. First, these three theoretical approaches were designed to analyze not environmental organizations but larger social systems. Although they have sometimes been extended in the direction of analyzing organizations—as in the cases of resource mobilization theory and studies of public interest organizations—each falls well short of offering a complete tool kit for studying organizations. Second, even when the work in these traditions does consider environmental organizations and others like them, it almost invariably conceptualizes them as operating in particular contexts using specific modes of operation. Unfortunately, these often depart significantly from the actual situations that have faced environmental organizations in various periods.

Interest group theories, for example, conceptualize environmental organizations as public interest groups seeking to advance political ends by working within pluralist or neocorporatist political systems. Some environmental organizations in some periods have, of course, functioned in this way. However, this model offers little insight into environmental organization in Nazi Germany or the GDR, the confrontational organizations of the 1970s and 1980s that insisted on working outside the system, or the WWF during the first quarter-century of its existence. Theories of social movements or civil society organizations might fit these organizations better, but they too fall far short of applying across the board. Indeed, a single organization may, at different points in its history, be reasonably conceptualized as a public interest group, an SMO, or part of civil society, and some environmental organizations in some periods function simultaneously in all three modes.

Organization theories, by contrast, provide the tools to analyze environmental organizations in virtually every situation. They focus attention on three key questions. First, what possibilities does the outside world offer such organizations for obtaining the funds, legitimacy, prestige, and influence they must have in order to acquire the labor, information and services they need to survive and make progress toward their general objectives? Second, how do environmental organizations tailor their specific goals, structures, and strategies to ensure a continued flow of resources, especially in complex environments where different external actors want different things of them? Examining the organizations from this perspective focuses attention on the key strategy dilemmas they face—the underlying contradictions that arise over and over and provoke anguish and controversy over specific decisions. Third, organization theory, especially the institutional approach, reminds us that history and culture matter. That is, decisions about goals, structures, and strategies are rarely resolved on grounds of efficiency or effectiveness alone. Particularly important in the analysis above were instances where supporters attached symbolic value to established goals and ways of doing things and resisted change as contrary to organizational tradition.

In short, general organizational theory provides a framework that illuminates important issues about every organization included here in every period. The fact that organization theory "works" in Germany, where changing social contexts have provided perhaps the widest variety of contextual conditions of any Western society, provides especially good evidence of its general applicability.

A drawback of organization theories is that they have seldom been applied to environmental organizations. Consequently, when I began this study, I found relatively few precedents to use as guidelines for identifying key issues for study, and I hope that one of this study's contributions has been to help identify issues that surface repeatedly and appropriate concepts for analyzing them. Existing studies of voluntary associations identified tendencies toward oligarchy and professionalization as key areas for exploration, and theories of social movements and interest groups suggested additional ones. Among those that proved most productive were discussions of the free rider problem, the importance of framing environmental issues to attract support and defuse opposition, the influence of the political opportunity structure on the kinds of goals and strategies environmental organizations choose, the distinction between pluralist and neocorporatist political structures, and the notion of recurring waves of social movement mobilization that cut across social movements. Although borrowed from diverse theoretical traditions, these ideas proved easy to integrate into organization theory, further demonstrating its utility as a *general* theory to which other concepts and hypotheses can be added. Investigators of environmental organizations in other nations might well profit from my effort to integrate these concepts into an analysis based on organization theory, but research on environmental organizations as organizations remains in its infancy, and researchers studying other societies will no doubt need to draw in additional concepts or develop new ones of their own.

The Future of German Environmental Organizations

The narrative of this book ends as the German organizations entered the twenty-first century. At the time of this writing (early 2007), relatively little had changed. A coalition of the two largest parties replaced the SPD/Green coalition government in 2005, and by 2007 there were signs that Germany's long-stagnant economy was reviving. Nevertheless, a reader of the environmental organizations' publications in 2006 would certainly have found a great deal of continuity with the situation described above.

One clear lesson of the history of German environmentalism, however, is the constancy of change, so I would be remiss to end without some reflections about the future. Interesting and important as it might be to speculate on how long-run, macro-level changes in German society may affect the organizations, my agenda here is more modest. I focus instead on the possible effects of rather concrete changes that are already visible.

The first of these changes is the "graying of the greens," the generation that reached early adulthood during the 1970s and 1980s. The analysis above suggests that this cohort has consistently displayed a relatively high level of environmental activism and concern. Activists drawn from it decisively shaped the environmental organizations during the heyday of the environmental movement, and they continued to play an important role at the beginning of the new century. Even though many have moderated their views, they have often been at the center of opposition to professionalization, centralization, mass fundraising, and excessive orientation toward working within the system. Other members of this same cohort are responsible for much of the criticism of these trends from outside.

If the aging out of these older activists over the next two decades forces the organizations to get by with fewer volunteers and raise a higher proportion of funds from less committed supporters, and if opposition to professionalization, centralization, mass fundraising, and cooperative strategies decreases, the organizations' tilt toward these strategy choices might well increase. In this scenario, they would come to resemble either highly institutionalized—and quite possibly weakened—interest groups working to protect and extend past gains, or civil society organizations focused on providing information, nature reserves, and a social outlet for their people interested in nature and the environment. In such a situation, the prominence of nature protection—the least controversial of the organizations' assignments and the one with the broadest public appeal—might be expected to rise (van Koppen and Markham, 2007). Indeed, some trends in this direction are already visible.

Whether this mix of goals and strategies can provide enough funds, donations, and influence for the organizations to remain effective is an open question. Focusing on nature protection does appeal to many supporters, but the success of the organizations may also hinge on whether their current emphasis on moving German society toward ecological modernization pays off in terms of visible, easily understandable outcomes—species and scenic vistas protected, fishable rivers, and reductions in greenhouse gas emissions—without requiring too much sacrifice from the public. While it is hard to imagine large numbers of social movement activists being recruited to a program based mainly on ecological modernization, this approach offers many opportunities for building alliances with government agencies and professional or business groups and attracting financial support from them. Lurking behind this scenario, however, are hard questions about whether ecological modernization can actually reduce the ecological footprint of a society where a growing consumer economy is viewed as the source of business profits and job creation (Schnaiberg and Gould, 1994; Blühdorn, 2006, 2007).

There remains the possibility that dramatic changes in the exogenous variables that have shaped the status quo might shock the organizations out of their current equilibrium. Such shocks might come, for example, from renewed social movement mobilization. Mobilization continues today on at least two themes potentially relevant to environmentalism. First, anti-nuclear protests continue to occur regularly in Germany. All of the environmental organizations align themselves with this movement by calling for an accelerated timetable for ending Germany's nuclear program, but they do not function as SMOs for it. And absent a reversal of the current government policy of gradual phaseout of nuclear power or a major nuclear accident, it is hard to imagine the anti-nuclear cause reenlivening the environmental organizations. The anti-globalization movement, too, has mobilized a passionate group of supporters, and environmentalism is closely related to some of its concerns. Should it take root and be embraced by broader segments of the population, it might offer the environmental organizations opportunities like those the BIs and anti-nuclear movement provided in the 1970s; however, as yet there is little persuasive evidence that this is occurring.

A more likely mobilization scenario is that global warming might produce a new wave of deep and widespread citizen concern that brings environmental issues back to the center of the political agenda. Alarming reports of faster-than-expected temperature increases and attention-catching weather events have great potential to attract media and public attention. However, much more than this is required to bring new legitimacy, influence, and support to the environmental organizations. Their success in capitalizing on the issue will depend, in part, on their ability to frame it in a way that makes the climate change problem seem serious but not hopeless, accords them a key role in solving it, and clearly demarcates what supporters can do to support their efforts. Like many other new environmental issues, climate change can be effectively addressed only at the international level, so environmental organizations that hope to be key actors in the fight against it will have overcome a whole set of obstacles to forging international alliances while attracting support at the national level. On the other hand, recalcitrant political and economic structures at the national or international level might ultimately provoke a wave of movement mobilization around climate change from which environmental organizations might benefit.

But however these scenarios—or others more difficult to foresee—play themselves out over the next decades, the persistence of environmental themes and the mutability and staying power of the German environmental organizations over the last century suggests that they are likely to be important players in the next.

References

AgrarBündnis. 2006. "Wir über Uns." Downloaded 5 December 2006. http://.kritischer-agrarbericht.de

Aldrich, Howard. 1999. *Organizations Evolving.* Thousand Oaks, CA: Sage.

Andersen, Arne. 1989. "Heimatschutz: Die Bürgerliche Naturschutzbewegung." In *Besiegte Natur: Geschichte der Umwelt im 19. und 20. Jahrhundert,* ed. Franz-Josef Brüggemeier and Thomas Rommelspacher. Munich: C.H. Beck. Pp. 143–158.

Andritzky, Walter. 1978. "Erstmals Wissenschaftlich Untersucht: Bürgerinitiativen." *Bild der Wissenschaft* 15, no. 2 (February): 85–94.

Andritzky, Walter, and Ulla Wahl-Terlinden. 1978. *Mitwirkung von Bürgerinitiativen an der Umweltpolitik.* Berlin: Erich Schmidt Verlag.

Anheier, Helmut K. 1992. "An Elaborate Network: Profiling the Third Sector in Germany." In *Government and the Third Sector: Emerging Relationships in Welfare States,* ed. Benjamin Gidron, Ralph M. Kramer, and Lester M. Salamon. San Francisco: Jossey-Bass Publishers. Pp. 31–56.

Anheier, Helmut K. 1998. "Der Dritte Sektor und der Staat." In *Dritter Sektor - Dritte Kraft: Versuch einer Standortbestimmung,* ed. Rupert Graf Strachwitz. Stuttgart: RAABE. Pp. 351–368.

Anheier, Helmut K., Eckhard Priller, Wolfgang Seibel, and Annette Zimmer, eds. 1997. *Der Dritte Sektor in Deutschland: Organisationen zwischen Staat und Markt im gesellschaftlichen Wandel.* Berlin: Edition Sigma.

Anheier, Helmut K., and Wolfgang Seibel. 1999. "Der Nonprofit Sektor in Deutschland." In *Handbuch der Nonprofit Organisation: Strukturen und Management,* ed. Christoph Badelt. Stuttgart: Schäffer-Poeschel. Pp. 19–41.

Arbeitsgemeinschaft für Umweltfragen. 2003. "Arbeitsgemeinschaft für Umweltfragen." Downloaded 12 July 2003. http://www.ag-umweltfragen.de

Aus der Arbeit der Natur- und Heimatfreunde im Kulturbund zur Demokratischen Erneuerung Deutschlands. 1954. "Organisationsplan der Natur- und Heimatfreunde." *Aus der Arbeit der Natur- und Heimatfreunde im Kulturbund zur Demokratischen Erneuerung Deutschlands,* no volume, no. 1 (February): 4–5.

Bagger, Wolfgang. 2002. "Naturfreunde Müssen Naturrevolutionäre Sein." In *Politische Landschaft: Die Andere Sicht auf die Natürliche Ordnung,* ed. Klaus-Peter Lorenz. Duisburg: Trikont. Pp. 30–40.

Ball, Allan R., and Frances Millard. 1987. *Pressure Politics in Industrial Societies.* Atlantic Highlands, NJ: Humanities Press.

Bammerlin, Ralf. 1998. *Umweltverbände in Deutschland: Herausforderung zum Wandel im Zeichen des Leitbildes Nachhaltiger Entwicklung,* Beiheft 24 to *Flora und Fauna in Rheinland-Pfalz.* Landau: Gesellschaft für Naturschutz und Ornithologie Rheinland-Pfalz.

Bauer, Ludwig. 2001. "Naturschutzarbeit der 1950er und 1960er Jahre in der Ehemaligen DDR." In *Natur im Sinn: Beiträge zur Geschichte des Naturschutzes,* ed. Stiftung Naturschutzgeschichte. Essen: Klartext. Pp. 47–61.

Baukloh, Anja, and Jochen Roose. 2002. "The Environmental Movement and Environmental Concern in Contemporary Germany." In *The Culture of German Environmentalism,* ed. Axel Goodbody. New York: Berghahn Books. Pp. 81–101.

B.A.U.M. 2007. "Unternehmer Sichern Zukunft." Downloaded 9 February 2007. http://www.baumev.de/baumev/home/index.php

Bauske, Berhhard. 2003. "Ökosponsoring als Gratwanderung: Bilanz der WWF-Kooperationen mit Wirtschaftsuntehmen." In *Das Große Buch des WWF. 40 Jahre Naturschutz für und mit den Menschen,* ed. Klaus-Henning Groth. Steinfurt: Tecklenborg Verlag. Pp. 201–211.

Beck, Ulrich. 1986. *Risikogesellschaft: Auf dem Weg in eine Andere Moderne.* Frankfurt: Suhrkamp.

Beck, Ulrich. 1993. "Abschied von der Abstraktionsidylle." *Politische Ökologie* 11, no. 31 (May/June): 20–25.

Becker, Carola. 1990. "Umweltgruppen in der DDR." In *DDR-Jugen: Politisches Bewusstsein und Lebensalltag,* ed. Barbara Hille and Walter Jaide. Opladen: Leske + Budrich. Pp. 216–247.

Behrens, Hermann. 1993. "Magere Mitgliederzahlen." *Politische Ökologie* 11, no. 31 (May/June): 44–47.

Behrens, Hermann. 2000. "Landschaftstage in der 'Seenplatte': Vorgeschichte, Themen, Ergebnisse." In *Landschaftsentwicklung und Landschaftsplanung in der Region 'Mecklenburgische Seenplatte,'* ed. Hermann Behrens. Neubrandenburg: Fachhochschule Neubrandenburg. Pp. 290–362.

Behrens, Hermann. 2001. "Die Ersten Jahre: Naturschutz und Landschaftspflege in der SBZ/DDR von 1945 bis Anfang der 60er Jahre." In *Naturschutz in den Neuen Bundesländern: Ein Rückblick,* ed. Institut für Umweltgeschichte und Regionalentwicklung. Berlin: Verlag für Wissenschaft und Forschung. Pp. 15–86.

Behrens, Hermann. 2003a. "Naturschutz in der DDR." In *Naturschutz Hat Geschichte,* ed. Stiftung Naturschutzgeschichte. Essen: Klartext. Pp. 189–253.

Behrens, Hermann. 2003b. "Naturschutz und Landeskultur in der Sowjetischen Besatzungszone und in der DDR: Ein Historischer Überblick." In *Die Veränderung der Kulturlandschaft,* ed. Günter Bayerl and Torsten Meyer. Münster: Waxmann. Pp. 213–271.

Behrens, Hermann. 2003c. "Naturschutz in der Deutschen Demokratischen Republik." In *Handbuch Naturschutz und Landschaftspflege* 9, ed. Werner Konold, R. Böcker, and U. Hampicke. Landsberg: Ecomed. Pp. 1–19.

Behrens, Hermann, and Ulrike Benkert. 1992. "Das Schicksal der Gesellschaft für Natur- und Umweltschutz im Kulturbund der DDR." *Zeitschrift für Umweltpolitik und Umweltrecht* 15, no. 3 (September): 353–371.

Behrens, Hermann, Ulrike Benkert, Jürgen Hopfmann, and Uwe Maechler. 1993. *Wurzeln der Umweltbewegung: Die 'Gesellschaft für Natur und Umwelt' (GNU) im Kulturbund der DDR.* Marburg: BdWi Verlag.

Benford, Robert D., and David A. Snow. 2000. "Framing Processes and Social Movements: An Overview and Assessment." *Annual Review of Sociology* 26: 611–639.

Bennie, Lynn G. 1998. "Brent Spar, Atlantic Oil, and Greenpeace." *Parliamentary Affairs* 51, no. 3 (July): 397–410.

Bensch, Margrit. 1999. "Blut oder Boden: Welche Natur Bestimmt den Rassismus?" In *Naturbilder in Naturschutz und Ökologie,* Landschaftsentwicklung und Umweltforschung, Schriftenreihe im Fachbereich Umwelt und Gesellschaft der TU Berlin, Nr. 111, ed. Stefan Korner et al. Berlin: Fachbereich Umwelt und Gesellschaft der Technische Universität Berlin. Pp. 37–45.

Bergmann, Klaus. 1970. *Agrarromantik und Großstadtfeindschaft.* Meisenheim am Glan: Verlag Anton Hain.

Bergstedt, Jörg. 1998. *Agenda, Expo, Sponsoring: Recherchen im Naturschutzfilz. Bd.1: Daten, Fakten, Historische und Aktuelle Hintergründe.* Frankfurt: IKO-Verlag für Interkulturelle Kommunikation.

Bergstedt, Jörg. 2002. *Reich oder Rechts: Umweltgruppen und NGOs im Filz mit Staat, Markt und Rechter Ideologie.* Frankfurt: IKO-Verlag für Interkulturelle Kommunikation.

Berliner Morgenpost. 1998. "Atom-Protest mit Wecker und Geigerzahler." *Berliner Morgenpost* 100, no. 161 (16 June): Politik, 2.

Berliner Zeitung. 2000a. "Ein Hoher Schutzstatus ist Nötig." *Berliner Zeitung* 56, no. 110 (12 May): Thema, 2.

Berliner Zeitung. 2000b. "Bauern Erhalten die Artenvielfalt." *Berliner Zeitung* 56, no. 110 (12 May): Thema, 2.

Berliner Zeitung. 2000c. "Deutschlands Atomkraftwerke Werden Langfristig Abgeschaltet." *Berliner Zeitung* 56, no. 138 (16 June): Tagesthema, 1.

Berliner Zeitung. 2000d. "Greenpeace: Unterstützung für Öko-Landwirtschaft." Downloaded 27 December 2000. http://www.BerlinOnline.de/aktuelles/tage...ma/.html/dpa_on160_1_2712_1227141257.html

Berliner Zeitung. 2001a. "18,200 Polizisten für Sechs Castor Behälter." *Berliner Zeitung* 57, no. 76 (30 March): Politik, 5.

Berliner Zeitung. 2001b. "Blockaden Verzögern Ankunft des Castor-Zuges um Einen Tag." *Berliner Zeitung* 57, no. 76 (30 March): Politik, 1.

Berliner Zeitung. 2001c. "Die Grünen Fordern von Ihren Parteichefs Stärkere Profilierung." Downloaded 10 March 2001. wysiwyg://8http://www.BerlineOnline.de/ak...berliner_zeitung/politik/.html/18818.html

Berliner Zeitung. 2001d. "Marie auf dem Gleis." *Berliner Zeitung* 57, no. 75 (29 March): Blickpunkt, 3.

Berliner Zeitung. 2001e. "Karriere mit Kopfkissen." *Berliner Zeitung* 57, no. 43 (20 February): Blickpunkt, 3.

Berliner Zeitung. 2004. "Verlorener Schwung: Die Umweltweisen Attestieren Rot-Grün Rückschritte im Naturschutz." *Berliner Zeitung* 60, no. 105 (6 May): Politik, 6.

Berry, Jeffrey M. 1984. *The Interest Group Society.* Boston: Little Brown.

Beucher, Friedhelm. 1995. "Ecological Modernization: Our Way Out of the Crisis: The SPD View." In *The Green Agenda: Environmental Politics and Policy in Germany,* ed. Ingolfur Blühdorn, Frank Krause, and Thomas Scharf. Keele: Keele University Press. Pp. 71–89.

bfp Analyse. 2000a. "Mut, Meer, Medien, und dazu Kompetent: Kernpunkte des Greenpeace-Image." In *Greenpeace auf der Wahrnehmungsmarkt,* ed. Christian Krüger and Matthias Müller-Hennig. Hamburg: LIT. Pp. 43–52.

bfp Analyse. 2000b. "Medienresonanzanalyse und Strategische Kommunikation: Am Beispiel der Greenpeace-Medienpräsenz." In *Greenpeace auf dem Wahrnehmungsmarkt,* ed. Christian Krüger and Matthias Müller-Hennig. Hamburg: LIT. Pp. 133–149.

bfp Analyse. 2000c. "Argumentationslinie zur Sozialen Wahrnehmung von Greenpeace als PR-Unternehmen." In *Greenpeace auf dem Wahrnehmungsmarkt,* ed. Christian Krüger and Matthias Müller-Hennig. Hamburg: LIT. Pp. 167–181.

bfp Analyse. 2000d. "Brent Spar: Eine Falschmeldung und ihre Karriere." In *Greenpeace auf dem Wahrnehmungsmarkt,* ed. Christian Krüger and Matthias Müller-Hennig. Hamburg: LIT. Pp. 205–221.

bfp Analyse. 2000e. "Weltmacht mit Schlauchboot und Schlips – oder: Sechs Dutzend Phrasen, Greenpeace zu Beschreiben." In *Greenpeace auf dem Wahrnehmungsmarkt,* ed. Christian Krüger and Matthias Müller–Hennig. Hamburg: LIT. Pp. 239–248.

Bick, Harmut, and Horst Obermann. 2000. "Stiefkind Naturschutz: Misere und Chancen des Naturschutzes in Deutschland." In *Umweltpolitik mit Augenmaß,* ed. Henning von Köller. Berlin: Erich Schmidt Verlag. Pp. 107–119.

Biehl, Janet. 1995. "'Ecology' and the Modernization of Fascism in the German Ultra-Right." In *Ecofascism: Lessons from the German Experience,* ed. Janet Biehl and Peter Staudenmaier. Edinburgh: AK Press. Pp. 31–73.

Biermann, Sabine, and Conny Böttger. 2000. "Lebende Bilder, Zeitungsbilder: Das Greenpeace-Foto in den Printmedien." In *Greenpeace auf der Wahrnehmungsmarkt,* ed. Christian Krüger and Matthias Müller-Hennig. Hamburg: LIT. Pp. 183–203.

Billig, Axel. 1994. *Ermittlung des Ökologischen Problembewußtseins der Bevölkerung.* Berlin: Umweltbundesamt.

Blackbourn, David. 2003. "Die Natur als Historisch zu Etablieren: Natur, Heimat und Landschaft in der Modernen Deutschen Geschichte." In *Naturschutz und Nationalsozialismus,* ed. Joachim Radkau and Frank Uekötter. Frankfurt: Campus. Pp. 65–76.

Blätter für Naturschutz und Naturpflege. 1931. "25 Jahre Naturschutz in Bayern." *Blätter für Naturschutz und Naturpflege* 14, no. 1 (May): 3–5.

Blätter für Naturschutz und Naturpflege. 1935. "Reichsnaturschutzgesetz vom 26. Juni 1935." *Blätter für Naturschutz und Naturpflege* 18, no. 2 (October): 97–98.

Blühdorn, Ingolfur. 1995. "Campaigning for Nature: Environmental Pressure Groups in Germany and Generational Change in the Ecology Movement." In *The Green Agenda: Environmental Politics and Policy in Germany,* ed. Ingolfur Blühdorn, Frank Krause, and Thomas Scharf. Keele: Keele University Press. Pp. 167–220.

Blühdorn, Ingolfur. 2000. *Post-Ecologist Politics: Social Theory and the Abdication of the Ecologist Paradigm.* London, New York: Routledge.

Blühdorn, Ingolfur. 2002. "Green Futures? A Future for the Greens?" In *The Culture of German Environmentalism,* ed. Axel Goodbody. New York, Oxford: Berghahn Books. Pp. 103–121.

Blühdorn, Ingolfur. 2006. "Self-Experience in the Theme Park of Radical Action? Social Movements and Political Articulation in the Late-Modern Condition." *European Journal of Social Theory* 9, no. 1 (February): 23–42.

Blühdorn, Ingolfur. 2007. "Self-Description, Self-Deception, Simulation. A Systems-theoretical Perspective on Contemporary Discourses of Radical Change." *Social Movement Studies* 6, no. 1 (May): 1–19.

Blühdorn, Ingolfur, Frank Krause, and Thomas Scharf ed. 1995. *The Green Agenda: Environmental Politics and Policy in Germany.* Keele: Keele University Press.

Blume, Johann-Friedrich, Alexander Schmidt, and Michael Zchiesche. 2001. *Verbandsklagen im Umwelt- und Naturschutz in Deutschland 1997–1999.* Berlin: Unabhängiges Institut für Umweltfragen.

Blumer, Herbert. 1969. "Social Movements." In *Studies in Social Movements: A Social Psychological Perspective,* ed. Barry McLaughlin. New York: Free Press.

Böckman, Hildegard, and Rosemarie Effers. 1985. "Verbandsideologie." In *Naturschutzverbände und Umweltpolitik,* ed. Institut für Landschaftspflege und Naturschutz und Institut für Landschaftsplanung und Raumforschung. Hanover: Institut für Landschaftspflege und Naturschutz and Institut für Landschaftsplanung und Raumforschung. Pp. 78–91.

Bode, Ingo. 2003. "Flexible Response in Changing Environments: The German Third Sector Model in Transition." *Nonprofit and Voluntary Sector Quarterly* 32, no. 2 (June): 190–210.

Bode, Thilo. 1996a. "Greenpeace und der Umbau der Gesellschaft: Sieben Thesen." In *Das Greenpeace Buch: Reflexionen und Aktionen,* ed. Greenpeace Deutschland. Munich: C.H. Beck. Pp. 12–19.

Bode, Thilo. 1996b. "Am Wendepunkt: Zur Stellung von Greenpeace in der Gesellschaft." In *Das Greenpeace Buch: Reflexionen und Aktionen,* ed. Greenpeace Deutschland. Munich: C.H. Beck. Pp. 254–265.

Bodenstein, Gerhard, Helmut Elbers, Achim Spiller, and Anke Zühlsdorf. 1998. *Umweltschützer als Zielgruppe des Ökologischen Innovationsmarketing: Ergebnisse einer Befragung von BUND-Mitgliedern.* Duisburg: Gerhard-Mercator-Universität-Duisberg, Fachbereich Wirtschaftswissenschaft.

Böker, Susanne. 2001. *Fundraising und Internet: Nutzungsmöglichkeiten des Internets als Kommunikationsmedium zwischen Nonprofit-Organisationen und Potentiellen und Existierenden Unterstützern.* Diplomarbeit aus dem Fachbereich Sozialwesen der Universität Gesamthochschule Kassel.

Börnecke, Stephan. 2003. "Naturschutz und Medien." In N*aturschutz in Deutschland: Eine Erfolgsstory?* ed. Deutscher Rat für Landespflege. Bonn: Deutscher Rat für Landespflege. Pp. 90–94.

Borsdorf-Ruhl, Barbara. 1973. *Bürgerinitiativen im Ruhrgebiet.* Essen: Siedlungsverband Ruhrgebiet.

Bosso, Christopher J. 1994. "After the Movement: Environmental Activism in the 1990s." In *Environmental Policy in the 1990s: Toward a New Agenda,* ed. Norman J. Vig and Michael E. Kraft. Washington, D.C.: Congressional Quarterly Press. Pp. 31–50.

Bosso, Christopher J. 1995. "The Color of Money: Environmental Groups and the Pathologies of Fundraising." In *Interest Group Politics,* ed. Allan J. Cigler and Burdett A. Loomis. Washington, D.C.: Congressional Quarterly Press. Pp. 101–130.

Bosso, Christopher J. 2005. *Environment, Inc.: From Grassroots to Beltway.* Lawrence: University of Kansas Press.

Boström, Magnus. 2001. *Miljörörelsens Mångfald.* Lund: Archiv Förlag.

Boström, Magnus. 2004. "Cognitive Practices and Collective Identities within a Heterogeneous Social Movement: The Swedish Environmental Movement." *Social Movement Studies* 3, no. 1 (April): 73–88.

Böttger, Conny. 1996. "Greenpeace Macht Bilder, Bilder Machen Greenpeace." In *Das Greenpeace Buch: Reflexionen und Aktionen,* ed. Greenpeace. Munich: C.H. Beck. Pp. 193–199.

Böttger, Conny. 2000. "Politik der Visualisation, Oder: Greenpeace Macht Bilder, Bilder Machen Greenpeace." In *Greenpeace auf dem Wahrnehmungsmarkt,* ed. Christian Krüger and Matthias Müller-Hennig. Hamburg: LIT. Pp. 35–42.

Bouman, Olav. 2003. "Gesunder Menschenverstand ist ein Wichtiger Erfolgsfaktor: Fundraising and Fördererpflege." In *Das Große Buch des WWF. 40 Jahre Naturschutz für und mit den Menschen,* ed. Klaus-Henning Groth. Steinfurt: Tecklenborg Verlag. Pp. 213–219.

Bozonnet, Jean-Paul. 2004. "De-institutionalising Environmentalism: The Shift from Civil Institutions to a Fake State Institutionalisation." Paper presented at the International

Sociological Association Research Committee 24 Conference on Local Institution Building for the Environment: Perspectives from East and West. Gorizia, Italy.

Bozonnet, Jean-Paul. 2005. "Unequal Environmentalism in Europe: Revisiting the Hypotheses of Affluence and Social Classes." Paper presented at the Biennial Meeting of the European Sociological Association. Torun, Poland.

Bramwell, Anna. 1985. *Blood and Soil: Richard Walther Darré and Hitler's "Green Party."* Abbotsbrook: Kensal Press.

Bramwell, Anna. 1994. *The Fading of the Greens: The Decline of Environmental Politics in the West.* New Haven, CT: Yale University Press.

Brand, Karl-Werner. 1987. "Kontinuität und Diskontinuität in den Neuen Sozialen Bewegungen." In *Neue Soziale Bewegungen in der Bundesrepublik Deutschland,* ed. Roland Roth and Dieter Rucht. Frankfurt: Campus. Pp. 19–29.

Brand, Karl-Werner. 1993. "Strukturveränderungen des Umweltdiskurses in Deutschland." *Forschungsjournal Neue Soziale Bewegungen* 6, no. 1 (March): 16–24.

Brand, Karl-Werner. 1995. "Der Ökologische Diskurs." In *Umweltbewußtsein und Massenmedien,* ed. Gerhard de Haan. Berlin: Akademie Verlag. Pp. 47–62.

Brand, Karl-Werner. 1997. "Environmental Consciousness and Behavior: The Greening of Lifestyles." In *The International Handbook of Environmental Sociology,* ed. Michael Redcliff and Graham Woodgate. Cheltenham: Edward Elgar. Pp. 204–217.

Brand, Karl-Werner. 1999a. "Dialectics of Institutionalisation: The Transformation of the Environmental Movement in Germany." In *Environmental Movements: Local, National, and Global,* ed. Christopher Rootes. London: Frank Cass. Pp. 35–58.

Brand, Karl-Werner. 1999b. "Transformationen der Ökologiebewegung." In *Neue Soziale Bewegungen: Impulse, Bilanzen und Perspektiven,* ed. Ansgar Klein, Hans-Josef Legrand, and Thomas Leif. Opladen: Westdeutscher Verlag. Pp. 237–256.

Brand, Karl-Werner. In press. "Umweltbewegung (inkl. Tierschutz)." In *Soziale Bewegungen in Deutschland seit 1945: Ein Handbuch,* ed. Roland Roth and Dieter Rucht. Frankfurt: Campus.

Brand, Karl-Werner, Detlef Büsser, and Dieter Rucht. 1986. *Aufbruch in eine Andere Gesellschaft: Neue Soziale Bewegungen in der Bundesrepublik.* Frankfurt: Campus.

Brand, Karl-Werner, Klaus Eder, and Angelika Poferl. 1997. *Ökologische Kommunikation in Deutschland.* Opladen: Westdeutscher Verlag.

Braun, Marie-Luise. 2003. *Umweltkommunikation im Lokalteil von Tageszeitungen: Eine Untersuchen zur Kritik am Umweltjournalismus.* Frankfurt: Peter Lang.

Brinkmann, Johannes, and Peter Lüdke-Höher. 1994. "Der Siemens Boykott: Ein Musterfall Ethnisch-politischen Protests." *Das Argument,* no volume, no. 206 (July/October): 785–798.

Broszat, Martin. 1987. *Hitler and the Collapse of Weimar Germany.* New York: Berg.

Brüggemeier, Franz-Josef. 1990. "Auf Kosten der Natur: Zu einer Geschichte der Umwelt 1880–1930." In *Jahrhundertwende: Der Aufbruch in die Moderne 1880–1930,* ed. August Nitschke et al. Reinbek bei Hamburg: Rowohlt. Pp. 75–91.

Brulle, Robert J. 1996. "Environmental Discourse and Social Movement Organizations: A Historical and Rhetorical Prospectus on the Development of US Environmental Organizations." *Sociological Inquiry* 66, no. 1 (February): 58–83.

Brulle, Robert J. 2000. *Agency, Democracy, and Nature: The US Environmental Movement from a Critical Theory Perspective.* Cambridge, MA: MIT Press.

Brulle, Robert J., and J. Craig Jenkins. 2005. "Foundations and the Environmental Movement: Priorities, Strategies, and Impacts." In *Foundations for Social Change: Critical Perspectives on Philanthropy and Popular Movements,* ed. Daniel Faber and Debra McCarthy. Lanham, NJ: Rowman and Littlefield. Pp. 151–173.

Bruns, Herbert. 1996. "Greenpeace Warenhaus." In *Das Greenpeace Buch: Reflexionen und Aktionen,* ed. Greenpeace Deutschland. Munich: C.H. Beck. Pp. 131–135.

Buechler, Steven M. 1993. "Beyond Resource Mobilization: Emerging Trends in Social Movement Theory." *The Sociological Quarterly* 34, no. 2 (May): 217–235.

Buechler, Steven M. 1995. "New Social Movement Theories." *The Sociological Quarterly* 36, no. 3 (June): 441–464.

Bulmer, Simon. 1989. "Territorial Government." In *Developments in West German Politics,* ed. Gordon Smith, William E. Paterson, and Peter H. Merkl. Durham, NC: Duke University Press. Pp. 40–59.

BUND. 1988. *Umwelt-Bilanz: Die Ökologische Lage der Bundesrepublik.* Hamburg: Rasch und Röhring Verlag.

BUND. 1995. *20 Jahre BUND.* Bonn: BUND.

BUND. 1998. *Das BUND-Jahr 1997: Luftsprünge im Gegenwind.* Bonn: BUND.

BUND. 2000. *Rückblick 99: Letztes Jahr in Bonn.* Berlin: BUND.

BUND. 2001. *2000 Jahresbericht.* Berlin: BUND.

BUND. 2002. *Der BUND Jahresbericht 2001.* Berlin: BUND.

BUND. 2004. *Der BUND Jahresbericht 2003.* Berlin: BUND.

BUND. 2005. *Der BUND Jahresbericht 2004.* Berlin: BUND.

BUND. 2007a. "Leitbild zur Verbandsentwicklung - BUND 2015." Downloaded 10 February 2007. http://www.bund.net/

BUND. 2007b. "Positionen." Downloaded 10 February 2007. http://www.bund.net/

BUND. 2007c. "Der BUND vor Ort." Downloaded 10 February 2007. http://www.bund.net/

BUND. 2007d. "6 plus 3: Die BUND-Spitze." Downloaded 10 February 2007. http://www.bund.net/

BUND. 2007e. "Satzung." Downloaded 10 February 2007. http://www.bund.net/

BUND. 2007f. "Hier Dreht sich die Arbeit nicht im Kreis." Downloaded 10 February 2007. http://www.bund.net/

BUND. No date. "BUND: Engagiert, Kompetent, in Ihrer Nähe" (pamphlet). Berlin: BUND.

Bundesamt für Naturschutz. 2007. "Rote Liste gefährdeter Tiere." Downloaded 26 January 2007. http://www.bfn.de/0322_tiere.html

Bundesministerium für Umwelt, Naturschutz und Reaktorsicherheit. 1996. *Umweltbewußtsein in Deutschland 1996.* Bonn: Bundesministerium für Umwelt, Naturschutz und Reaktorsicherheit.

Bundesministerium für Umwelt, Naturschutz und Reaktorsicherheit. 1998. *Umweltbewußtsein in Deutschland 1998.* Bonn: Bundesministerium für Umwelt, Naturschutz und Reaktorsicherheit.

Bundesministerium für Umwelt, Naturschutz und Reaktorsicherheit. 2000. *Umweltbewusstsein in Deutschland 2000.* Berlin: Bundesministerium für Umwelt, Naturschutz und Reaktorsicherheit.

Bundesministerium für Umwelt, Naturschutz und Reaktorsicherheit. 2002. *Umweltbewusstsein in Deutschland 2002.* Berlin: Bundesministerium für Umwelt, Naturschutz und Reaktorsicherheit.

Bundesministerium für Umwelt, Naturschutz und Reaktorsicherheit. 2004. *Umweltbewusstsein in Deutschland 2004.* Berlin: Bundesministerium für Umwelt, Naturschutz und Reaktorsicherheit.

Bundesverband Bürgerinitiativen Umweltschutz. 1980. "Orientierungspapier des Bundesverbandes Bürgerinitiativen e. V. (BBU)." In *Bürgerinitiativen in der Gesellschaft: Politische Dimensionen und Reaktionen,* ed. Ottheim Rammstedt. Villingen: Neckar-Verlag. Pp. 341–350.

Bundesverband Bürgerinitiativen Umweltschutz. 2005. "Tätigkeitsbericht." Downloaded 21 April 2005. http://www.bbu-online.de/Arbeitsbereiche/Taetigkeitsbericht.htm

Bundesverband Bürgerinitiativen Umweltschutz. 2007. "Willkommen beim BBU." Downloaded 2 January 2007. http://bbu-online.de/index.html.

Bundesverband Bürgerinitiativen Umweltschutz. No date. "Bürgerinnen und Bürger Initiieren Zukunft" (pamphlet). Bonn: Bundesverband Bürgerinitiativen Umweltschutz.

Bund für Vogelschutz. 1914. *Jahresbericht 1914*. Stuttgart: Bund für Vogelschutz.

Bund für Vogelschutz. 1921. "Jahresbericht 1920 des Bundes für Vogelschutz." *Zeitschrift für Vogelschutz und Andere Gebiete des Naturschutzes* 2, no. 1 (January): 10–16.

Bund für Vogelschutz. 1925. *Jahresbericht des Bundes für Vogelschutz, 1921–1924*. Stuttgart: Bund für Vogelschutz.

Bund Heimat und Umwelt in Deutschland. 2007. "Index." Downloaded 29 January 2007. http://www.bhu.de

BUNDMagazin. 1998a. "Natur in der Schußlinie." *BUNDMagazin* 2, no. 1 (February): 20–21.

BUNDMagazin. 1998b. "Jeder Baum ist Anders." *BUNDMagazin* 2, no. 2 (May): 42–3.

BUNDMagazin. 1998c. "P.S.: BUNDMagazin-Berichte und Was Daraus Wurde." *BUNDMagazin* 2, no. 2 (May): 54.

BUNDMagazin. 1998d. "Radeln für die Umwelt." *BUNDMagazin* 2, no. 3 (August): 36.

BUNDMagazin. 1998e. "Mediation Hilft Nicht Immer." *BUNDMagazin* 2, no. 3 (August): 11.

BUNDMagazin. 1999a. "Dünger für's Reförmchen." *BUNDMagazin* 3, no. 1 (February): 22–24.

BUNDMagazin. 1999b. "Erbe für den Naturschutz." *BUNDMagazin* 3, no. 1 (February): 47.

BUNDMagazin. 1999c. "Strategien Künftiger BUND-Politik." *BUNDMagazin* 3, no. 1 (February): 28.

BUNDMagazin. 2000a. "Die Zukunft Braucht Uns." *BUNDMagazin* 4, no. 2 (May): 6–7.

BUNDMagazin. 2000b. "25 Jahre Bund für Umwelt und Naturschutz." *BUNDMagazin* 4, no. 2 (May): 12–16, 18–20, 22.

BUNDMagazin. 2000c. "Friends of the Earth International." *BUNDMagazin* 4, no. 2 (May): 32.

BUNDMagazin. 2000d. "Die Erde Braucht Freunde - und Sie Hat Welche." *BUNDMagazin* 4, no. 2 (May): 32.

BUNDMagazin. 2000e. "BUND: 'Atom-Konsens' Ist kein Ausstieg." *BUNDMagazin* 4, no. 3 (August): 7.

BUNDMagazin. 2000f. "Dosen vor die Brauerei Gekippt." *BUNDMagazin* 4, no. 3 (August): 27.

BUNDMagazin. 2000g. "Dank im Grünen." *BUNDMagazin* 4, no. 3 (August): 30.

BUNDMagazin. 2001a. "Wie Grün ist Grün 2020?" *BUNDMagazin* 5, no. 4 (November): 12.

BUNDMagazin. 2001b. "Ausstieg aus dem Ausstieg." *BUNDMagazin* 5, no. 4 (November): 30.

BUNDMagazin. 2001c. "Dreikönigstreffen von BUND und NABU." *BUNDMagazin* 5, no. 4 (November): 32.

BUNDMagazin. 2002a. "Neuer Bundesvorstand." *BUNDMagazin* 6, no. 1 (February): 27.

BUNDMagazin. 2002b. "Überparteilich, Nicht Unparteiisch." *BUNDMagazin* 6, no. 3 (August): 10.

BUNDMagazin. 2002c. "Wahlprogramme unter der Lupe." *BUNDMagazin* 6, no. 3 (August): 14–15.

BUNDMagazin. 2003a. "Wer Regiert die Republik?" *BUNDMagazin* 7, no. 1 (February): 10.

BUNDMagazin. 2003b. "Umwelt in der Globalisierungsfalle." *BUNDMagazin* 7, no. 2 (May): 26.

BUNDMagazin. 2003c. "Windpark Bedroht Meere." *BUNDMagazin* 7, no. 2 (May): 31.

BUNDMagazin. 2003d. "Unwiderruflicher Verlust." *BUNDMagazin* 7, no. 3 (August): 14–15.

BUNDMagazin. 2003e. "Gutes Tun." *BUNDMagazin* 7, no. 4 (November): 6.

BUNDMagazin. 2004a. "Klimaoasen Treiben Vielfältige Blüten." *BUNDMagazin* 8, no. 2 (May): 37.

BUNDMagazin. 2004b. "Naturschutz im BUND." *BUNDMagazin* 8, no. 3 (August): 12–14.

BUNDMagazin. 2004c. "Natur Nachhaltig Schützen." *BUNDMagazin* 8, no. 3 (August): 15–17.

BUNDMagazin. 2004d. "Rot-Grün die Zweite." *BUNDMagazin* 8, no. 4 (November): 10.

BUNDMagazin. 2005a. "Große BUND-Mobilitätsaktion für Mitglieder und Förderer." *BUNDMagazin* 9, no. 1 (February): 21.

BUNDMagazin. 2005b. "Mitglieder Werben Mitglieder." *BUNDMagazin* 9, no. 1 (February): 23.

BUNDMagazin. 2005c. "Liebe Leserinnen and Leser." *BUNDMagazin* 9, no. 2 (May): 3.

BUNDMagazin. 2005d. "30 Jahre BUND." *BUNDMagazin* 9, no. 2 (May): 15–20.

BUNDMagazin. 2005e. "Der Braunzaun im Kopf." *BUNDMagazin* 9, no. 2 (May): 22–23.

BUNDMagazin. 2005f. "Zuversichtlich in die Zukunft." *BUNDMagazin* 9, no. 2 (May): 30–33.

BUNDMagazin. 2005g. "Als BUNDmitglied Günstiger Telefonieren." *BUNDMagazin* 9, no. 3 (August): 5.

BUNDMagazin. 2005h. "Wir Sind Gefragt." *BUNDMagazin* 9, no. 3 (August): 11.

BUNDMagazin. 2005i. "Zehn Mal Zukunft." *BUNDMagazin* 9, no. 3 (August): 12–13.

BUNDMagazin. 2005j. "Ein Bock als Gärtner." *BUNDMagazin* 9, no. 3 (August): 16.

BUNDMagazin. 2005k. "Testamente Stiften Zukunft." *BUNDMagazin* 9, no. 3 (August): 35.

BUNDMagazin. 2005l. "Martin Rocholl Verlässt Brüssel." *BUNDMagazin* 9, no. 3 (August): 39.

BUND/Misereor. 1996. *Zukunftsfähiges Deutschland: Ein Beitrag zu einer Global Nachhaltigen Entwicklung.* Basel: Birkhäuser.

BUND/Misereor. 2002. *Wegweiser für ein Zukunftsfähiges Deutschland.* Munich: Riemann Verlag.

Bund Naturschutz in Bayern. 1925. "Sonderheft für den Ersten Deutschen Naturschutztag in Munich, 26., 27. and 28. Juni 1925." *Blätter für Naturschutz und Naturpflege* 8 (July).

Bund Naturschutz Service. 2007. "Bund-Reisen." Downloaded 7 February 2007. http://www.bund-reisen.de/

Bündnis 90/Die Grünen. 2007. "Die Zukunft ist Grün: Grundsatzprogramm von Bündnis 90/Die Grünen." Downloaded 27 January 2007. http://www.gruene.de/cms/files/dokbin/68/68425.grundsatzprogramm_die_zukunft_ist_gruen.pdf

BUNDschau. 2003a. "Wer Sind die BUNDaktiven?" *BUNDschau,* no volume, no. 3 (July): 3.

BUNDschau. 2003b. "Kyoto-Abkommen Jetzt!" *BUNDschau,* no volume, no. 4 (October): 1.

BUNDschau. 2003c. "Klimakiller Flugverkehr im Visier." *BUNDschau,* no volume, no. 4 (October): 20–21.

BUNDschau. 2004a. "BUND-Jahrbuch 2004." *BUNDschau,* no volume, no. 1 (January): 27.

BUNDschau. 2004b. "Ökotips." *BUNDschau,* no volume, no. 1 (January): 28.

BUNDschau. 2004c. "Fossile ins Museum." *BUNDschau,* no volume, no. 3 (July): 1.

BUNDschau. 2004d. "BUND/NABU-Workshop Erfolgreich." *BUNDschau,* no volume, no. 3 (July): 19.

BUNDschau. 2004e. "Ab 2005 Elektronisch." *BUNDschau,* no volume, no. 4 (October): 3.

BUNDschau. 2004f. "Neues aus den Gremien." *BUNDschau,* no volume, no. 4 (October): 3.

Bunz, Axel R. 1973. *Umweltpolitisches Bewußtsein 1972.* Berlin: Erich Schmidt Verlag.

Burstein, Paul. 1998. "Interest Organizations, Political Parties, and the Study of Democratic Politics." In *Social Movements and American Political Institutions,* ed. Anne N. Costain and Andrew S. McFarland. Lanham, MD: Rowman and Littlefield. Pp. 39–55.

Buttel, Frederick H. 2003. "Environmental Sociology and the Explanation of Environmental Reform." *Organization and Environment* 16, no. 3 (September): 306–344.

Caniglia, Beth Schaeffer, and JoAnn Carmin. 2005. "Scholarship on Social Movement Organizations: Classic Views and Emerging Trends." *Mobilization: An International Journal* 10, no. 2 (June): 201–212.

Central Intelligence Agency. 2007. *The World Factbook, 2007.* Pittsburgh, PA: Superintendent of Documents.

Chaney, Sandra. 1996. *Visions and Revisions of Nature: From the Protection of Nature to the Invention of the Environment in the Federal Republic of Germany, 1945–1975.* Unpublished doctoral dissertation, Department of History, University of North Carolina at Chapel Hill.

Chaney, Sandra. 2005. "Protecting Nature in a Divided Nation: Conservation in the Two Germanys, 1945–1972." In *Germany's Nature: Cultural Landscapes and Environmental History,* ed. Thomas Lekan and Thomas Zeller. New Brunswick, NJ: Rutgers University Press. Pp. 207–243.

Christlich Demokratische Union Deutschlands. 2007. "Umweltpolitik." Downloaded 27 January 2007. http://www.cdu.de/politikaz/umwelt.php

Christmann, Gabriela B. 1996. "Zur 'Ökologischen Moral' im Wandel der Zeiten." *Forschungsjournal Neue Soziale Bewegungen* 9, no. 4 (December): 66–75.

Clausen, Jens. 2002. "Wie Sieht die Umweltberichterstattung von Unternehmen aus?" In *Umweltkommunikation - Vom Wissen zum Handeln,* ed. Fritz Brickwedde and Ulrike Peters. Berlin: Erich Schmidt Verlag. Pp. 87–101.

Commission of the European Communities. 1992. *Europeans and the Environment 1992.* Brussels: Commission of the European Communities, Directorate General Environment, Nuclear Safety, and Civil Protection.

Cooper, Belinda. 1995. "Die Arche Berlin-Brandenburg (West): Hilfe vom Klassenfeind." In *Arche Nova: Opposition in der DDR,* ed. Carlo Jordan and Hans Michael Kloth. Berlin: BasisDruck. Pp. 99–111.

Cornelsen, Dirk. 1991. *Anwälte der Natur. Umweltschutzverbände in Deutschland.* Munich: C.H. Beck.

CSU. 2007. "Umweltpolitik." Downloaded 27 January 2007. http://www.csu.de/home/Display/Politik/Themen/Umwelt/Umwelt?Thema=Umwelt&Unterthema=Umweltpolitik

Curtis, Russell, and Louis A. Zurcher. 1974. "Social Movements: An Analytical Exploration of Organizational Forms." *Social Problems* 21, no. 3: 356–370.

Cyert, Richard M., and James G. March. 1963. *A Behavioral Theory of the Firm.* Englewood Cliffs, NJ: Prentice Hall.

Dahl, Robert A. 1961. *Who Governs?* New Haven, CT: Yale University Press.

Dalton, Russell. 1994. *The Green Rainbow: Environmental Groups in Western Europe.* New Haven, CT: Yale University Press.

Dalton, Russell. 2005. "The Greening of the Globe? Cross-national Levels of Environmental Group Membership." *Environmental Politics* 14, no. 4 (August): 441–459.

Dannenbaum, Thomas. 2005. "'Atom-Staat' oder 'Unregierbarkeit'? Wahrnehmungsmuster im Westdeutschen Atomkonflikt der Siebziger Jahre." In *Natur und Umweltschutz nach 1945: Konzepte, Konflikte, Kompetenzen,* ed. Franz-Josef Brüggemeier and Jens Ivo Engels. Frankfurt: Campus. Pp. 268–286.

Davis, Gerald F., et al. eds. 2005. *Social Movements and Organization Theory.* Cambridge: Cambridge University Press.

de Haan, Gerhard. 1995. "Umweltbewußtsein und Massenmedien: Der Stand der Debatte." In *Umweltbewußtsein und Massenmedien,* ed. Gerhard de Haan. Berlin: Akademie Verlag. Pp. 17–34.

Deakin, Nicholas. 2001. *In Search of Civil Society.* Hampshire: Palgrave.

Degenhardt, Wolfgang. 2003. "Wo die Eigentliche Arbeit Gemacht Wird. Das Aktivitätenprofil des BUND in NRW." In *Keine Berufsprotestierer oder Schornsteinkletterer,* ed. Stiftung Naturschutzgeschichte. Essen: Klartext. Pp. 159–179.

della Porta, Donatella, and Mario Diani. 1999. *Social Movements: An Introduction.* Oxford: Blackwell.

della Porta, Donatella, and Hanspeter Kriesi. 1999. "Social Movements in a Globalizing World: An Introduction." In *Social Movements in a Globalizing World,* ed. Donatella della Porta, Hanspeter Kriesi, and Dieter Rucht. New York: St. Martin's Press. Pp. 3–22.

della Porta, Donatella, and Dieter Rucht. 2002. "The Dynamics of Environmental Campaigns." *Mobilization* 7, no. 1 (February): 1–14.

Der Spiegel. 1991a. "McDonald's der Umweltszene." *Der Spiegel* 45, no. 38 (16 September): 84–99, 103–105.

Der Spiegel. 1991b. "Naturschutz mit Betroffenheit." *Der Spiegel* 45, no. 38 (16 September): 102.

Der Spiegel. 1995a. "Ich Kann ohne Wale Leben." *Der Spiegel* 49, no. 38 (18 September): 50–54.

Der Spiegel. 1995b. "Lust auf Langsamkeit." *Der Spiegel* 49, no. 43 (23 October): 214–217.

Der Spiegel. 1997a. "Ein Spiel mit dem Feuer." *Der Spiegel* 51, no. 12 (17 March): 82–85.

Der Spiegel. 1997b. "Der Sieg des Geldes." *Der Spiegel* 51, no. 14 (31 March): 58–59.

Der Spiegel. 1998. "Ich Finde Missionare Eklig." *Der Spiegel* 52, no. 52 (21 December): 44–45.

Deutsche Bundesstiftung Umwelt. 2006. *Jahresbericht 2005.* Onsnabrück: Deutsche Bundesstiftung Umwelt.

Deutscher Bundestag. 2006. *Bekanntmachung von Rechenschaftsberichten Politischer Parteien für das Kalenderjahr 2004 (1. Teil: Bundestagsparteien).* Berlin: Deutscher Bundestag.

Deutscher Naturschutzring. 1976. *25 Jahre Deutscher Naturschutzring.* Bonn: Deutscher Naturschutzring.

Deutscher Naturschutzring. 1998. "Der Deutsche Naturschutzring: Selbstdarstellung." In *Jahrbuch Ökologie,* ed. Günter Altner, Barbara Mettler-von Meibom, Udo E. Simonis, and Ernst U. von Weizsäcker. Munich: C.H. Beck. P. 275.

Deutscher Naturschutzring. 2000. *50 Jahre Lobbyismus für Natur und Umwelt.* Bonn: Deutscher Naturschutzring.

Deutscher Naturschutzring. 2002. "EU-Rundschreiben." Downloaded 11 December 2002. http://www.dnr.de/publikationen/eur/index.php

Deutscher Naturschutzring. 2007. "Mitglieder." Downloaded 30 January 2007. http://www.dnr.de/dnr/verbaende/index.php

Deutscher Rat für Vogelschutz. 2006. "Der DRV Stellt sich vor." Downloaded 5 December 2006. http://www.drv-web.de/start_d.htm

Deutsche Umwelthilfe. 2005. *Tatigkeitsbericht 2004.* Radolfzell: Deutsche Umwelthilfe.

Deutscher Verband für Landschaftspflege. 2005. "Wir über Uns." Downloaded 13 July 2005. http://www.lpv.de

DGB. 2007. "Gewerkschaften." Downloaded 25 January 2007. http://www.dgb.de/dgb/gewerkschaften/

Diani, Mario. 1992. "The Concept of Social Movement." *The Sociological Review* 40, no. 1 (February): 1–25.

Diani, Mario, and Paolo R. Donati. 1999. "Organizational Change in Western European Environmental Groups: A Framework for Analysis." *Environmental Politics* 8, no. 1 (spring): 13–34.

Diani, Mario and Doug McAdam, eds. 2003. *Social Movements and Networks: Relational Approaches to Collective Action.* New York: Oxford University Press.

Diefenbacher, Hans. 1994. "Umweltbewußtsein in der Bundesrepublik: Folgenarme Betroffenheit?" *Blätter für Deutsche und Internationale Politik* 39, no. 6 (June): 767–770.

Diekmann, Andreas, and Peter Preisendörfer. 1998. "Umweltbewußtsein und Umweltverhalten in Low- und High-Cost-Situationen." *Zeitschrift für Soziologie* 27, no. 6 (December): 438–453.

Die Linke.PDS. 2007. "Ökologische Plattform bei der Linkspartei.PDS." Downloaded 27 January 2007. http://sozialisten.de/partei/strukturen/agigs/oekologische_plattform/index.htm

Dierkes, Meinolf, and Hans-Joachim Fietkau. 1988. *Umweltbewußtsein - Umweltverhalten.* Karlsruhe: Kohlhammer.

Die Zeit. 1991. "Greenpeace in Seenot." *Die Zeit* 46, no. 33 (9 August): Wissenschaft, 55.

Die Zeit. 1995a. "Triumph der Aktivisten." *Die Zeit* 50, no. 26 (23 June): Wirtschaft, 17.

Die Zeit. 1995b. "Wohlfeil aber Willkommen." *Die Zeit* 50, no. 26 (23 June): Wirtschaft, 20.

Die Zeit. 1995c. "Ruch der Kumpanei." *Die Zeit* 50, no. 28 (7 July): Wirtschaft, 25.

Die Zeit. 1995d. "Manager unter dem Regenbogen." *Die Zeit* 50, no. 29 (14 July): Dossier, 9.

Die Zeit. 1995e. "Grün in die Regale." *Die Zeit* 50, no. 34 (18 August): Wirtschaft, 15.

Die Zeit. 1995f. "Wir Machen Politische Kunst." *Die Zeit* 50, no. 36 (1 September): Politik, 7.

Die Zeit. 1995g. "Sparen mit Vollgas." *Die Zeit* 50, no. 38 (15 September): Wissen, 46.

Die Zeit. 1995h. "Bosch: Auch Sonntags in die Fabrik. Italien: Prominenz im Zwielicht. Mineralwasser: Plastik in Kommen." *Die Zeit* 50, no. 42 (13 October): Wirtschaft, 26.

Die Zeit. 1995i. "Zensur gegen Umweltschützer." *Die Zeit* 50, no. 49 (1 December): Wirtschaft, 26.

Die Zeit. 1996a. "Die Unternehmensberatung McKinsey Trimmt Betriebe und Verwaltung Fit für den Wettbewerb - und Verändert im Stillen unser Ganzes Leben." *Die Zeit* 51, no. 3 (12 January): Dossier, 9.

Die Zeit. 1996b. "Pioneer Sucht Partner." *Die Zeit* 51, no. 9 (23 February): Wirtschaft, 26.

Die Zeit. 1996c. "Benzindurst." *Die Zeit* 51, no. 29 (12 July): Wissen, 30.

Die Zeit. 1996d. "Die Politik Allein Ist Überfordert." *Die Zeit* 51, no. 32 (2 August): Wirtschaft, 19.

Die Zeit. 1996e. "Die Protestmaschine." *Die Zeit* 51, no. 37 (6 September): Dossier, 9.

Die Zeit. 1996f. "Signal für die Umwelt." *Die Zeit* 51, no. 38 (13 September): Wirtschaft, 20.

Die Zeit. 1997a. "Joghurtbecher Bringen's Nicht." *Die Zeit* 52, no. 25 (13 June): Dossier, 12.

Die Zeit. 1997b. "Öko-Krieger zu TV-Kämpfern." Die Zeit 52, no. 37 (September, 5): Medienseite, 67.

Die Zeit. 1997c. "Die Appetitverderber." *Die Zeit* 52, no. 40 (26 September): Wirtschaft, 25.

Die Zeit. 1997d. "Ausgerechnet Hannover." *Die Zeit* 52, no. 43 (17 October): Reise, 65.

Die Zeit. 1998. "Die Grünen auf die Finger Sehen." *Die Zeit* 53, no. 38 (10 September): Politik, 15.

Die Zeit. 1999a. "Umwelt Ist Uncool." *Die Zeit* 54, no. 11 (3 March): Wirtschaft, 17.

Die Zeit. 1999b. "Alles Geschmackssache." *Die Zeit* 54, no. 48 (11 November): Dossier, 18.

Die Zeit. 1999c. "Neues Vorbild: Manager und Märkte Rohstoff Holz." *Die Zeit* 54, no. 49 (12 December): Wirtschaft, 37.
Die Zeit. 1999d. "Ärger: Manager und Märkte Elbe." *Die Zeit* 54, no. 51 (16 December): Wirtschaft, 33.
DiMaggio, Paul J., and Walter W. Powell. 1991a. "Introduction." In *The New Institutionalism in Organizational Analysis,* ed. Walter W. Powell and Paul J. DiMaggio. Chicago: University of Chicago Press. Pp. 1–38.
DiMaggio, Paul J., and Walter W. Powell. 1991b. "The Iron Cage Revisited: Institutional Isomorphism and Collective Rationality." In *The New Institutionalism in Organizational Analysis,* ed. Walter W. Powell and Paul J. DiMaggio. Chicago: University of Chicago Press. Pp. 63–82.
Directorate General Environment. 2005. "The Attitudes of European Citizens toward Environment." (Special Eurobarometer 217/Wave 62.1). Brussels: Directorate General Environment, European Union.
Ditt, Karl. 2003. "Die Anfänge der Naturschutzgesetzgebung in Deutschland und England 1935/49." In *Naturschutz und Nationalsozialismus,* ed. Joachim Radkau and Frank Uekötter. Frankfurt: Campus. Pp. 107–144.
Dix, Andreas. 2003. "Nach dem Ende der 'Tausend Jahre': Landschaftsplanung in der Sowjetischen Besatzungszone und Frühen DDR." In *Naturschutz und Nationalsozialismus,* ed. Joachim Radkau and Frank Uekötter. Frankfurt: Campus. Pp. 331–362.
DNR Deutschland Rundbrief. 2000a. "Umweltverbände mit 6,8 Millionene Mark Gefördert." *DNR Deutschland Rundbrief* 6, no. 4 (April): 10.
DNR Deutschland Rundbrief. 2000b. "Schlecht Gedämmte Gebäude sind Bremsklotz für Klimapolitik." *DNR Deutschland Rundbrief* 6, no. 4 (April): 39.
DNR Deutschland Rundbrief. 2000c. "BUND: Zivildienst in Sozialen und Ökologischen Bereichen Muss trotz Bundeswehrreform Ausgebaut Werden." *DNR Deutschland Rundbrief* 6, no. 6/7 (June/July): 36.
DNR Deutschland Rundbrief. 2000d. "Wunderwelt der Vielfalt." *DNR Deutschland Rundbrief* 6, no. 6/7 (June/July): 41.
DNR Deutschland Rundbrief. 2000e. "Glückwünsche an den BUND zum 25. Jubiläum." *DNR Deutschland Rundbrief* 6, no. 6/7 (June/July): 42.
DNR Deutschland Rundbrief. 2000f. "Freiburg Solar Erfahren." *DNR Deutschland Rundbrief* 6, no. 6/7 (June/July): 46.
DNR Deutschland Rundbrief. 2000g. "Umweltbewegung als Gesellschaftspolitische Kraft: Chance für den Deutschen Naturschutzring." *DNR Deutschland Rundbrief* 6, no. 8/9 (August/September): 15.
DNR Deutschland Rundbrief. 2000h. "Greenpeace Verstärkt Weltweit Lösungskampagnen." *DNR Deutschland Rundbrief* 6, no. 8/9 (August/September): 48.
DNR Deutschland Rundbrief. 2000i. "Nationales Klimaschutzprogramm der Bundesregierung." *DNR Deutschland Rundbrief* 6, no. 10 (October): 7–9.
DNR Deutschland Rundbrief. 2000j. "Hiwi statt Zivil." *DNR Deutschland Rundbrief* 6, no. 10 (October): 49.
DNR Deutschland Rundbrief. 2000k. "Fast so Alt wie die Bundesrepublik...." *DNR Deutschland Rundbrief* 6, no. 11/12 (November/December): 19–21.
DNR Deutschland Rundbrief. 2000l. "Biopatentrecht: Enquête-Kommission und Bundesrat Kritisch gegenüber Umweltzung der EU-Richtlinie." DNR *Deutschland Rundbrief* 6, no. 11/12 (November/December): 36.
DNR Deutschland Rundbrief. 2001a. "Bürgern Schmeckt die Agrarwende." *DNR Deutschland Rundbrief* 7, no. 1/2 (January/February): 5.

DNR Deutschland Rundbrief. 2001b. "WWF und NABU zur Agrar- und Umweltministerkonferenz." *DNR Deutschland Rundbrief* 7, no. 6 (June): 6.
DNR Deutschland Rundbrief. 2001c. "Neues Antragsverfahren für Förderprojekte des BMU." *DNR Deutschland Rundbrief* 7, no. 7/8 (July/August): 13.
DNR Deutschland Rundbrief. 2001d. "Umweltschutz Macht Arbeit." *DNR Deutschland Rundbrief* 7, no. 7/8 (July/August): 34.
DNR Deutschland Rundbrief. 2001e. "BUND und B.KWK: Bundeswirtschaftminister Torpediert Effektiven Klimaschutz." *DNR Deutschland Rundbrief* 7, no. 9 (September): 9.
DNR Deutschland Rundbrief. 2001f. "Mit Abfallkatalog Klarheit Schaffen." *DNR Deutschland Rundbrief* 7, no. 9 (September): 13.
DNR Deutschland Rundbrief. 2001g. "B.A.U.M. - Unternehmen Engagiert im Klimaschutz." *DNR Deutschland Rundbrief* 7, no. 9 (September): 19.
DNR Deutschland Rundbrief. 2001h. "Kampagnen-Erfahrungen hinter den Kulissen." *DNR Deutschland Rundbrief* 7, no. 9 (September): 21.
DNR Deutschland Rundbrief. 2001i. "IÖW-Gutachten zu Elbausbau." *DNR Deutschland Rundbrief* 7, no. 9 (September): 27.
DNR Deutschland Rundbrief. 2001j. "Sonderkonferenz d. HELCOM-Staaten in Kopenhagen." *DNR Deutschland Rundbrief* 7, no. 9 (September): 32.
DNR Deutschland Rundbrief. 2001k. "Rückkehr zur Gentechnikfreien Landwirtschaft Sei Unmöglich." *DNR Deutschland Rundbrief* 7, no. 10 (October): 6.
DNR Deutschland Rundbrief. 2001l. "'Stilllegung der Atomkraftwerke Ist Atom- und Verfassungsrechtlich Zwingend." *DNR Deutschland Rundbrief* 7, no. 10 (October): 15.
DNR Deutschland Rundbrief. 2002a. "Castor-Gegner Sollen Zahlen." *DNR Deutschland Rundbrief* 8, no. 3 (March): 11.
DNR Deutschland Rundbrief. 2002b. "Industrie Sammelt Information über 'Gegner.'" *DNR Deutschland Rundbrief* 8, no. 4 (April): 30.
DNR Deutschland Rundbrief. 2002c. "Umweltminister und Verbände zu Wichtigen Themen Einig." *DNR Deutschland Rundbrief* 8, no. 6/7 (June/July): 18.
DNR Deutschland Rundbrief. 2002d. "Umweltverbände: 'Tafelsilber der Einheit' Jetzt Sichern." *DNR Deutschland Rundbrief* 8, no. 11/12 (November/December): 28.
DNR Deutschland Rundbrief. 2003a. "Überwiegand Positive: Umweltbilanz des Jahres 2002." *DNR Deutschland Rundbrief* 9, no. 1/2 (January/February): 26.
DNR Deutschland Rundbrief. 2003b. "Warnung vor Umweltpolitischem Rollback." *DNR Deutschland Rundbrief* 9, no. 3 (March): 5.
DNR Deutschland Rundbrief. 2003c. "Umwelt: Kein Thema in den Medien Mehr?" *DNR Deutschland Rundbrief* 9, no. 3 (March): 22.
DNR Deutschland Rundbrief. 2003d. "Spitzentreffen des DNR mit CDU-Chefin Angela Merkel." *DNR Deutschland Rundbrief* 9, no. 3 (March): 41.
DNR Deutschland Rundbrief. 2003e. "Neuer NABU-Präsident." *DNR Deutschland Rundbrief* 9, no. 6 (June): 38.
DNR Deutschland Rundbrief. 2003f. "Diesel Weiter Billig." *DNR Deutschland Rundbrief* 9, no. 7 (July): 22.
DNR Deutschland Rundbrief. 2004a. "Jahresbilanz der Umweltverbände." *DNR Deutschland Rundbrief* 10, no. 1/2 (January/February): 22.
DNR Deutschland Rundbrief. 2004b. "2001–2004: Das DNR Präsidium Zieht Balanz." *DNR Deutschland Rundbrief* 10, no. 10 (October): 4–10.
DNR Deutschland Rundbrief. 2004c. "Halbzeit bei Rot-Grün unter der Ökologischen Lupe." *DNR Deutschland Rundbrief* 10, no. 10 (October): 11.
DNR Deutschland Rundbrief. 2005a. "Kernforderung der Umweltverbände zur Wahl." *DNR Deutschland Rundbrief* 11, no. 9 (September): 5.

DNR Deutschland Rundbrief. 2005b. "Industrie Will keinen Verbindlichen Klimaschutz." *DNR Deutschland Rundbrief* 11, no. 9 (September): 16.

Doherty, Brian. 1994. "The Fundi-Realo Controversy: An Analysis of Four European Green Parties." *Environmental Politics* 1, no. 1 (spring): 95–120

Dölle, Wilgard, and Reiner Lüginbühl. 1985. "Öffentlichkeitsarbeit der Naturschutzverbände." In *Naturschutzverbände und Umweltpolitik,* ed. Institut für Landschaftspflege und Naturschutz und Institut für Landschaftsplanung und Raumforschung. Hanover: Institut für Landschaftspflege und Naturschutz und Institut für Landschaftsplanung und Raumforschung. Pp. 34–46.

Dominick, Raymond H. III. 1986. "Nascent Environmental Protection in the Second Empire." *German Studies Review* 9, no. 2 (May): 257–291.

Dominick, Raymond H. III. 1987. "The Nazis and the Nature Conservationists." *The Historian* 49, no. 4 (August): 508–537.

Dominick, Raymond H. III. 1988. "The Roots of the Green Movement in the United States and West Germany." *Environmental Review* 12, no. 3 (fall): 1–30.

Dominick, Raymond H. III. 1992. *The Environmental Movement in Germany.* Bloomington: Indiana University Press.

Dowie, Mark. 1995. *Losing Ground: American Environmentalism at the Close of the Twentieth Century.* Cambridge, MA: MIT Press.

Drews, Hans-Peter. 1995. "Umweltsponsoring: Chancen einer Neuen Kommunikation." In *Umweltbewußtsein und Massenmedien,* ed. Gerhard de Haan. Berlin: Akademie Verlag. Pp. 177–184.

Dryzek, John S. et al. 2003. *Green States and Social Movements.* New York: Oxford University Press.

Dulk, Christiane. 1984. "Neubeginn." In *Mit Uns Zieht die Neue Zeit: Die Naturfreunde,* ed. Jochen Zimmer. Cologne: Pahl-Rugenstein. Pp. 118–140.

Dulk, Christiane, and Jochen Zimmer. 1984. "Die Auflösung des Touristenvereins 'Die Naturfreunde' nach dem März 1933." In *Mit Uns Zieht die Neue Zeit: Die Naturfreunde,* ed. Jochen Zimmer. Cologne: Pahl-Rugenstein. Pp. 112–117.

Dyson, Kenneth. 1982. "Party Government and Party State." In *Party Government and Political Culture in West Germany,* ed. Herbert Döring and Gordon Smith. New York: St. Martin's Press. Pp. 77–100.

Ebert, Theodor. 1980. "Konfliktformation im Wandel: Von den Bürgerinitiativen zur Ökologiebewegung." In *Bürgerinitiativen in der Gesellschaft: Politische Dimensionen und Reaktionen,* ed. Otthein Rammstedt. Villingen: Neckar-Verlag. Pp. 351–371.

Eckersley, Robin. 1989. "Green Politics and the New Class: Selfishness or Virtue?" *Political Studies* 37, no. 2 (June): 205–223.

Eder, Klaus. 1985. "The New Social Movements: Moral Crusades, Political Pressure Groups, or Social Movements?" *Social Research* 52, no. 4 (winter): 869–890.

Eder, Klaus. 1996. "The Institutionalization of Environmentalism." In *Risk, Modernity, and Environment: Towards a New Ecology,* ed. S. B. Lash, N. Szerszinski, and B. Wynee. London: Sage. Pp. 203–223.

Edwards, Bob, and John D. McCarthy. 2004. "Resource and Social Movement Mobilization." In *The Blackwell Companion to Social Movements,* ed. David A. Snow, Sarah A. Soule, and Hanspeter Kriesi. Malden, MA: Blackwell. Pp. 116–152.

Edwards, Michael. 2004. *Civil Society.* Cambridge: Polity Press.

Ehlers, Jörn. 1995. "Kokettieren mit dem Kartoffeldruck: Außendarstellungen von Non-Profit- und Anderen Unternehmen." In *Umweltbewußtsein und Massenmedien,* ed. Gerhard de Haan. Berlin: Akademie Verlag. Pp. 167–175.

Eitner, Kerstin. 1996. "Greenpeace Chronik 1971 bis 1996." In *Das Greenpeace Buch: Reflexionen und Aktionen*, ed. Greenpeace Deutschland. Munich: C.H. Beck. Pp. 296–307.

Ellwein, Thomas. 1984. "Bürgerinitiativen und Verbände." In *Bürgerinitiativen und Repräsentatives System*, ed. Bernd Guggenberger and Udo Kempf. Opladen: Westdeutscher Verlag. Pp. 239–256.

EMNID Institute. 1993. "Der WWF in der Sicht der Bevölkerung 1993." Bielefeld: EMNID Institute.

Enders, Gisela. 1995. "German Unification and the Chances for the Environment: A BUND Perspective." In *The Green Agenda: Environmental Politics and Policy in Germany*, ed. Ingolfur Blühdorn, Frank Krause, and Thomas Scharf. Keele: Keele University Press. Pp. 227–234.

Engels, Jens Ivo. 2003. "'Hohe Zeit' und 'Dicker Strich': Vergangenheitsdeutung und -bewahrung im Westdeutschem Naturschutz nach dem Zweiten Weltkrieg." In *Naturschutz und Nationalsozialismus*, ed. Joachim Radkau and Frank Uekötter. Frankfurt: Campus. Pp. 363–404.

Engelhardt, Wolfgang. 2000. "Die Rolle der Umweltverbände." In *Umweltpolitik mit Augenmaß*, ed. Henning von Köller. Berlin: Erich Schmidt Verlag. Pp. 87–97.

Erdmann, Wulf. 1991. "Mit dem Wandern Fing Es an: Kurze Geschichte der Naturfreunde." In *Hundert Jahre Kampf um Freie Natur*, ed. Wulf Erdmann and Jochen Zimmer. Essen: Klartext. Pp. 10–36.

Erz, Wolfgang. 1987. "Naturschutz im Wandel der Zeit: Eine Bewertung." *Geographische Rundschau* 39, no. 6 (June): 307–315.

European Commission. 1999. *What Do Europeans Think about the Environment?* Luxembourg: Office for Official Publications of the European Communities.

European Commission. 2005. *Eurobarometer 63: Public Opinion in the European Union.* Brussels: European Commission.

European Environmental Bureau. 2007. "How the EEB Works." Downloaded 29 January 2007. http://www.eeb.org/how_the_EEB_works/Index.htm

European Opinion Research Group. 2002. *Attitudes of Europeans toward the Environment.* (Eurobarometer 58.0). Brussels: Directorate-General Environment, European Union.

Ewringmann, Dieter, and Klaus Zimmermann. 1978. "Umweltpolitische Interessenanalyse der Unternehmen, Gewerkschaften, und Gemeinden." In *Umweltpolitik: Beiträge zur Politologie des Umweltschutzes*, ed. Martin Jänicke. Opladen: Leske + Budrich. Pp. 66–100.

Eyerman, Ron, and Andrew Jamison. 1989. "Environmental Knowledge as an Organizational Weapon: The Case of Greenpeace." *Social Science Information* 28, no. 1 (March): 99–119.

Faber, Daniel. 1998. "The Political Ecology of American Capitalism: New Challenges for the Environmental Justice Movement." In *The Struggle for Ecological Democracy: Environmental Justice Movements in the United States*, ed. Daniel Faber. New York: Guilford Press. Pp. 27–59.

FDP. 2007. "Arbeit har Vorfahrt: Deutschlandprogramm 2005." Downloaded 27 January 2007. http://files.liberale.de/fdp-wahlprogramm.pdf

Felbinger, Doris. 2003. "Umweltschützer als Strategen: Strategischer Umweltschutz?" In *Mission Impossible? Strategien in Nonprofitorganisationen*, ed. Arbeitskreis Nonprofit-Organisationen. Frankfurt: Eigenverlag des Deutschen Vereins für Öffentliche und Private Fürsorge. Pp. 69–95.

Felbinger, Doris. 2005. *Ego oder Öko: Spenden and Umweltschutzorganisationen.* Berlin: Logos Verlag.

Fisher, Dana R. 2006. *Activism Inc.: How the Outsourcing of Grassroots Campaigns Is Strangling Progressive Politics in America.* Stanford, CA: Stanford University Press.

Fischer, Helmut. 1994. *90 Jahre für Umwelt und Naturschutz: Geschichte eines Programms.* Bonn: Deutscher Heimatbund.

Flasbarth, Jochen. 1992. "Strategiedebatte und Krise der Umweltpublizistik." *Ökologische Briefe,* no volume, no. 38 (September): 13–15.

Flasbarth, Jochen. 1993. "Nur Gemeinsam eine Chance." *Politische Ökologie* 11, no. 31 (May/June): 61–62.

Flasbarth, Jochen. 1995. "Ziele und Probleme des Umwelt-Sponsoring aus Sicht von Umweltverbänden." In *Ökosponsoring: Werbestrategie oder Selbstverpflichtung?* ed. Bayrische Akademie für Naturschutz und Landschaftspflege und Norddeutsche Naturschutzakademie. Schneeverdingen: Norddeutsche Naturschutzakademie. Pp. 51–53.

Flechner, Ralf. 1999. "Die Greenpeace-Gruppen vor Ort." In *Perspektiven Gesellschaftlichen Zusammenhalts: Empirische Befunde, Praxiserfahrungen, Meßkonzepte,* ed. Ernst Kister, Heinz-Herbert Noll, and Eckhard Priller. Berlin: Edition Sigma. Pp. 371–376.

Foley, Michael, and Bob Edwards. 1996. "The Paradox of Civil Society." *Journal of Democracy* 7, no. 3 (July): 38–52.

Foljanty-Jost, Gesine. 2004. "Environmental NGO's in Germany: The Question of Power and Influence." In *The Environmental Challenges for Japan and Germany,* ed. György Széll and Ken'ichi Tominaga. Frankfurt: Peter Lang. Pp. 121–136.

Forum Umwelt und Entwicklung. 2006. "Über das Forum." Downloaded 5 December 2006. http://www.forumue.de

Foster, John Bellamy. 1999. "Marx's Theory of Metabolic Rift: Classical Foundations of Environmental Sociology." *American Journal of Sociology* 105, no. 2 (September): 366–405.

Frankfurter Allgemeine Zeitung. 1997a. "Öko-Realismus." *Frankfurter Allgemeine Zeitung* 48, no. 1 (2 January): Zeitgeschehen, 8.

Frankfurter Allgemeine Zeitung. 1997b. "Strahlung von Castor Nicht Nachweisbar." *Frankfurter Allgemeine Zeitung* 48, no. 105 (7 May): Natur und Wissenschaft, 41.

Frankfurter Allgemeine Zeitung. 1997c. "Im Schweinsgallop von Smart zu Smile." *Frankfurter Allgemeine Zeitung* 48, no. 218 (10 September): Wirtschaft, 25.

Frankfurter Allgemeine Zeitung. 1998a. "Castor-Blockade War Straftat." *Frankfurter Allgemeine Zeitung* 49, no. 38 (14 February): Politik, 4.

Frankfurter Allgemeine Zeitung. 1998b. "Erster Schritt zum Grünen PC." *Frankfurter Allgemeine Zeitung* 49, no. 34 (10 February): Technik und Motor, T1.

Frankfurter Allgemeine Zeitung. 1998c. "Aus dem Frühjahrsprogramm der Verlage." *Frankfurter Allgemeine Zeitung* 49, no. 69 (23 March): Feuilleton, 46.

Frankfurter Allgemeine Zeitung. 1998d. "Naturschutz und Eigentum." *Frankfurter Allgemeine Zeitung* 49, no. 70 (24 March): Die Gegenwart, 10.

Frankfurter Allgemeine Zeitung. 1998e. "Immer Mehr Privatpersonen Fördern die Arbeit des WWF Deutschland." *Frankfurter Allgemeine Zeitung* 49, no. 110 (13 May): Wirtschaft, 18.

Frankfurter Allgemeine Zeitung. 1998f. "Kritik an 'Pakt der Halbherzigen'." *Frankfurter Allgemeine Zeitung* 49, no. 243 (20 October): Politik, 4.

Frankfurter Allgemeine Zeitung. 1998g. "Naturschutzbund Fordert eine Korrektur des Ökosteuer-Gesetzentwurfs." *Frankfurter Allgemeine Zeitung* 49, no. 287 (10 December): Wirtschaft, 18.

Frankfurter Allgemeine Zeitung. 1998h. "Agrar- und Regionalpolitik der EU Aufeinander Abstimmen." *Frankfurter Allgemeine Zeitung* 49, no. 302 (12 December): Politik, 4.

Frankfurter Allgemeine Zeitung. 1999a. "Greenpeace-Chef Homolka Entlassen." *Frankfurter Allgemeine Zeitung* 50, no. 32 (2 February): Wirtschaft, 20.

Frankfurter Allgemeine Zeitung. 1999b. "Schröder Brüskiert Naturschützer." *Frankfurter Allgemeine Zeitung* 50, no. 43 (20 February): Politik, 1.

Frankfurter Allgemeine Zeitung. 1999c. "Kritik an den Agrarplänen der Europäischen Union." *Frankfurter Allgemeine Zeitung* 50, no. 43 (20 February): Wirtschaft, 14.

Frankfurter Allgemeine Zeitung. 1999d. "Die Aufbruchstimmung auf dem Markt für Ökostrom Wird durch Komplizierte Durchleitungsregelungen Getrübt." *Frankfurter Allgemeine Zeitung* 50, no. 59 (11 March): Wirtschaft, 20.

Frankfurter Allgemeine Zeitung. 1999e. "Hilger Unterliegt beim Bundesverfassungsgericht." *Frankfurter Allgemeine Zeitung* 50, no. 129 (6 June): Wirtschaft, 22.

Frankfurter Allgemeine Zeitung. 1999f. "Die Elbe Ist eine Ausnahme in Deutschland." *Frankfurter Allgemeine Zeitung* 50, no. 161 (15 July): Reiseblatt, R6.

Frankfurter Allgemeine Zeitung. 1999g. "BUND Kritisiert Rot-Grüne Politik." *Frankfurter Allgemeine Zeitung* 50, no. 168 (23 July): Politik, 4.

Frankfurter Allgemeine Zeitung. 1999h. "Mit Naiven Umwelt-Spezialisten Kann Keiner Etwas Anfangen." *Frankfurter Allgemeine Zeitung* 50, no. 175 (31 July): Beruf und Chance, 59.

Frankfurter Allgemeine Zeitung. 1999i. "Greenpeace Will Strom Anbieten." *Frankfurter Allgemeine Zeitung* 50, no. 261 (9 November): Wirtschaft, 31.

Frankfurter Allgemeine Zeitung. 1999j. "IG Bau für Wärmedämmung." *Frankfurter Allgemeine Zeitung* 50, no. 277 (21 November): Wirtschaft, 14.

Frankfurter Rundschau. 1991. "Von Weltverbesserern zu Marktstrategen." *Frankfurter Rundschau* 47, no. 269 (19/20 November): Berichte, 6.

Frankfurter Rundschau. 1996. "BUND Verzichtet zunächst auf Klinikenputzer." *Frankfurter Rundschau* 52, no. 94 (22 April): Hessen, 18.

Frankfurter Rundschau. 1999. "Fast wie beim Denver Clan." *Frankfurter Rundschau* 55, no. 113 (18 May): Umwelt und Wissenschaft, 6.

Frankfurter Rundschau. 2000. "Die Zerstörerische Kraft des Kapitalismus: Ist die Natur noch zu Retten?" *Frankfurter Rundschau* 56, no. 16 (20 January): Dokumentation, 9.

Frankland, E. Gene. 1995. "The Rise and Fall of Die Grünen." In *The Green Challenge,* ed. Dirk Richardson and Chris Rootes. London: Routledge. Pp. 23–44.

Frederichs, Günter. 1980. "Ursachen und Entwicklungstendenzen der Opposition gegen die Kernenergie." *Zeitschrift für Umweltpolitik* 3, no. 3 (September): 681–705.

Friedman, Debra, and Doug McAdam. 1992. "Collective Identity and Activism: Networks, Choices and the Life of a Social Movement." In *Frontiers in Social Movement Theory,* ed. Aldon D. Morris and Carol McClurg Mueller. New Haven, CT: Yale University Press. Pp. 156–173.

Friends of the Earth International. 2001. *Sparks of Hope, Fires of Resistance: FOEI Celebrates the Sustainable Path Forward.* Amsterdam: Friends of the Earth International Secretariat.

Friends of the Earth International. 2005. *Annual Report 2004: Friends of the Earth International.* Brussels: Friends of the Earth International.

Fritzler, Marc. 1997. *Ökologie und Umweltpolitik.* Bonn: Bundeszentrale für Politische Bildung.

Fromson, Sandra Bender. 2003. "Mission versus Maintenance: Social Movement Organizations' Multiple Forms." *Sociological Focus* 36, no. 3 (August): 257–273.

Fuchs, Dieter, and Dieter Rucht. 1994. "Support for New Social Movements in Five Western European Countries." In *Social Change and Political Transformations,* ed. Chris Rootes and Howard Davis. London: UCC Press. Pp. 86–111.

Fulbrook, Mary. 1995. *Anatomy of a Dictatorship: Inside the GDR, 1949–1989.* Oxford: Oxford University Press.

Fulbrook, Mary. 2005. *The People's State: East German Society from Hitler to Honecker.* New Haven, CT: Yale University Press.

Fung, Archon. 2003. "Associations and Democracy: Between Hopes and Realities." *Annual Review of Sociology* 29: 515–539.

Gamson, William A. 1991. "Commitment and Agency in Social Movements." *Sociological Forum* 6, no. 1 (March): 27–50.

Gamson, William A. 1992. "The Social Psychology of Collective Action." In *Frontiers in Social Movement Theory,* ed. Aldon D. Morris and Carol McClurg Mueller. New Haven, CT: Yale University Press. Pp. 53–76.

Gamson, William A. 1995. "Constructing Social Protest." In *Social Movements and Culture,* ed. Hank Johnston and Bert Klandermans. Minneapolis: University of Minnesota Press.

Gamson, William A., and Gadi Wolfsfeld. 1993. "Movements and Media as Interacting Systems." *Annals of the American Academy of Political and Social Science* 528, no. 1 (July): 114–125.

Garner, Curt. 1995. "Public Service Personnel in West Germany in the 1950s: Controversial Policy Decisions and their Effects on Social Composition, Gender Structure, and the Role of Former Nazis." *Journal of Social History* 29, no. 1 (fall): 25–80.

Gebers, Betty, Ralf Jülich, Peter Küppers, and Gerhard Roller. 1996. *Bürgerrechte im Umweltschutz: Impulse für ein Konzept zur Stärkung der Beteiligungsrechte im Umweltverfahren.* Darmstadt: Öko-Institut.

Geden, Oliver. 1999. *Rechte Ökologie: Umweltschutz zwischen Emanzipation und Faschismus.* Berlin: Elefanten Press.

Gensichen, Hans-Peter. 1991. "Kritisches Umweltengagement in den Kirchen." In *Zur Freiheit Berufen: Die Kirche in der DDR als Schutzraum der Opposition 1981–1989,* ed. Jürgen Israel. Berlin: Aufbau Taschenbuch Verlag. Pp. 146–170.

Gensichen, Hans-Peter. 1994. "Das Umweltengagement in den Evangelischen Kirchen in der DDR." In *Umweltgeschichte: Wissenschaft und Praxis,* ed. Hermann Behrens and Horst Paucke. Marburg: Verlag des Bundes Demokratischer Wissenschaftlerinnen und Wissenschaftler. Pp. 65–83.

Gensichen, Hans-Peter. 1998. "Zur Geschichte der Kirchlichen Umweltbewegung in der DDR." In *Umweltschutz in Ostdeutschland und Osteuropa: Bilanz und Perspektiven,* ed. Fritz Brickwedde. Osnabrück: Steinbacher. Pp. 181–191.

Gensichen, Hans-Peter. 2005. "Umweltverantwortung in einer Betonierten Gesellschaft: Anmerkungen zur Kirchlichen Umweltarbeit in der DDR 1970 bis 1990." In *Natur und Umweltschutz nach 1945: Konzepte, Konflikte, Kompetenzen,* ed. Franz-Josef Brüggemeier and Jens Ivo Engels. Frankfurt: Campus. Pp. 287–304.

Gensicke, Thomas, Sibylle Picot, and Sabine Geiss. 2005. *Freiwilliges Engagement in Deutschland 1999–2004: Ergibnisse der Repräsentativen Trenderhebung zu Ehrenamt, Freiwilligenarbeit und Bürgerschaftlichem Engagement.* Munich: TNS Infratest Sozialforschung.

George, A. A. 1955. "Fünfzig Jahre Naturfreundebewegung in Deutschland." In *Touristenverein "Die Naturfreunde": Denkschrift zum Sechzigjährigen Bestehen 1895–1955,* ed. Zentralausschuß der Naturfreunde-Internationale. Zurich: Zentralausschuß der Naturfreunde-Internationale. Pp. 73–78.

Georgi-Valtin, Albert. 1955a. "Von 1906–1933." In *Touristenverein "Die Naturfreunde": Denkschrift zum Sechzigjährigen Bestehen 1895–1955,* ed. Zentralausschuß der Naturfreunde-Internationale. Zurich: Zentralausschuß der Naturfreunde-Internationale. Pp. 31–33.

Georgi-Valtin, Albert. 1955b. "Die Naturkundliche Arbeit." In *Touristenverein "Die Naturfreunde": Denkschrift zum Sechzigjährigen Bestehen 1895–1955,* ed. Zentralausschuß der Naturfreunde-Internationale. Zurich: Zentralausschuß der Naturfreunde-Internationale. Pp. 144–146.

Gerhard, Gesine. 2003. "Richard Walther Darré: Naturschützer oder 'Rassenzüchter'?" In *Naturschutz und Nationalsozialismus,* ed. Joachim Radkau and Frank Uekötter. Frankfurt: Campus. Pp. 257–272.

German Federal Foreign Office. 2006. *Facts about Germany.* Frankfurt: Societäts-Verlag.

German Watch. 2006. "German Watch." Downloaded 27 December 2006. http://www.germanwatch.org

Gilsenbach, Reimar. 1989. "Der Minister Blieb, die Grünen Kommen." In *Aufbruch in eine Andere DDR,* ed. Hubertus Knabe. Reinbek bei Hamburg: Rowohlt. Pp. 107–117.

Godwin, R. Kenneth. 1988. *One Billion Dollars of Influence.* Chatham, NJ: Chatham House.

Greenpeace Deutschland. 1995. *Greenpeace 1994: Jahresrückblick: Kampagnen, Finanzen, Hintergründe.* Hamburg: Greenpeace Deutschland.

Greenpeace Deutschland, ed. 1996. *Das Greenpeace Buch: Reflexionen und Aktionen.* Munich: C.H. Beck.

Greenpeace Deutschland. 1997. *Jahresrückblick 1996.* Hamburg: Greenpeace Deutschland.

Greenpeace Deutschland. 1999. *Jahresrückblick 1998.* Hamburg: Greenpeace Deutschland.

Greenpeace Deutschland. 2000. *Jahresüberblick, 1999.* Hamburg: Greenpeace Deutschland.

Greenpeace Deutschland. 2002. *Jahresrückblick 2001.* Hamburg: Greenpeace Deutschland.

Greenpeace Deutschland. 2003. *Jahresrückblick 2002.* Hamburg: Greenpeace Deutschland.

Greenpeace Deutschland. 2004. *Jahresrückblick 2003.* Hamburg: Greenpeace Deutschland.

Greenpeace Deutschland. 2005a. *Jahresrückblick 2004.* Hamburg: Greenpeace Deutschland.

Greenpeace Deutschland. 2005b. "Greenpeace: Die Wichtigsten Fakten." Downloaded 21 May 2005 http://www.greenpeace.de/fileadmin/gpd/user_upload/wir_ueber_uns/GP_wichtigste_fakten.pdf#search=%22Greenpeace%3A%20%20Die%20Wichtigsten%20Fakten%22

Greenpeace Deutschland. 2005c. "Wahlkompass Umweltpolitik" (brochure). Hamburg: Greenpeace Deutschland.

Greenpeace Deutschland. 2007a. "Themen." Downloaded 14 February 2007. http://www.greenpeace.de/themen

Greenpeace Deutschland. 2007b. "Tierpark Arche Warder." Downloaded 14 February 2007. http://www.greenpeace.de/themen/landwirtschaft/alternativen/artikel/tierpark_arche_warder/ansicht/bild/

Greenpeace Deutschland. No date. "Umweltstiftung Greenpeace" (brochure). Hamburg: Greenpeace Deutschland.

Greenpeace International. 2006a. *Greenpeace Annual Report 2005.* Amsterdam: Greenpeace International.

Greenpeace International. 2006b. "Greenpeace International." Downloaded 9 June 2006. http://www.greenpeace.org/international

Greenpeace Magazin. 1996a. "Ein Tag im Leben von GREENPEACE." *Greenpeace Magazin,* no volume, no. 3 (May/June): Heft zum Greenpeace-Jubiläum 1971–1996: 4–11.

Greenpeace Magazin. 1996b. "Greenpeace Verändert das Denken." *Greenpeace Magazin,* no volume, no. 3 (May/June): Heft zum Greenpeace-Jubiläum 1971–1996, 12.

Greenpeace Magazin. 1996c. "25 Jahre Greenpeace." *Greenpeace Magazin,* no volume, no. 3 (May/June): Heft zum Greenpeace-Jubiläum 1971–1996, 33–48.

Greenpeace Magazin. 1996d. "Nachher ist man ein Bißchen Klüger." *Greenpeace Magazin,* no volume, no. 3 (May/June): Heft zum Greenpeace-Jubiläum 1971–1996, 58–59.

Greenpeace Nachrichten für Alle Förderinnen und Förderer. 1998. "Unser Engagement Bringt Spaß und Hält Jung." *Greenpeace Nachrichten für Alle Förderinnen und Förderer,* no volume, no. 4 (November–January): 7.

Greenpeace Nachrichten für Alle Förderinnen und Förderer. 1999a. "See-Seminare." *Greenpeace Nachrichten für Alle Förderinnen und Förderer,* no volume, no. 2 (February–May): 3.

Greenpeace Nachrichten für Alle Förderinnen und Förderer. 1999b. "Atom." *Greenpeace Nachrichten für Alle Förderinnen und Förderer,* no volume, no. 2 (February–May): 4.

Greenpeace Nachrichten für Alle Förderinnen und Förderer. 1999c. "Kreativer Aufbruch." *Greenpeace Nachrichten für Alle Förderinnen und Förderer,* no volume, no. 3 (no month): 2.

Greenpeace Nachrichten für Alle Förderinnen und Förderer. 1999d. "Umweltschutz Soll Laune Machen." *Greenpeace Nachrichten für Alle Förderinnen und Förderer* no volume, no. 4 (no month): 5.

Greenpeace Nachrichten für Alle Förderinnen und Förderer. 1999e. "Von Wegen Geheim: Greenpeace Weist die Vorwürfe Spendengelder zu Verschleudern Entscheiden Züruck." *Greenpeace Nachrichten für Alle Förderinnen und Förderer,* no volume, no. 4 (no month): 6.

Greenpeace Nachrichten für Alle Förderinnen und Förderer. 1999f. "Weniger Gremium, mehr Mitbestimmung." *Greenpeace Nachrichten für Alle Förderinnen und Förderer,* no volume, no. 4 (no month): 6.

Greenpeace Nachrichten für Alle Förderinnen und Förderer. 2000a. "430.000 Neue Jobs Dank Klimaschutz." *Greenpeace Nachrichten für Alle Förderinnen und Förderer,* no volume, no. 1 (February–April): 3.

Greenpeace Nachrichten für Alle Förderinnen und Förderer. 2000b. "Viel zu Schade zum Verschrotten." *Greenpeace Nachrichten für Alle Förderinnen und Förderer,* no volume, no. 1 (February–April): 6.

Greenpeace Nachrichten für Alle Förderinnen und Förderer. 2000c. "Die Aufmüpfigen aus der Chausseestraße 131." *Greenpeace Nachrichten für Alle Förderinnen und Förderer,* no volume, no. 1 (February–April): 7.

Greenpeace Nachrichten für Alle Förderinnen und Förderer. 2000d. "Mogel-Expo." *Greenpeace Nachrichten für Alle Förderinnen und Förderer,* no volume, no. 2 (May–July): 5.

Greenpeace Nachrichten für Alle Förderinnen und Förderer. 2000e. "Preise." *Greenpeace Nachrichten für Alle Förderinnen und Förderer,* no volume, no. 3 (August–October): 5.

Greenpeace Nachrichten für Alle Förderinnen und Förderer. 2001a. "Jahresbeschluss 2000." *Greenpeace Nachrichten für Alle Förderinnen und Förderer,* no volume, no. 3 (August–October): 6.

Greenpeace Nachrichten für Alle Förderinnen und Förderer. 2001b. "Sauberer Strom zu Fairen Preis." *Greenpeace Nachrichten für Alle Förderinnen und Förderer,* no volume, no. 4 (November–January): 2.

Greenpeace Nachrichten für Alle Förderinnen und Förderer. 2002a. "Schmutz-Konzern Esso." *Greenpeace Nachrichten für Alle Förderinnen und Förderer,* no volume, no. 2 (February–April): 1.

Greenpeace Nachrichten für Alle Förderinnen und Förderer. 2002b. "Starker Rückhalt." *Greenpeace Nachrichten für Alle Förderinnen und Förderer,* no volume, no. 2 (February–April): 6.

Greenpeace Nachrichten für Alle Förderinnen und Förderer. 2002c. "Votum für Mehr Ökologie." *Greenpeace Nachrichten für Alle Förderinnen und Förderer,* no volume, no. 4 (November–January): 9.

Greenpeace Nachrichten für Alle Förderinnen und Förderer. 2002d. "Jeder Euro Macht Uns Stärker." *Greenpeace Nachrichten für Alle Förderinnen und Förderer,* no volume, no. 4 (November–January): 3.

Greenpeace Nachrichten für Alle Förderinnen und Förderer. 2003a. "Schutz vor Kriminalisierung." *Greenpeace Nachrichten für Alle Förderinnen und Förderer,* no volume, no. 2 (February–April): 7.

Greenpeace Nachrichten für Alle Förderinnen und Förderer. 2003b. "Botschafter auf der Straße." *Greenpeace Nachrichten für Alle Förderinnen und Förderer,* no volume, no. 4 (November–January): 9.

Greenpeace Nachrichten für Alle Förderinnen und Förderer. 2004a. "Maulkorb für Greenpeace." *Greenpeace Nachrichten für Alle Förderinnen und Förderer,* no volume, no. 4 (August–October): 3–4.

Greenpeace Nachrichten für Alle Förderinnen und Förderer. 2004b. "Kein Club von Millionären." *Greenpeace Nachrichten für Alle Förderinnen und Förderer,* no volume, no. 4 (November–January): 6.

Greenpeace Nachrichten für Alle Förderinnen und Förderer. 2004c. "Ausrufezeichen am Lebensende." *Greenpeace Nachrichten für Alle Förderinnen und Förderer,* no volume, no. 4 (November–January): 9.

Greenpeace Nachrichten für Alle Förderinnen und Förderer. 2005a. "Wir Bleiben Unbequem." *Greenpeace Nachrichten für Alle Förderinnen und Förderer,* no volume, no. 1 (February–April): 1.

Greenpeace Nachrichten für Alle Förderinnen und Förderer. 2005b. "Florett statt Keule." *Greenpeace Nachrichten für Alle Förderinnen und Förderer,* no volume, no. 1 (February–April): 6.

Greenpeace Nachrichten für Alle Förderinnen und Förderer. 2005c. "Vertrauen in Greenpeace." *Greenpeace Nachrichten für Alle Förderinnen und Förderer,* no volume, no. 2 (May–July): 5.

Greenpeace Nachrichten für Alle Förderinnen und Förderer. 2005d. "Gemeinsam Stark!" *Greenpeace Nachrichten für Alle Förderinnen und Förderer,* no volume, no. 3 (August–October): 3.

Greenpeace Nachrichten für Alle Förderinnen und Förderer. 2005e. "Liebe Greenpeace-Föderinnen und –Förderer." *Greenpeace Nachrichten für Alle Förderinnen und Förderer,* no volume, no. 4 (November–January): 1.

Gröning, Gert. 1996. "Naturschutz und Nationalsozialismus." *Grüner Weg 31a* 10, no. 4 (December): 5–25.

Gröning, Gert, and Joachim Wolschke. 1986. "Soziale Praxis statt Ökologischer Ethik: Zum Gesellschafts- und Naturverständnis der Arbeiterbewegung." *Jahrbuch des Archivs der Deutschen Jugendbewegung,* vol. 15: 201–252.

Gröning, Gert, and Joachim Wolschke-Bulmahn. 1987. "Politics, Planning, and the Protection of Nature: Political Abuse of Early Ecological Ideas in Germany, 1933–45." *Planning Perspectives* 2, no. 2 (May): 127–148.

Gröning, Gert, and Joachim Wolschke-Bulmahn. 1993. "Ganz Deutschland ein Großer Garten: Landespflege und Stadtplanung im Nationalsozialismus." *Kursbuch,* no volume, no. 113 (June): 29–46.

Gröning, Gert, and Joachim Wolschke-Bulmahn. 1998. "Landschafts- und Naturschutz." In *Handbuch der Deutschen Reformbewegung 1880–1933,* ed. Diethart Kerbs and Jürgen Reulecke. Wuppertal: Peter Hammer Verlag. Pp. 23–34.

Groß, Matthias. 2001. *Die Natur der Gesellschaft: Eine Geschichte der Umweltsoziologie.* Weinheim: Juventa.

Großer, Karl Heinz. 1990/1991. "Naturschutz in Brandenburg 1945 bis 1990: Ein Rückblick im Zeitgeschehen," *Naturschutzarbeit in Berlin und Brandenburg,* vol. 26: 17–26.

Groth, Klaus-Henning. 2003. "Kommunikation ist nicht Alles, aber ohne Kommunikation ist Alles Nichts: Die Öffentlichkeitsarbeit des WWF-Deutschland." In *Das Große Buch des WWF: 40 Jahre Naturschutz für und mit den Menschen,* ed. Klaus-Henning Groth. Steinfurt: Tecklenborg Verlag. Pp. 179–189.

Grüger, Christine, and Irmgard Usersfeld. 1986. *Erfolgskontrolle der Verbandsbeteiligung in Nordrhein-Westfalen.* Dortmund: Diplomarbeit, Abteilung Raumplanung, Universität Dortmund.

Gruhl, Herbert. 1987. *Überleben Ist Alles: Erinnerungen des Autors von 'Ein Planet Wird Geplündert.'* Munich: F. A. Herbig.

Gruhn, Werner. 1972. "Umweltschutz in der DDR." *Deutschland Archiv* 5, no. 10 (October): 1038–1050.

Grüne Liga. 1999. *Tätigkeitsbericht 1999.* Berlin: Grüne Liga.

Grüne Liga. 2003. *Tätigkeitsbericht 2003.* Berlin: Grüne Liga.

Grüne Liga. 2005. *Konsequent für Natur und Umwelt, Mensch!: Tätigkeitsbericht 2005.* Berlin: Grüne Liga.

Grüne Liga Berlin. 2001. *Jahresbericht 2000/2001.* Berlin: Grüne Liga Berlin.

Grunenberg, Heiko, and Udo Kuckartz. 2003. *Umweltbewusstsein im Wandel.* Opladen: Leske + Budrich.

Grüßer, Birgit. 1995. "Ökosponsoring: Welche Kriterien Spielen für Unternehmer eine Rolle." In *Ökosponsoring: Werbestrategie oder Selbstverpflichtung?* ed. Bayrische Akademie für Naturschutz und Landschaftspflege und Norddeutsche Naturschutzakademie. Schneeverdingen: Norddeutsche Naturschutzakademie. Pp. 34–36.

Guggenberger, Bernd. 1984. "Von der Bürgerinitiativbewegung zur Umweltpartei: Zur Parteipolitischen Formierung des Grünen Protests." In *Bürgerinitiativen und Repräsentatives System,* ed. Bernd Guggenberger and Udo Kempf. Opladen: Westdeutscher Verlag. Pp. 376–403.

Günther, Michael. 1996. "Greenpeace und das Recht." In *Das Greenpeace Buch: Reflexionen und Aktionen,* ed. Greenpeace Deutschland. Munich: C.H. Beck. Pp. 65–81.

Haag, Dietrich W. 1986. "Die Umweltstiftung WWF-Deutschland." In *Lebensbilder Deutscher Stiftungen 5,* ed. Rolf Hauer et al. Tübingen: J.C.B. Mohr. Pp. 471–482.

Hagedorn, Friedrich, Peter Stawowy, and Heinz H. Meyer. 2002. "Umweltkommunikation im Hörfunk: Zwischen Quote und Kröte." In *Umweltkommunikation - Vom Wissen zum Handeln,* ed. Fritz Brickwedde and Ulrike Peters. Berlin: Erich Schmidt Verlag. Pp. 343–350.

Haibach, Marita. 1998a. *Handbuch Fundraising: Spenden, Sponsoring, Stiftungen in der Praxis.* Frankfurt: Campus.

Haibach, Marita. 1998b. "Spezifika der Finanzierung des Dritten Sektors." In *Dritter Sektor - Dritte Kraft: Versuch einer Standortbestimmung,* ed. Rupert Graf Strachwitz. Stuttgart: RAABE. Pp. 475–490.

Haibach, Marita. 2006. *Handbuch Fundraising: Spenden, Sponsoring, Stiftungen in der Praxis.* 2nd ed. Frankfurt: Campus.

Halbrock, Christian. 1995. "Störfaktor Jugend: Die Anfänge der Unabhängigen Umweltbewegung in der DDR." In *Arche Nova: Opposition in der DDR,* ed. Carlo Jordan and Hans Michael Kloth. Berlin: BasisDruck. Pp. 13–32.

Halcour, Florian. 1995. "Umweltsponsoring: Empfehlungen an Umweltschutz-Organisationen." In *Ökosponsoring: Werbestrategie oder Selbstverpflichtung?* ed. Bayrische Akademie für Naturschutz und Landschaftspflege und Norddeutsche Naturschutzakademie. Schneeverdingen: Norddeutsche Naturschutzakademie. Pp. 39–44.

Hall, Richard H. 2002. *Organizations: Structure, Processes, and Outcomes.* Upper Saddle River, NJ: Prentice-Hall.

Halperin, Samuel W. 1946. *Germany Tried Democracy: A Political History of the Reich from 1918 to 1933.* New York: Crowell.

Hampele, Anne. 1991. "Das Wahlbündnis 'Die Grünen/Bündnis 90 – BürgerInnenbewegung.'" In *Von der Illegalität ins Parlament: Werdegang und Konzept der Neuen Bürgerbewegungen,* ed. Helmut Müller-Enbergs, Marianne Schultz, and Jan Wielgohs. Berlin: LinksDruck. Pp. 307–341.

Hanemann, Horst, and Jürgen M. Simon. 1987. *Deutscher Bund für Vogelschutz e.V.: Die Chronik eines Naturschutzverbandes von 1899–1984.* Wiesbaden: Wirtschaftsverlag.

Hannan, Michael T., and John Freeman. 1977. "The Population Ecology of Organizations." *American Journal of Sociology* 82, no. 5 (March): 929–964.

Hannigan, John. 1995. *Environmental Sociology: A Social Constructionist Perspective.* London: Routledge.

Hansen, Anders. 1993. "Greenpeace and Press Coverage of Environmental Issues." In *The Mass Media and Environmental Issues,* ed. Anders Hansen. Leicester: Leicester University Press. Pp. 150–177.

Hart, Klaus. 2001. "Offen miteinander Reden." *Der Rabe Ralf* 12, no. 100 (October/November): 6.

Haßler, Robert. 1993. "Umweltschutz als Werbegag?" *Politische Ökologie* 11, no. 31 (May/June): 93–96.

Hayes, Michael T. 1983. "Interest Groups: Pluralism or Mass Society?" In *Interest Group Politics,* ed. Allan J. Ciglar and Burdett A. Loomis. Washington: Congressional Quarterly Press. Pp. 110–125.

Held, David. 1995. *Democracy and the Global Order.* Cambridge, MA: Polity Press.

Held, David. 2004. *Global Covenant: The Social Democratic Alternative to the Washington Consensus.* Cambridge, MA: Polity Press.

Hempel, Richard. 1930. "Heimatschutz in der Landwirtschaft." In *Der Deutsche Heimatschutz: Ein Rückblick und Ausblick,* ed. Gesellschaft der Freunde des Deutschen Heimatschutzes. Munich: Kastner & Callwey. Pp. 158–173.

Hermand, Jost. 1991. *Grüne Utopien in Deutschland: Zur Geschichte des Ökologischen Bewußtseins.* Frankfurt: Fischer Taschenbuch.

Hermand, Jost. 1993. "'Erst die Bäume, dann Wir!': Proteste gegen das Abholzen der Deutschen Wälder 1780–1950." In *Mit den Bäumen Sterben die Menschen,* ed. Jost Hermand. Cologne: Böhlau Verlag. Pp. 1–23.

Heuser, Stefan. 1994. "Das Ehrenamtliche Engagement in Umweltverbänden als Naturschutzpolitische Größe: Der Stand einer Ökologiebewegung am Beispiel der Umweltverbände." Unpublished Diplomarbeit im Fachbereich Landschaftsarchitektur und Umweltentwicklung, Institut für Landschaftspflege und Naturschutz. Hanover: Universität Hanover. Pp. 12–23.

Hey, Christian. 1994. *Umweltpolitik in Europa: Fehler, Risiken, Chancen.* Munich: C.H. Beck.

Hey, Christian. 2004. "Die Europäische Politik im Europa der 25." In *Jahrbuch Ökologie 2005,* ed. Günter Altner et al. Munich: C.H. Beck. Pp. 11–25.

Hey, Christian, and Uwe Brendle. 1994. *Umweltverbände und EG: Strategien, Politische Kulturen und Organisationsformen.* Opladen: Westdeutscher Verlag.

Hirche, Walter. 1998. "Umweltschutz in den Neuen Ländern: Bilanz und Perspektive." In *Umweltschutz in Ostdeutschland und Osteuropa: Bilanz und Perspektiven,* ed. Fritz Brickwedde. Osnabrück: Steinbacher. Pp. 13–30.

Hoffman, Jürgen. 2002. "From Cooperation to Confrontation: The Greens and the Ecology Movement in Germany." In *The Culture of German Environmentalism,* ed. Axel Goodbody. New York: Berghahn Books. Pp. 63–79.

Hoffman, Ludwig. 1998. "Umweltschutz als ein Motiv der Bürgerrechtsbewegung in der DDR: Erfüllte Erwartungen oder Enttäuschte Hoffnungen?" In *Umweltschutz in Ostdeutschland und Osteuropa: Bilanz und Perspektiven,* ed. Fritz Brickwedde. Osnabrück: Steinbacher. Pp. 160–170.

Hömberg, Walter. 1993. "Ökologie: Ein Schwieriges Medienthema." In *Krieg, Aids, Katastrophen: Gegenwartsprobleme als Herausforderung für die Publizistikwissenschaft,* ed. Heinz Bonfadelli and Werner A. Meier. Konstanz: Universitätsverlag. Pp. 81–94.

Hopfenbeck, Waldemar, and Peter Roth. 1994. *Öko-Kommunikation: Wege zu einer Neuen Kommunikationskultur.* Landsberg/Lech: Moderne Industrie.

Hoplitschek, Ernst. 1984. "Der Bund Naturschutz in Bayern: Traditioneller Naturschutzverband oder Teil der Neuen Sozialen Bewegung?" Unpublished Doctoral Dissertation, Berlin: Fachbereich Politische Wissenschaften, Freie Universität.

Hucke, Jochen. 1990. "Umweltpolitik: Die Entwicklung eines Neuen Politikfeldes." In *Politik in der Bundesrepublik Deutschland,* ed. Klaus von Beyme and Manfred G. Schmidt. Opladen: Westdeutscher Verlag. Pp. 382–398.

Hünemörder, Kai F. 2005. "Epochenschwelle der Umweltgeschichte?" In *Natur und Umweltschutz nach 1945: Konzepte, Konflikte, Kompetenzen,* ed. Franz-Josef Brüggemeier and Jens Ivo Engels. Frankfurt: Campus. Pp. 124–144.

Hunold, Christian. 2001a. "Nuclear Waste in Germany: Environmentalists between State and Society." *Environmental Politics* 10, no. 3 (fall): 127–133.

Hunold, Christian. 2001b. "Environmentalists, Nuclear Waste, and the Politics of Passive Exclusion in Germany." *German Politics and Society* 19, no. 4 (winter): 43–63.

Illig, Hubert, Helmut Donath, and Ralf Donat. 2001. "Biologischer Arbeitskreis 'Alwin Arndt' Luckau: 30 Jahre Naturkundliche Heimatforschung und Naturschutzarbeit in der Niederlausitz." In *Naturschutz in den Neuen Bundesländern: Ein Rückblick,* ed. Institut für Umweltgeschichte und Regionalentwicklung. Berlin: Verlag für Wissenschaft und Forschung. Pp. 281–289.

Inglehart, Ronald. 1990. "Values, Ideology, and Cognitive Mobilization in New Social Movements." In *Challenging the Political Order: New Social Movements in Western Democracies,* ed. Russell J. Dalton and Manfred Keuchler. New York: Oxford University Press. Pp. 43–66.

Ingram, Helen M., David H. Colnic, and Dean E. Mann. 1995. "Interest Groups and Environmental Policy." In *Environmental Politics and Policy: Theories and Evidence,* ed. James P. Lester. Durham, NC: Duke University Press. Pp. 115–145.

INRA (Europe) – ECO. 1995. "Europeans and the Environment in 1995" (Eurobarometer 43.1 BIS). Brussels: Directorate General Environment, Nuclear Safety and Civil Protection, European Union.

Institut der Deutschen Wirtschaft Köln. 2007. "Institut der Deutschen Wirtschaft Köln." Downloaded 11 January 2007. http://www.iwkoeln.de

Jäger, Wolfgang. 1984. "Bürgerinitiativen - Verbände - Parteien: Thesen zu einer Funktionalen Analyse." In *Bürgerinitiativen und Repräsentatives System,* ed. Bernd Guggenberger and Udo Kempf. Opladen: Westdeutscher Verlag. Pp. 217–232.

Jahn, Thomas, and Peter Wehling. 1990. *Ökologie von Rechts: Nationalismus und Umweltschutz bei der Neuen Rechten und den 'Republikanern.'* Frankfurt: Campus.

Jänicke, Martin, Philip Kunig, and Michael Stitzel. 1999. *Umweltpolitik.* Bonn: Dietz.

Jänicke, Martin, and Helmut Weidner. 1996. "Germany." In *National Environmental Policies: A Comparative Study of Capacity-Building,* ed. Martin Jänicke and Helmut Weidner. Berlin: Springer. Pp. 133–155.

Jannasch, Alexander. 2001. "Die Verbandsklage: Einige Bemerkungen aus der Sicht des Bundesverwaltungsgerichts." In *Die Verbandsklage: Ein Wirksames Instrument für den Natur- und Umweltschutz,* ed. Michael Zschiesche. Berlin: Unabhängiges Institut für Umweltfragen. Pp. 6–12.

Jaschke, Hans-Gerd. 1990. "Modernisierung von Rechts: Anmerkung zur Historisch-politischen Dynamik des Rechten Lagers." In *Ökologie von Rechts: Nationalismus und Umweltschutz bei der Neuen Rechten und den 'Republikanern,'* ed. Thomas Jahn and Peter Wehling. Frankfurt: Campus. Pp. 167–180.

Jeffries, Matthew. 1997. "Heimatschutz: Environmental Activism in Wilhelmine Germany." In

Green Thought in German Culture, ed. Colin Riordan. Cardiff: University of Wales. Pp. 42–53.

Jenkins, C. Craig. 1983. "Resource Mobilization Theory and the Study of Social Movements." *Annual Review of Sociology* 9: 527–553.

Johnson, Hank, Enrique Laraña, and Joseph R. Gusfield. 1994. "Identities, Grievances, and New Social Movements." In *New Social Movements: From Ideology to Identity,* ed. Enrique Laraña, Hank Johnston, and Joseph R. Gusfield. Philadelphia: Temple University Press.

Jones, Merrill E. 1993. "Origins of the East German Environmental Movement." *German Studies Review* 16, no. 2 (May): 235–264.

Joppke, Christian. 1993. *Mobilizing against Nuclear Energy: A Comparison of Germany and the United States.* Berkeley: University of California Press.

Jordan, Andrew. 2001. "The European Union: An Evolving System of Multilevel Governance . . . or Government?" *Policy and Politics* 29, no. 2: 193–208.

Jordan, Carlo. 1993. "Im Wandel: Ökologiebewegung und Grüne im Osten." In *Die Bürgerbewegung in der DDR und in den Ostdeutschen Ländern,* ed. Gerda Haufe and Karl Bruckmeier. Opladen: Westdeutscher Verlag. Pp. 240–260.

Jordan, Carlo. 1995. "Akteure und Aktionen der Arche." In *Arche Nova: Opposition in der DDR,* ed. Carlo Jordan and Hans Michael Kloth. Berlin: BasisDruck. Pp. 37–70.

Jordan, Grant, and William Maloney. 1997. *The Protest Business.* Manchester: Manchester University Press.

Kaase, Max. 1986. "Die Entwicklung des Umweltbewußtseins in der Bundesrepublik Deutschland." In *Umwelt, Wirtschaft, Gesellschaft: Wege zu einem Neuen Grundverständnis: Kongreß der Landesregierung "Zukunftschancen eines Industrielandes" Dezember 1985,* ed. Rudolf Wildenmann. Stuttgart: Staatsministerium Baden-Württemberg. Pp. 289–314.

Katz, Daniel, and Robert L. Kahn. 1978. *The Social Psychology of Organizations.* New York: Wiley.

Katz, Linda Sobel, Sarah Orrick, and Robert Honig. 1993. *Environmental Profiles: A Global Guide to Projects and People.* New York: Garland.

Katzenstein, Peter J. 1987. *Policy and Politics in West Germany: The Growth of a Semi-Sovereign State.* Philadelphia: Temple University Press.

Kazcor, Markus. 1986. "Der Bundesverband Bürgerinitiativen Umweltschutz: Geschichte einer Bewegungsorganisation unter dem Aspekt des Ziel- und Strategiewandels." Unpublished Halbjahresarbeit gemäß der Prüfungsordnung für Diplom-Politologen, Institut für Politische Wissenschaft, Universität Hamburg.

Kazcor, Markus. 1989. "Institutionen in der Umweltpolitik: Erfolg der Ökologiebewegung?" *Forschungsjournal Neue Soziale Bewegungen* 2, no. 3/4 (September): 47–62.

Kazcor, Markus. 1990. "Der Bundesverband Bürgerinitiativen Umweltschutz." In *Institutionalisierungsprozesse Sozialer Bewegungen,* ed. Stiftung Mitarbeit. Bonn: Stiftung Mitarbeit.

Kazcor, Markus. 1992. "Anmerkungen zum Führungspersonal Deutscher Umweltverbände." In *Die Politische Klasse in Deutschland: Eliten auf dem Prüfstand,* ed. Thomas Leif, Hans-Josef Legrande, and Ansgar Klein. Bonn: Bouvier Verlag. Pp. 339–361.

Kempf, Udo. 1984. "Der Bundesverband Bürgerinitiativen Umweltschutz." In *Bürgerinitiativen und Repräsentatives System,* ed. Bernd Guggenberger and Udo Kempf. Opladen: Westdeutscher Verlag. Pp. 404–423.

Kerbs, Diethart, and Jürgen Reulecke, eds. 1998. *Handbuch der Deutschen Reformbewegung 1880–1933.* Wuppertal: Peter Hammer Verlag.

Kielbowicz, Richard B., and Clifford Scheer. 1986. "The Role of the Press in the Dynamics of Social Movements." In *Research in Social Movements, Conflict, and Change,* vol. 9, ed. Kurt Lang and Gladys Engel Lang. Greenwich, CT: JAI Press. Pp. 71–96.

King, David, and Jack L. Walker. 1991. "The Origin and Maintenance of Groups." In *Mobilizing Interest Groups in America,* ed. Jack L. Walker. Ann Arbor: University of Michigan Press. Pp. 75–103.

Kirschey, Tom. 2001. "Greenpeace Enterbt." *Der Rabe Ralf* 12, no. 100 (October/November): 12.

Kitschelt, Herbert P. 1986. "Political Opportunity Structures and Political Protest: Anti-Nuclear Movements in Four Democracies." *British Journal of Political Science* 16, no. 1 (January): 57–85.

Klandermans, Bert. 1986. "New Social Movements and Resource Mobilization: The European and the American Approach." *International Journal of Mass Emergencies and Disasters* 4, no. 2 (August): 13–37.

Klandermans, Bert. 1988. "The Formation and Mobilization of Consensus." In *International Social Movement Research,* vol. 1, ed. Bert Klandermans, Hanspeter Kriesi, and Sidney Tarrow. Greenwich, CT: JAI Press. Pp. 173–196.

Klandermans, Bert. 2001. "Why Social Movements Come into Being and Why People Join Them." In *Blackwell Handbook of Sociology,* ed. Judith R. Blau. London: Blackwell. Pp. 268–281.

Klausa, Udo. 1979. "75 Jahre Deutscher Heimatbund: Rückblick und Ausblick." In *75 Jahre Deutscher Heimatbund,* ed. Deutscher Heimatbund. Siegburg: Deutscher Heimatbund. Pp. 7–12.

Klein, Ansgar. 1996. "Die Legitimität von Greenpeace und die Risiken Symbolischer Politik: Konturen und Probleme einer Medialen Stellvertreterpolitik im Bewegungssektor." *Forschungsjournal Neue Soziale Bewegungen* 9, no. 1 (March): 11–14.

Kleinert, Hubert. 1996. "Bündnis 90/Die Grünen: Die Neue Dritte Kraft?" *Aus Politik und Zeitgeschichte* 46, no. 6 (February 2): 36–44.

Kline, Benjamin. 2000. *First Along the River.* Lanham, MD: Acada Books.

Klingemann, Hans-Dieter. 1991. "Bürger Mischen sich ein: Die Entwicklung der Unkonventionellen Politischen Beteiligung in Berlin 1981–1990." In *Politische Klasse und Politische Institutionen,* ed. Hans-Dieter Klingemann, Richard Stöss, and Bernard Weßels. Opladen: Westdeutscher Verlag. Pp. 375–405.

Kloth, Hans Michael. 1995. "Grüne Bewegung, Grünes Netzwerk, Grüne Partei: Ein Politologischer Versuch." In *Arche Nova: Opposition in der DDR,* ed. Carlo Jordan and Hans Michael Kloth. Berlin: BasisDruck. Pp. 145– 179.

Klueting, Edeltraud. 2003. "Die Gesetzlichen Regelungen der Nationalsozialistischen Reichsregierung für den Tierschutz, den Naturschutz und den Umweltschutz." In *Naturschutz und Nationalsozialismus,* ed. Joachim Radkau and Frank Uekötter. Frankfurt: Campus. Pp. 77–106.

Kluth, Sabine, and Dietrich Kraetzchmer. 1985. "Die Binnenstruktur der Naturschutzverbände." In *Naturschutzverbände und Umweltpolitik,* ed. Institut für Landschaftspflege und Naturschutz und Institut für Landschaftsplanung und Raumforschung. Hannover: Institut für Landschaftspflege und Naturschutz und Institut für Landschaftsplanung und Raumforschung. Pp. 107–118.

Knabe, Hubertus. 1985. "Gesellschaftlicher Dissens im Wandel: Ökologische Diskussion und Umweltengagement in der DDR." In *Umweltprobleme und Umweltbewußtsein in der DDR,* ed. Werner Gruhn. Cologne: Verlag Wissenschaft und Politik. Pp. 169–197.

Knabe, Hubertus. 1988. "Neue Soziale Bewegungen im Sozialismus: Zur Genesis Alternativer Politischer Orientierungen in der DDR." *Kölner Zeitschrift für Soziologie und Sozialpsychologie* 40, no. 3: 551–569.

Knapp, Hans Dieter. 2003. "Knapp Braucht Dringend ein Auto, um Beweglich zu Sein: Wie der WWF in der Wendezeit in der DDR Fuß Fasste." In *Das Große Buch des WWF: 40 Jahre Naturschutz für und mit den Menschen,* ed. Klaus-Henning Groth. Steinfurt: Tecklenborg Verlag. Pp. 244–247.

Knaut, Andreas. 1991. "Ernst Rudorff und die Anfänge der Deutschen Heimatbewegung." In *Antimodernismus und Reform: Zur Geschichte der Deutschen Heimatbewegung,* ed. Edeltraud Klueting. Darmstadt: Wissenschaftliche Buchgesellschaft. Pp. 20–49.

Knaut, Andreas. 1993. *Zurück zur Natur! Die Wurzeln der Ökologiebewegung.* Bonn: Arbeitsgemeinschaft Beruflicher und Ehrenamtlicher Naturschutz.

Kneitz, Gerhard, and Christa Kley. 1986. "Zur Geschichte des Deutschen Natur- und Umweltschutzes." In *Was Wir Wollen: Ja zum Leben, Mut zum Handeln. Die Natur- und Umweltverbände auf dem 1. Deutschen Umwelttag,* ed. Reinhard Sander and Gerhard Kneitz. Bonn: Deutscher Naturschutzring. Pp. 17–36.

Knoke, David. 1990. *Organizing for Collective Action: The Political Economics of Associations.* New York: Aldine de Gruyter.

Knoke, David, and James R. Wood. 1981. *Organized for Action: Committment in Voluntary Organizations.* New Brunswick, NJ: Rutgers University Press.

Kohl, Andrea. 2003. "Parteilichkeit für die Natur: Ideale, Strategien, Realitäten des WWF." In *Das Große Buch des WWF: 40 Jahre Naturschutz für und mit den* Menschen, ed. Klaus-Henning Groth. Steinfurt: Tecklenborg Verlag. Pp. 49–57.

Kolb, Felix. 1997. "Der Kastor-Konflikt: Das Comeback der Anti-AKW Bewegung." *Forschungsjournal Neue Soziale Bewegungen* 10, no. 3 (September): 16–29.

Konferenz der Umweltminister des Bundes und der Länder. 2007. "Umweltministerkonferenz." Downloaded 25 January 2007. http://www.umweltministerkonferenz.de/

Koopmans, Ruud. 1995. *Democracy from Below: New Social Movements and the Political System in West Germany.* Boulder, CO: Westview.

Koopmans, Ruud. 2004. "Protest in Time and Space: The Evolution of Waves of Contention." In *The Blackwell Companion to Social Movements,* ed. David A. Snow, Sarah A. Soule, and Hanspeter Kriesi. Malden, MA: Blackwell. Pp. 19–46.

Körner, Stefan. 2005. "Die Entwicklung des Naturschutzes und der Landschaftsplanung nach dem Zweiten Weltkrieg." In *Natur und Umweltschutz nach 1945: Konzepte, Konflikte, Kompetenzen,* ed. Franz-Josef Brüggemeier and Jens Ivo Engels. Frankfurt: Campus. Pp. 87–102.

Koshar, Rudy. 1998. *Germany's Transient Pasts: Preservation and National Memory in the Twentieth Century.* Chapel Hill: University of North Carolina Press.

Kramer, Dieter. 1984. "Arbeiter als Touristen: Ein Privileg Wird Gebrochen." In *Mit Uns Zieht die Neue Zeit: Die Naturfreunde,* ed. Jochen Zimmer. Cologne: Pahl-Rugenstein. Pp. 31–65.

Kretschmer, Winfried, and Dieter Rucht. 1991. "Beispiel Wackersdorf: Die Protestbewegung gegen die Wiederaufarbeitungsanlage. Gruppen, Organisationen, Netzwerke." In *Neue Soziale Bewegungen in der Bundesrepublik Deutschland,* ed. Roland Roth and Dieter Rucht. Bonn: Bundeszentrale für Politische Bildung. Pp. 181–212.

Kriesi, Hanspeter. 2004. "Political Context and Opportunity." In *The Blackwell Companion to Social Movements,* ed. David A. Snow, Sarah A. Soule, and Hanspeter Kriesi. Malden, MA: Blackwell. Pp. 67–90.

Kriesi, Hanspeter, and Marco G. Giugni. 1996. "Ökologische Bewegungen im Internationalen Vergleich: Zwischen Konflikt und Kooperation." In *Umweltsoziologie,* ed. Andreas Diekmann and Carlo C. Jaeger. Opladen: Westdeutscher Verlag. Pp. 324–349.

Krüger, Christian. 1996. "Greenpeace: Politik der Symbolischen Konfrontation." *Forschungsjournal Neue Soziale Bewegungen* 9, no. 3 (September): 39–47.

Krüger, Christian. 2000a. "Kommunikation der Aktion: Grundzüge der Kommunikationspolitik, nach der Praxis Entworfen." In *Greenpeace auf der Wahrnehmungsmarkt,* ed. Christian Krüger and Matthias Müller-Hennig. Hamburg: LIT. Pp. 19–33.

Krüger, Christian. 2000b. "Der Grüne Kanal: Ein Pioneerproject?" In *Greenpeace auf der Wahrnehmungsmarkt,* ed. Christian Krüger and Matthias Müller-Hennig. Hamburg: LIT. Pp. 95–115.

Krüger, Christian, and Matthias Müller-Hennig. 2000. "Wahrnehmungsprozesse, Kommunikationspolitik: Greenpeace als Unikum und Exempel." In *Greenpeace auf der Wahrnehmungsmarkt,* ed. Christian Krüger and Matthias Müller-Hennig. Hamburg: LIT. Pp. 9–15.

Krüger, Sabine. 1998. "Gewerkschaften und NGOs: Spannungsfelder im Rio-Nachfolge-Prozeß." *Forschungsjournal Neue Soziale Bewegungen* 11, no. 3 (September): 101–107.

Krüger, Sabine. 2000. "Arbeit und Umwelt Verbinden: Probleme der Interaktion zwischen Gewerkschaften und Nicht-Regierungsorganisationen." Berlin: Wissenschaftszentrum Berlin für Sozialforschung.

Kuckartz, Udo, and Anke Rheingans-Heintze. 2006. *Trends im Umweltbewusstsein: Umweltgerechtigkeit, Lebensqualität und Persönliches Engagement.* Wiesbaden: Verlag für Sozialwissenschaften.

Kuhn, Axel. 1971. "Herrschaftsstruktur und Ideologie des Nationalsozialismus." *Neue Politische Literatur* 16, no. 3 (July–September): 395–406.

Kühnel, Wolfgang, and Carola Sallmon-Metzner. 1991. "Grüne Partei und Grüne Liga." In *Von der Illegalität ins Parlament: Werdegang und Konzept der Neuen Bürgerbewegungen,* ed. Helmut Müller-Enbergs, Marianne Schultz, and Jan Wielgohs. Berlin: LinksDruck. Pp. 166–220.

Kunz, Hildegard. 1989. "Die Öffentlichkeitsarbeit der Umweltorganisation Greenpeace." *Forschungsjournal Neue Soziale Bewegungen* 2, no. 1 (June): 27–37.

Küppers, Günter, Peter Lundgreen, and Peter Weingart. 1978. *Umweltforschung: Die Gesteuerte Wissenschaft? Eine Empirische Studie zum Verhältnis von Wissenschaftsentwicklung und Wissenschaftspolitik.* Frankfurt: Suhrkamp.

Lahusen, Christian. 1998. "Der Dritte Sektor als Lobby: Umweltverbände im Räderwerk der Nationalen Politik." In *Dritter Sektor - Dritte Kraft: Versuch einer Standortbestimmung,* ed. Rupert Graf Strachwitz. Stuttgart: RAABE. Pp. 411–436.

Lake, Beate. 1998. "NPO im Spannungsfeld von Solidarität und Wettbewerb." In *Dritter Sektor - Dritte Kraft: Versuch einer Standortbestimmung,* ed. Rupert Graf Strachwitz. Stuttgart: RAABE. Pp. 447–462.

Lange, Hellmuth. 2004. "Rapid Change in Agricultural Policies: The BSE Crisis in Germany (2000–2001)." Artec Working Paper Nr. 119. Bremen: Artec, Forschungszentrum Nachhaltigkeit, University of Bremen.

Lange, Rolf-Peter. 1973. "Zur Rolle und Funktion von Bürgerinitiativen in der Bundesrepublik und West-Berlin: Analyse von 61 Bürgerinitiativen." *Zeitschrift für Parlamentsfragen* 4, no. 2 (June): 247–286.

Lange, Ulrich. 2000. "Rahmenbedingungen des Naturschutzes in den USA und in Deutschland: Ein Vergleich." In *Naturschutz im Vereinigten Deutschland,* ed. Use Wegener and Hermann Behrens. Berlin: Verlag für Wissenschaft und Forschung. Pp. 23–37.

Lausch, Wolfgang. 1986. "Umweltschutz als Immanenter Bestandteil der Wissenschaftlich-technischen Revolution im Sozialismus." *Natur und Umwelt,* no volume, no. 2: 3–8.

Lebrecht, Jeschke. 2003. "Naturschutz der Wendezeit in der DDR." In *Naturschutz Hat Geschichte,* ed. Stiftung Naturschutz. Essen: Klartext. Pp. 245–253.

Lehmbruch, Gerhard. 1982. "Introduction: Neocorporatism in Comparative Perspective." In *Patterns of Corporatist Policy Making,* ed. Gerhard Lehmbruch and Phillippe C. Schmitter. Beverly Hills, CA: Sage. Pp. 1–28.

Leif, Thomas. 1993. "Greenpeace vor einer Großen Strategiediskussion." *Forschungsjournal Neue Soziale Bewegungen* 6, no. 2 (June): 115–117.

Lekan, Thomas M. 2003. "Organische Raumordnung: Landschaftspflege und die Durchführung des Reichsnaturschutzgesetzes im Rheinland und in Westfalen." In *Naturschutz und Nationalsozialismus,* ed. Joachim Radkau and Frank Uekötter. Frankfurt: Campus. Pp. 145–168.

Lense, Fritz. 1973. "60 Jahre Bund Naturschutz." *Blätter für Natur- und Umweltschutz* 53, no 2 (April): 42–44.

Leonhard, Martin. 1986. *Umweltverbände: Zur Organisation von Umweltinteressen in der Bundesrepublik Deutschland.* Opladen: Westdeutscher Verlag.

Linse, Ulrich. 1986. *Ökopax und Anarchie.* Nördlingen: Deutscher Taschenbuch Verlag.

Linse, Ulrich. 2002. "Frauen und Mädels: Schließt die Reihen." In *Politische Landschaft: Die Andere Sicht auf die Natürliche Ordnung,* ed. Klaus-Peter Lorenz. Duisburg: Trikont. Pp. 109–114.

Lipsky, Michael. 1968. "Protest as a Political Resource." *American Political Science Review* 62, no. 4 (December): 1144–1158.

Lofland, John. 1996. *Social Movement Organizations: Guide to Research on Insurgent Realities.* New York: Aldine de Gruyter.

Löhndorf, Joachim, and Eckard von Lühmann. 1996. "Betriebsrat bei Greenpeace: Eine Gratwanderung." In *Das Greenpeace Buch: Reflexionen und Aktionen,* ed. Greenpeace Deutschland. Munich: C.H. Beck. Pp. 233–237.

Long, Tony. 1998. "The Environmental Lobby." In *British Environmental Policy,* ed. Philip Lowe and Stephen Ward. London: Routledge. Pp. 105–118.

Loos, Anneliese. 1995. "Entwicklungsstationen der Umweltpolitik in der Bundesrepublik Deutschland: Von der Symptombekämpfung zur Präventiven Umweltpolitik?" In *Umwelt und Gesellschaft: Eine Einführung in die Sozialwissenschaftliche Umweltforschung,* ed. Wolfgang Joußen and Armin G. Hessler. Berlin: Akademie Verlag. Pp. 143–169.

Lorenz, Klaus-Peter. 1996. "Früher Deutscher Heimatschutz. Oder: Aus der Perspektive der Ästhetischen Revolte Sind Alle Politischen Katzen Grau." *Grüner Weg 31a* 10, no. 3 (September): 3–31.

Lorenz, Klaus-Peter. 2002. "Zur Sonne, zur Freiheit, zur Freizeit? Die Zeitgenossen der Naturfreunde." In *Politische Landschaft: Die Andere Sicht auf die Natürliche Ordnung,* ed. Klaus-Peter Lorenz. Duisburg: Trikont. Pp. 91–107.

Louis, Hans Walter. 2003. "Von der Polizeiverordnung zum Komplexen Naturschutzrecht." In *Naturschutz in Deutschland: Eine Erfolgsstory,* ed. Deutscher Rat für Landespflege. Bonn: Deutscher Rat für Landespflege. Pp. 39–42.

Lugger, Beatrice. 1994. "Woher Nehmen? Weniger Geld für Umweltverbände." *Politische Ökologie* 12, no. 39 (November/December): 71.

Lurisch, Joachim. 1995. "Einheit der Vielfalt." *Politische Ökologie* 13, no. 42 (July/August): 27–30.

Luthe, Detlef. 1997. *Fundraising: Fundraising als Beziehungsorientiertes Marketing: Entwicklungsaufgaben für Nonprofit-Organisationen.* Augsburg: Maro Verlag.

Malunat, Bernd M. 1994. "Die Umweltpolitik der Bundesrepublik Deutschland." *Aus Politik und Zeitgeschichte* 44, no. 49 (9 December): 3–12.

Markham, William T. 1999. "Fundraising Strategien von Umweltorganisationen: World-Wide-Web-Präsentationen Deutscher und Amerikanischer Umweltverbände." *BSM - Newsletter: Informationsbulletin der Bundesarbeitsgemeinschaft Sozialmarketing e. V.,* no volume, no. 1 (February): 54–57.

Markham, William T. 2005a. "German Environmental Organizations as Symbolic Actors: Myth, Ritual, and the Imperatives of Fundraising in Mainstream Environmental Organizations." Paper Presented at the Conference on Double Standards: Symbolism, Rhetoric and Irony in Eco-Politics of the Research Committee on Environment and Society, International Sociological Association. Bath, England.

Markham, William T. 2005b. "Networking Local Environmental Groups in Germany." *Environmental Politics* 14, no. 5 (November): 667–685.

Markham, William T., and Sabrine Broselow. 2001. "Trolling for Members and Donations: The Internet Pages of German and US Environmental Organizations." Paper presented at the Meetings of the American Sociological Association. Anaheim, California.

Markham, William T., and C.S.A. (Kris) van Koppen. 2007. "Nature Protection in Nine Countries: A Framework for Analysis." In *Protecting Nature: Organizations and Networks in Europe and the USA,* ed. C.S.A. (Kris) van Koppen and William T. Markham. Cheltenham: Edward Elgar.

Marks, Gary, Lisbeth Hooghe, and Kermit Blank. 1996. "European Integration from the 1980s: State-centric v. Multi-level Governance." *Journal of Common Market Studies* 34, no. 3 (September): 341–378.

Martindale, Don. 1981. *The Nature and Types of Sociological Theory.* Boston: Houghton-Mifflin.

Maxim, Wilfried, and Wolfgang Degenhardt. 2003. "'Keine Berufsprotestierer oder Schornsteinkletterer': Die Geschichte des BUND in Nordrhein-Westfalen." In *Keine Berufsprotestierer oder Schornsteinkletterer,* ed. Stiftung Naturschutzgeschichte. Essen: Klartext. Pp. 97–158.

May, Helge. 1999. *100 Jahre NABU: Ein Historischer Abriss.* Bonn: NABU.

Mayer-Tasch, Cornelius. 1985. *Die Bürgerinitiativbewegung: Der Aktive Bürger als Rechts- und Politikwissenschaftliches Problem.* Reinbek bei Hamburg: Rowohlt.

Mazey, Sonia, and Jeremy Richardson. 1992. "Environmental Groups and the EC: Challenges and Opportunities." *Environmental Politics* 1, no. 1 (spring): 109–128.

McAdam, Doug. 1994. "Taktiken von Protestbewegungen: Das Framing der Amerikanischen Bürgerrechtsbewegung." In *Öffentlichkeit, Öffentliche Meinung, Soziale Bewegungen,* ed. Friedhelm Neidhart. Opladen: Westdeutscher Verlag. Pp. 393–411.

McAdam, Doug, John D. McCarthy, and Mayer N. Zald. 1988. "Social Movements." In *Handbook of Sociology,* ed. Neil Smelser. Newbury Park, CA: Sage. Pp. 695–737.

McAdam, Doug, and W. Richard Scott. 2005. "Organizations and Movements." In *Social Movements and Organizational Theory,* ed. Gerald F. Davis et al. Cambridge: Cambridge University Press. Pp. 4–40.

McAdam, Doug, Sidney Tarrow, and Charles Tilly. 1996. "To Map Contentious Politics." *Mobilization* 1, no. 1 (September): 17–34.

McCarthy, John D. 2005. "Persistence and Change among Nationally Federated Social Movements." In *Social Movements and Organization Theory,* ed. Gerald F. Davis et al. Cambridge: Cambridge University Press. Pp. 193–225.

McCarthy, John D., and Mayer N. Zald. 1977. "Resource Mobilization and Social Movements: A Partial Theory." *American Journal of Sociology* 82, no. 6 (May): 1212–1241.

Melucci, Alberto. 1980. "The New Social Movements: A Theoretical Approach." *Social Science Information* 19, no. 2 (May): 199–226.

Melucci, Alberto. 1985. "The Symbolic Character of Contemporary Social Movement." *Social Research* 52, no. 4 (winter): 789–816.

Meroth, Peter, and Konrad von Moltke. 1987. *Umwelt und Umweltpolitik in der Bundesrepublik Deutschland.* Munich: Iudicium.

Meyer, Christian. 1991. "Fundraising für Greenpeace: Erfolgreich durch den Dialog." In *Handbuch Direct Marketing,* ed. Heinz Dallmer. Wiesbaden: Gabler. Pp. 844–852.

Meyer, David S. 2004. "Protest and Political Opportunities." *Annual Review of Sociology* 26: 125–145.

Meyer, David S., and Nancy Whittier. 1994. "Social Movement Spillover." *Social Problems* 41, no. 2 (May): 277–298.

Meyer, Hermann. 1996. "Wer Arbeitet bei Greenpeace? Zur Unternehmenssoziologie." In *Das Greenpeace Buch: Reflexionen und Aktionen,* ed. Greenpeace Deutschland. Munich: C.H. Beck. Pp. 239–245.

Meyer, John W., and Brian Rowan. 1977. "Institutionalized Organizations: Formal Structure as Myth and Ceremony." *American Journal of Sociology* 83, no. 2 (September): 340–363.

Mez, Lutz. 1987. "Von den Bürgerinitiativen zu den Grünen: Zur Entstehungsgeschichte der Wahlalternativen in der Bundesrepublik." In *Neue Soziale Bewegungen in der Bundesrepublik Deutschland,* ed. Roland Roth and Dieter Rucht. Bonn: Bundeszentrale für Politische Bildung. Pp. 379–391.

Michels, Robert. 1949. *Political Parties: A Sociological Study of Oligarchical Tendencies in Modern Democracy.* Glencoe, IL: Free Press.

Michelsen, Gerd, Lars Degenhardt, Jasmin Godemann, and Heike Molitor. 2001. *Umweltengagement von Kindern und Jugendlichen in der Außerschulischen Umweltbildung: Ergebnisse - Bedingungen - Perspektiven.* Frankfurt: Peter Lang.

Milnik, Albrecht. 2003. "Hugo Conwentz: Zur Geschichte der Staatlichen Stelle für Naturdenkmalpflege in Preußen." In *Naturschutz Hat Geschichte,* ed. Stiftung Naturschutzgeschichte. Essen: Klartext. Pp. 131–143.

Miner, John B. 1982. *Theories of Organizational Structure and Process.* Chicago: Dryden Press.

Minkoff, Debra. 1997. "Producing Social Capital: National Social Movements in Civil Society." *American Behavioral Scientist* 40, no. 5 (March/April): 606–619.

Minkoff, Debra C. 2001. "Social Movement Politics and Organization." In *Blackwell Handbook of Sociology,* ed. Judith R. Blau. London: Blackwell. Pp. 282–294.

Mitchell, Robert Cameron. 1989. "From Conservation to Environmental Movement: The Development of Modern Environmental Lobbies." In *Government and Economic Politics: Essays on Historical Developments since World War II,* ed. Michael J. Lacey. Washington, D.C.: Woodrow Wilson Center Press. Pp. 82–113.

Mitchell, Robert Cameron, Angela G. Mertig, and Riley E. Dunlap. 1992. "Twenty Years of Environmental Mobilization: Trends among National Environmental Organizations." In *American Environmentalism,* ed. Riley G. Dunlap and Angela G. Mertig. Philadelphia: Taylor and Francis. Pp. 11–26.

Mitlacher, Günter, and Ralf Schulte. 2005. *Steigerung des Ehrenamtlichen Engagements in Naturschutzverbänden.* Bonn: Bundesamt für Naturschutz.

Moe, Terry M. 1980. *The Organization of Interest Groups.* Chicago: University of Chicago Press.

Moeller, Robert G., ed. 1997. *West Germany Under Construction: Politics, Society, and Culture in the Adenauer Era.* Ann Arbor: University of Michigan Press.

Mol, Arthur P. J. 2000. "The Environmental Movement in an Era of Ecological Modernisation." *Geoforum* 31, no. 1 (February): 45–57.

Mol, Arthur. P. J. 2006. "From Environmental Sociologies to Environmental Sociology? A Comparison of US and European Environmental Sociology." *Organization and Environment* 19, no. 1: 5–27.

Möller, Christel. 1993. "Heraus aus der Oppositionsrolle." *Politische Ökologie* 11, no. 31 (May/June): 64–65 and 67–69.

Möller, Matthias. 1993. "Mehr Mut zur Professionalität." *Politische Ökologie* 11, no. 31 (May/June): 58–60.

Moritz, Torsten. 1997. "Die Entwicklung von DDR-Oppositionsgruppen nach 1989: Das Beispiel Umweltbibliothek Berlin." In *Zwischen Verweigerung und Opposition: Politischer Protest in der DDR 1970–1989,* ed. Detlef Pollack and Dieter Rink. Frankfurt: Campus. Pp. 208–235.

Müller, Edda. 1986. *Innenwelt der Umweltpolitik: Sozial-liberale Umweltpolitik: (Ohn)macht durch Organisation?* Opladen: Westdeutscher Verlag.

Müller, Edda. 2001. "Die Beziehung von Umwelt- und Naturschutz in den 1970er Jahren." In *Natur im Sinn: Beiträge zur Geschichte des Naturschutzes,* ed. Stiftung Naturschutzgeschichte. Essen: Klartext. Pp. 31–45.

Müller-Henning, Matthias. 2000. "Der Mythos von der Allmacht der Öffentlichkeitsarbeit: Ergebnisse der Nachrichten- und Informationsflußanalyse zur Informationsquelle Greenpeace." In *Greenpeace auf der Wahrnehmungsmarkt,* ed. Christian Krüger and Matthias Müller-Hennig. Hamburg: LIT. Pp. 55–69.

Müller-Rommel, Ferdinand. 1993. *Grüne Parteien in Westeuropa.* Opladen: Westdeutscher Verlag.

Mundo, Philip A. 1992. "Introduction." In *Interest Groups,* ed. Philip A. Mundo. Chicago: Nelson Hall. Pp. 2–17.

Muthesium, Stefan. 1981. "The Origins of the German Conservation Movement." In *Planning for Conservation,* ed. Roger Kain. New York: St. Martins Press. Pp. 37–49.

NABU. 1997. *Jahresbericht '96.* Bonn: NABU.

NABU. 1999. *Jahresbericht 1998.* Bonn: NABU.

NABU. 2000a. *NABU - Grundsatzprogramm.* Bonn: NABU.

NABU. 2000b. *Jahresbericht 1999.* Bonn: NABU.

NABU. 2001. *Jahresbericht 2000.* Bonn: NABU.

NABU. 2002. *Jahresbericht 2001.* Bonn: NABU.

NABU. 2003. *Jahresbericht 2002.* Bonn: NABU.

NABU. 2004a. *Jahresbericht 2003.* Bonn: NABU.

NABU. 2004b. "Umwelt-Halbzeit bei Rot-Grün." NABU Pressedienst, NABU, Released 16 September 2004. Downloaded 12 November 2004. http://www.nabu.de/modules/presseservice/index.php?show=340&db

NABU. 2005a. *Jahresbericht 2004.* Bonn: NABU.

NABU. 2005b. "NABU-Naturschutzzentren." Downloaded 15 April 2005. http://www.nabu.de/m03/m03_03/index.html

NABU. 2007. "Wir Wollen Menschen für die Natur Begeistern." Downloaded 10 February 2007. http://www.nabu.de/m09/m09_01/

NABU-Stiftung. 2002. *Jahresbericht 2001.* Bonn: NABU.

NABU-Stiftung. 2005. *Jahresbericht 2004.* Bonn: NABU.

Naturfreunde Deutschlands. 2007. "Naturfreunde Deutschlands." Downloaded 29 January 2007. http://www.naturfreunde.de/

NaturschutzForum Deutschland. 2007. "NaturschutzForum Deutschland." Downloaded 30 January 2007. http://www.nafor.de/

Naturschutz Heute. 1994. "Gesamtnote 'Gut': Die Ergebnisse der NH-Leserumfrage." *Naturschutz Heute* 26, no. 4 (October/November/December): 52–53.

Naturschutz Heute. 1998a. "Kulturlandschaft statt Agrarsteppe." *Naturschutz Heute* 30, no. 1 (January): 10–18.

Naturschutz Heute. 1998b. "Uwe Hüser Interview." *Naturschutz Heute* 30, no. 2 (April): 9.

Naturschutz Heute. 1998c. "Wenig Versprochen, Nichts Gehalten." *Naturschutz Heute* 30, no. 3 (August): 10–14.

Naturschutz Heute. 1998d. "Umweltbewusstes Einkaufen." *Naturschutz Heute* 30, no. 3 (August): 17.

Naturschutz Heute. 1998e. "NABU Kompetent aber Machtlos." *Naturschutz Heute* 30, no. 3 (August): 18.

Naturschutz Heute. 1998f. "Naturschutz an der Ladentheke." *Naturschutz Heute* 30, no. 4 (November): 40.

Naturschutz Heute. 1999a. "Hundert Jahre für Mensch und Natur." *Naturschutz Heute* 31, no. 1 (February): 8–13.
Naturschutz Heute. 1999b. "So Funktioniert der NABU." *Naturschutz Heute* 31, no. 1 (February): 20.
Naturschutz Heute. 1999c. "Schutz für Tengis und Kaspisches Meer," *Naturschutz Heute* 31, no. 2 (May): 32.
Naturschutz Heute. 2000a. "Dinosaurier für Erwin Teufel." *Naturschutz Heute* 32, no. 1 (February): 6.
Naturschutz Heute. 2000b. "Geflügelte Zugpferde: Die Aktion Vogel des Jahres Wird 30 Jahre Alt." *Naturschutz Heute* 32, no. 1 (February): 10–13.
Naturschutz Heute. 2000c. "Das Wallnau-Gefühl." *Naturschutz Heute* 32, no. 1 (February): 38–39.
Naturschutz Heute. 2000d. "Vogelparadies im Wattenmeer." *Naturschutz Heute* 32, no. 1 (February): 40–41.
Naturschutz Heute. 2000e. "Der Natur eine Chance: Der NABU-Club Hilft, Naturschutzprojekte Dauerhaft zu Sichern." *Naturschutz Heute* 32, no. 1 (February): 45.
Naturschutz Heute. 2000f. "Erfolg Braucht Größe." *Naturschutz Heute* 32, no. 2 (April): 3.
Naturschutz Heute. 2000g. "100,- DM Startguthaben für Mitglieder des Naturschutzbundes." *Naturschutz Heute* 32, no. 2 (April): 10.
Naturschutz Heute. 2000h. "Wir Haben uns Durchchecken Lassen und Wurden Zertifiziert." *Naturschutz Heute* 32, no. 2 (April): 19.
Naturschutz Heute. 2000i. "Heute Mal Nix Billiges." *Naturschutz Heute* 32, no. 2 (April): 21.
Naturschutz Heute. 2000j. "Strom Muss Grün Sein." *Naturschutz Heute* 32, no. 2 (April): 20.
Naturschutz Heute. 2000k. "Abgabe statt Zwangspfand." *Naturschutz Heute* 32, no. 2 (April): 22.
Naturschutz Heute. 2000l. "Zwei Millionen Unterschriften." *Naturschutz Heute* 32, no. 2 (April): 36.
Naturschutz Heute. 2000m. "Anspruch und Wirklichkeit." *Naturschutz Heute* 32, no. 3 (July): 44–45.
Naturschutz Heute. 2000n. "Die Große Beliebigkeit." *Naturschutz Heute* 32, no. 4 (November): 28–29.
Naturschutz Heute. 2001a. "Auf dem Ersten Blick." *Naturschutz Heute* 33, no. 4 (October): 9.
Naturschutz Heute. 2001b. "Nebenwirkungen Beachten." *Naturschutz Heute* 33, no. 4 (October): 39–41.
Naturschutz Heute. 2002a. "Erfolgreiche Modernisierung." *Naturschutz Heute* 34, no. 2 (April): 6–7.
Naturschutz Heute. 2002b. "Die Natur Braucht Freunde: Wie Steht's mit Ihren?" *Naturschutz Heute* 34, no. 2 (April): 27.
Naturschutz Heute. 2002c. "NABU-Club: Projekte Gezielt Unterstützen und Direkt Erleben." *Naturschutz Heute* 34, no. 2 (April): 33.
Naturschutz Heute. 2002d. "Virtuelle Welt: Der Neue Internet-Auftritt des NABU." *Naturschutz Heute* 34, no. 3 (July): 6–7.
Naturschutz Heute. 2002e. "Bilanz und Ausblick: Die Deutsche Umweltpolitik Zwei Monate vor der Bundestagswahl 2002." *Naturschutz Heute* 34, no. 3 (July): 12–16.
Naturschutz Heute. 2003a. "Liebe Freunde!" *Naturschutz Heute* 35, no. 2 (April): 3.
Naturschutz Heute. 2003b. "Vom Wald Leben und Ihn Schützen," *Naturschutz Heute* 35, no. 2 (April): 48.
Naturschutz Heute. 2003c. "NABU-Akademie Stellt Seminarbetrieb ein" *Naturschutz Heute* 35, no. 3 (July): 16.

Naturschutz Heute. 2003d. "Unsere Welt: Start der Lotterie für Mensch und Natur," *Naturschutz Heute* 35, no. 4 (October): 28.

Naturschutz Heute. 2003e. "Der Frosch im Kleid," *Naturschutz Heute* 35, no. 4 (October): 48–49.

Naturschutz Heute. 2005a. "Vogelbeobachtung für die Wissenschaft." *Naturschutz Heute* 37, no. 1 (January): 25.

Naturschutz Heute. 2005b. "Die Naturparadiese Wachsen." *Naturschutz Heute* 37, no. 1 (January): 31.

Naturschutz Heute. 2005c. "Die NABU VISA Card: Die Karte für den Naturschutz." *Naturschutz Heute* 37, no. 1 (January): 49.

Naturschutz Heute. 2005d. "NABU-Reisen Informiert." *Naturschutz Heute* 37, no. 2 (April): 67.

Naturschutz Heute. 2005e. "NABU Homepage: Natur Erleben im Web." *Naturschutz Heute* 37, no. 2 (April): 6.

Naturschutz Heute. 2005f. "66 Mal Nature Erleben: Die NABU- und LBV-Zentren im Überblick." *Naturschutz Heute* 37, no. 2 (April): 9.

Naturschutz Heute. 2005g. "Sparen und die Natur Unterstützen: Die NABU Bahn-Card." *Naturschutz Heute* 37, no. 3 (July): 21.

Naturschutz Heute. 2005h. "Zum Leben zu Wenig: Versuch einer Umweltpolitischen Bilanz von Rot-Grün." *Naturschutz Heute* 37, no. 3 (July): 24–25.

Naumann, Jörg. 1996. "Von der Umweltbewegung der DDR zu Greenpeace-Ost." In *Das Greenpeace Buch: Reflexionen und Aktionen,* ed. Greenpeace Deutschland. Munich: C.H. Beck. Pp. 51–63.

Neubert, Ehrhardt. 1997. *Geschichte der Opposition in der DDR 1949–1989.* Bonn: Bundzentrale für Politische Bildung.

Neumann, Ulrich. 1995. "Was War, War Wenig und Viel: Die Anfänge der Arche." In *Arche Nova: Opposition in der DDR,* ed. Carlo Jordan and Hans Michael Kloth. Berlin: Basis-Druck. Pp. 81–92.

Newig, Jens. 2003. *Symbolische Umweltgesetzgebung: Rechtssoziologische Untersuchungen am Beispiel des Ozongesetzes, des Kreislaufwirtschaft- und Abfallgesetzes sowie der Großfeueranlageverordnung.* Berlin: Dunker und Humblot.

Niermann, Inga. 1996. "Öko- und Sozialsponsoring: Neue Chance für Gemeinnützige Organisationen?" *Forschungsjournal Neue Soziale Bewegungen* 9, no. 4 (December): 84–89.

Noelle-Neumann, Elisabeth. 1976. *Allensbacher Jahrbuch der Demoskopie 1974–1976.* Vienna: Verlag Fritz Molden.

Noelle-Neumann, Elisabeth. 1977. *Allensbacher Jahrbuch der Demoskopie 1976–1977.* Vienna: Verlag Fritz Molden.

Noelle-Neumann, Elisabeth. 1983. *Allensbacher Jahrbuch der Demoskopie 1978–1983.* Munich: K.G. Saur.

Noelle-Neumann, Elisabeth, and Renate Köcher. 1993. *Allensbacher Jahrbuch der Demoskopie 1984–1992.* Munich: K.G. Saur.

Noelle-Neumann, Elisabeth, and Renate Köcher. 1997. *Allensbacher Jahrbuch der Demoskopie 1993–1997.* Munich: K.G. Saur.

Nölting, Benjamin. 2002. *Strategien und Handlungsspielräume Lokaler Umweltgruppen in Brandenburg und Ostberlin 1980–2000.* Frankfurt: Peter Lang.

Oberkrome, Willi. 2003a. "'Liberos' auf Altlasten." In *Keine Berufsprotestierer oder Schornsteinkletterer,* ed. Stiftung Naturschutzgeschichte. Essen: Klartext. Pp. 23–96.

Oberkrome, Willi. 2003b. "Hans Klose, Walther Schoenichen und der Erlass des Reichsnaturschutzgesetzes." In *Naturschutz Hat Geschichte,* ed. Stiftung Naturschutzgeschichte. Essen: Klartext. Pp. 145–162.

Oberkrome, Willi. 2005. "Kontinuität und Wandel im Deutschen Naturschutz 1930 bis 1970: Bemerkungen und Thesen." In *Natur und Umweltschutz nach 1945: Konzepte, Konflikte, Kompetenzen,* ed. Franz-Josef Brüggemeier and Jens Ivo Engels. Frankfurt: Campus. Pp. 23–37.

Offe, Claus. 1981. "The Attribution of Public Status to Interest Groups: Observations on the West German Case." In *Organizing Interests in Western Europe: Pluralism, Corporatism, and the Transformation of Politics,* ed. Suzanne Berger. Cambridge, MA: Cambridge University Press. Pp. 123–158.

Offe, Claus. 1985. "New Social Movements: Challenging the Boundaries of Institutional Politics." *Social Research* 52, no. 4 (winter): 817–868.

Oliver, Pamela E., and Gerald Marwell. 1992. "Mobilizing Technologies for Collective Action." In *Frontiers in Social Movement Theory,* ed. Aldon D. Morris and Carol McClurg Mueller. New Haven, CT: Yale University Press. Pp. 251–272.

Olsen, Jonathan. 1999. *Nature and Nationalism: Right Wing Ecology and the Politics of Identity in Contemporary Germany.* New York: St. Martin's Press.

Olson, Mancur Jr. 1965. *The Logic of Collective Action.* Cambridge, MA: Harvard University Press.

O'Neill, Michael. 1997. *Green Parties in Contemporary Europe: New Politics. Old Predicaments.* Aldershot: Ashgate.

Opp, Karl-Dieter. 1996. "Aufstieg und Niedergang der Ökologiebewegung in der Bundesrepublik." In *Umweltsoziologie,* ed. Andreas Diekmann and Carlo C. Jaeger. Opladen: Westdeutscher Verlag. Pp. 350–379.

Osti, Giorgio. 2007. "Nature Protection Organisations in Italy: From Elitist Fervour to Confluence with Environmentalism." In *Protecting Nature: Organizations and Networks in Europe and the USA,* ed. C.S.A. (Kris) van Koppen and William T. Markham. Cheltenham: Edward Elgar.

Oswald von Nell Breuning Institut für Wissenschafts- und Gesellschaftsethik der Philosophisch-theologischen Hochschule Sankt Georgen. 1996. *Die Rolle der Umweltverbände in den Demokratischen und Umweltethischen Lernprozessen der Gesellschaft.* Stuttgart: Metzler-Poeschel.

Padgett, Stephen. 1989. "The Party System." In *Developments in West German Politics,* ed. Gordon Smith, William E. Paterson, and Peter H. Merkl. Durham, NC: Duke University Press. Pp. 122–146.

Parsons, Talcott. 1970. "Some Problems of General Theory." In *Theoretical Sociology: Perspectives and Developments,* ed. J. C. McKinney and E. A. Tiryakian. New York: Appleton-Century-Crofts. Pp. 28–68.

Paterson, William E. 1989. "Environmental Politics." In *Developments in West German Politics,* ed. Gordon M. Smith, William E. Paterson, and Peter Merkl. Durham, NC: Duke University Press. Pp. 267–288.

Pehle, Heinrich. 1997. "Germany: Domestic Obstacles to an International Frontrunner." In *European Environmental Policy: The Pioneers,* ed. Michael Skou Andersen and Duncan Liefferink. Manchester: Manchester University Press. Pp. 161–209.

Pehle, Heinrich. 1998. *Das Bundesministerium für Umwelt, Naturschutz und Reaktorsicherheit: Ausgegrenzt statt Integriert?* Wiesbaden: Deutscher Universitätsverlag.

Perrow, Charles. 1961. "The Analysis of Goals in Complex Organizations." *American Sociological Review* 26, no. 6 (December): 854–865.

Perrow, Charles. 1993. *Complex Organizations: A Critical Essay.* New York: McGraw Hill.

Petracca, Mark P. 1992. "The Rediscovery of Interest Group Politics." In *The Politics of Interests,* ed. Mark P. Petracca. Boulder, CO: Westview. Pp. 3–31.

Pfeffer, Jeffrey. 1997. *New Directions for Organization Theory: Problems and Prospects.* New York: Oxford University Press.

Pfeffer, Jeffrey, and Gerald R. Salancik. 1978. *The External Control of Organizations: A Resource Dependence Perspective.* New York: Harper and Row.

Pichardo, Nelson A. 1997. "New Social Movements: A Critical Review." *Annual Review of Sociology* 23: 411–430.

Pietschmann, Manfred. 1996. "Engagement im Abo: Greenpeace Magazin und Nachrichten." In *Das Greenpeace Buch: Reflexionen und Aktionen,* ed. Greenpeace Deutschland. Munich: C.H. Beck. Pp. 125–129.

Poguntke, Thomas. 1998. "Alliance 90/The Greens in East Germany: From Vanguard to Insignificance." *Party Politics* 4, no. 1: 33–55.

Preisendörfer, Peter. 1999. *Umwelteinstellungen und Umweltverhalten in Deutschland.* Opladen: Leske + Budrich.

Preisendörfer, Peter. 2005. *Organisationssoziologie: Grundlagen, Theorien und Problemstellungen.* Wiesbaden: Verlag für Sozialwissenschaften.

Preisendörfer, Peter, and Axel Franzen. 1996. "Der schöne Schein des Umweltbewusstseins: Zu den Ursachen und Konsequenzen von Umwelteinstellung in der Bevölkerung." In *Umweltsoziologie,* ed. Andreas Diekmann and Carlo C. Jaeger. Opladen: Westdeutscher Verlag. Pp. 219–224.

Price, James L., and Charles W. Mueller. 1986. *Handbook of Organizational Measurement.* Boston: Pitman.

Priller, Eckhard, and Jana Sommerfeld. 2005. "Wer Spendet in Deutschland? Eine Sozialstrukturelle Analyse." Berlin: Wissenschaftszentrum Berlin für Sozialforschung.

Putnam, Robert D. 2000. *Bowling Alone: The Collapse and Revival of American Community.* New York: Simon and Schuster.

Radkau, Joachim. 2003. "Naturschutz und Nationalsozialismus: Wo ist das Problem?" In *Naturschutz und Nationalsozialismus,* ed. Joachim Radkau and Frank Uekötter. Frankfurt: Campus. Pp. 41–54.

Radkau, Joachim, and Frank Uekötter. 2003. "Ernst Rudorff und die Moderne." In *Naturschutz Hat Geschichte,* ed. Stiftung Naturschutzgeschichte. Essen: Klartext. Pp. 89–99.

Radow, Birgit, and Christian Krüger. 1996. "Öffentlichkeit Herstellen." In *Das Greenpeace Buch: Reflexionen und Aktionen,* ed. Greenpeace Deutschland. Munich: C.H. Beck. Pp. 211–221.

Ramthun, Susanne. 2000. "Aktionsraum Nordsee: Campagnenpolitik an Drei Fallbeispielen." In *Greenpeace auf der Wahrnehmungsmarkt,* ed. Christian Krüger and Matthias Müller-Hennig. Hamburg: LIT. Pp. 119–130.

Raschke, Joachim. 1987. "Zum Begriff der Sozialen Bewegung." In *Neue Soziale Bewegungen in der Bundesrepublik Deutschland,* ed. Roland Roth and Dieter Rucht. Frankfurt: Campus. Pp. 19–29.

Rat von Sachverständigen für Umweltfragen. 1996. *Umweltgutachten 1996: Zur Umsetzung einer Dauerhaft Umweltgerechten Entwicklung.* Stuttgart: Metzler-Poeschel.

Rauprich, Nina. 1985. *Erst Wenn der Letzte Baum Gestorben Ist! Alternative Organisationen im Umweltschutz.* Frankfurt: Fischer Taschenbuch.

Rawcliffe, Peter. 1998. *Environmental Pressure Groups in Transition.* Manchester: Manchester University Press.

Reichert, Sandra, and Holger Schmied. 1995. "Greenpeace Strategien." In *Macht der Zeichen - Zeichen der Macht,* ed. Sigrid Baringhorst, Bianca Müller, and Holger Schmied. Frankfurt: Peter Lang. Pp. 143–154.

Reichsbund für Vogelschutz. 1936. *Jahresbericht 1936.* Giengen: Reichsbund für Vogelschutz.

Reichsbund für Vogelschutz. 1944. *Jahresbericht 1943.* Stuttgart: Reichsverband für Vogelschutz.

Reiss, Jochen. 1988. *Greenpeace: Der Umweltmulti - Sein Apparat, Seine Aktionen.* Rheda-Wiedenbrück: Daedalus Verlag.

Renn, Ortwin. 1985. "Die Alternative Bewegung: Eine Historisch-soziologische Analyse des Protests gegen die Industriegesellschaft." *Zeitschrift für Politik* 32, no. 2: 153–194.

Reuß, Johann. 1926. "Mitgliederversammlung des Bundes Naturschutz in Bayern." *Blätter für Naturschutz und Naturpflege* 9, no. 1 (May): 2–6.

Reuß, Johann. 1933. "Die Neue Zeit und Wir." *Blätter für Naturschutz und Naturpflege* 16, no. 2 (November): 97–104.

Reuß, Johann. 1935. "Aus der Arbeit des Bundes." *Blätter für Naturschutz und Naturpflege* 18, no. 2 (October): 136–140.

Reusswig, Fritz. 2004. "Naturschutz und Naturbilder in Verschiedenen Lebensstilgruppen." In *Land - Natur - Konsum,* ed. Wolfgang Serbser, Heide Inhetveen, and Fritz Reusswig. Munich: oekom. Pp. 143–176.

Reutter, Werner. 2001. "Deutschland." In *Verbände und Verbandssysteme in Westeuropa,* ed. Werner Reutter and Peter Rütters. Opladen: Leske + Budrich. Pp. 75–101.

Rieder, Barbara. 1980. *Der Bundesverband Bürgerinitiativen Umweltschutz: Geschichte, Struktur und Aktionsformen einer Dachorganisation der Ökologiebewegung.* Wissenschaftliche Hausarbeit im Rahmen der Ersten (Wissenschaftlichen) Staatsprüfung für das Amt des Studienrats. Berlin: Freie Universität.

Riesenberger, Dieter. 1991. "Heimatgedanke und Heimatschutz in der DDR." In *Antimodernismus und Reform: Zur Geschichte der Deutschen Heimatbewegung,* ed. Edeltraud Klueting. Darmstadt: Wissenschaftliche Buchgesellschaft. Pp. 320–343.

Rink, Dieter. 1991. "Soziale Bewegungen in der DDR: Die Entwicklung bis Mai 1990." In *Neue Soziale Bewegungen in der Bundesrepublik Deutschland,* ed. Roland Roth and Dieter Rucht. Bonn: Bundeszentrale für Politische Bildung. Pp. 54–70.

Rink, Dieter. 1999. "Mobilisierungsschwäche, Latenz, Transformation oder Auflösung?" In *Neue Soziale Bewegungen: Impulse, Bilanzen und Perspektiven,* ed. Ansgar Klein, Hans-Josef Legrand, and Thomas Leif. Opladen: Westdeutscher Verlag. Pp. 180–195.

Rink, Dieter. 2001. "Institutionalization Instead of Mobilization: The Environmental Movement in Eastern Germany." In *Pink, Purple, Green: Women's, Religious, Environmental and Gay/Lesbian Movements in Central Europe Today,* ed. Helena Flamm. Boulder, CO: East European Monographs. Pp. 120–131.

Rink, Dieter. 2002. "Environmental Policy and the Environmental Movements in East Germany." *Socialism - Nature - Capitalism* 13, no. 3: 73–91.

Rink, Dieter, and Saskia Gerber. No date. "Die Ostdeutsche Umweltbewegung: Von der Opposition zur Neuen Lebensreform?" Unpublished Manuscript.

Riordan, Colin. 1997. "Green Ideas in Germany: A Historical Survey." In *Green Thought in German Culture: Historical and Contemporary Perspectives,* ed. Colin Riordan. Cardiff: University of Wales Press. Pp. 3–41.

Robin Wood. 2003a. "Robin Wood: Aktiv für die Umwelt." Downloaded 12 July 12 2003. http://www.umwelt.org/robin-wood/german/index-allgemei.htm

Robin Wood. 2003b. "Robin Wood Magazin." Downloaded 12 July 2003. http://www.umwelt.org/robin-wood/german/magazin/neu/index-abo.htm

Robin Wood. 2007a. "ROBIN WOOD e.V. - Auch in Ihrer Nähe?" Downloaded 2 January 2007. http://www.umwelt.org/robin-wood/german/index-regional.htm

Robin Wood. 2007b. "Jahresrechnung 2004." Downloaded 2 January 2007. http://www.umwelt.org/robin-wood/german/rechenschaftsbericht/2004/index.htm

Robin Wood. 2007c. "Robin Wood." Downloaded 29 January 2007. http://www.robinwood.de/

Rogall, Holger. 2003. *Akteure der Nachhaltigen Entwicklung.* Munich: oekom.
Rohkrämer, Thomas. 2002. "Contemporary Environmentalism and Its Links with the German Past." In *The Culture of German Environmentalism,* ed. Axel Goodbody. New York: Berghahn Books. Pp. 47–61.
Rohrschneider, Robert. 1991. "Public Opinion toward Environmental Groups in Western Europe: One Movement or Two?" *Social Science Quarterly* 72, no. 2 (June): 251–266.
Rollins, William. 1993. "Bund Heimatschutz: Zur Integration von Ästhetik und Politik." In *Mit den Bäumen Sterben die Menschen: Zur Kulturgeschichte der Ökologie,* ed. Jost Hermand. Cologne: Böhlau Verlag. Pp. 97–104.
Rollins, William. 1997. *A Greener Vision of Home: Cultural, Political, and Environmental Reform in the German Heimatschutz Movement, 1904–1918.* Ann Arbor: University of Michigan Press.
Roose, Jochen. 2002. *Die Europäisierung von Umweltorganisationen: Die Umweltbewegung auf dem Langen Weg nach Brüssel.* Opladen: Westdeutscher Verlag.
Rootes, Christopher. 1999a. "The Transformation of Environmental Activism: Activists, Organizations, and Policy-Making." *Innovation: The European Journal of Social Sciences* 12, no. 2 (June): 155–173.
Rootes, Christopher. 1999b. "Environmental Movements: From Global to Local." In *Environmental Movements: Local, National, and Global,* ed. Christopher Rootes. London: Frank Cass. Pp. 1–12.
Rootes, Christopher. 1999c. "Political Opportunity Structure: Promise, Problems, and Prospects." *La Lette de la Maison Francais d'Oxford,* no volume, no. 10: 75–97.
Rootes, Christopher. 2002. "Environmentalism Transformed? Environmental Protest in Western Europe, 1988–1997." Paper presented at the World Congress of the International Sociological Association. Brisbane, Australia.
Rootes, Christopher. 2004a. "Is There a European Environmental Movement?" In *Europe, Globalization, and the Challenge of Sustainability,* ed. John Barry, Brian Baxter, and Richard Dunphy. London: Routledge. Pp. 47–72.
Rootes, Christopher. 2004b. "Environmental Movements." In *The Blackwell Companion to Social Movements,* ed. David A. Snow, Sarah A. Soule, and Hanspeter Kriesi. Malden, MA: Blackwell. Pp. 608–640.
Röscheisen, Helmut. 2006. *Der Deutsche Naturschutzring: Geschichte, Interessenvielfalt, Organisationsstruktur und Perspektiven.* Munich: oekom.
Rosegger, Hans, Helga Schneider, and Hans-Josef Hönig. 2000. *Database Fundraising: Wie Sie Ihr Fundraising zum Erfolg Führen.* Ettlingen: IM Marketing Forum GmbH Fachverlag.
Rösler, Markus, Elisabeth Schwab, and Markus Lambrecht. 1990. *Naturschutz in der DDR.* Bonn: Economica Verlag.
Rossman, Torsten. 1993. "Öffentlichkeitsarbeit und Ihr Einfluß auf die Medien: Das Beispiel Greenpeace." *Media Perspektiven,* no volume, no. 2 (February): 85–94.
Roth, Karin, ed. 1992. *Ökologische Reform der Wirtschaft.* Cologne: BUND Verlag.
Rucht, Dieter. 1980. *Von Whyl nach Gorleben: Bürger gegen Atomprogramm und Nukleare Entsorgung.* Munich: C.H. Beck.
Rucht, Dieter. 1984. *Flughafenprojekte als Politikum: Die Konflikte in Stuttgart, Munich und Frankfurt.* Frankfurt: Campus.
Rucht, Dieter. 1989. "Environmental Movement Organizations in West Germany and France: Structures and Interorganizational Relations." In *International Social Movement Research: A Research Annual,* ed. Bert Klandermans. Greenwich, CT: JAI Press. Pp. 61–94.
Rucht, Dieter. 1990. "Campaigns, Skirmishes, and Battles: Anti-Nuclear Movements in the USA, France, and West Germany." *Industrial Crisis Quarterly* 4, no. 3 (fall): 193–222.

Rucht, Dieter. 1991a. "Von der Bewegung zur Institution?" In *Neue Soziale Bewegungen in der Bundesrepublik Deutschland,* ed. Roland Roth and Dieter Rucht. Frankfurt: Campus. Pp. 334–358.

Rucht, Dieter. 1991b. "The Study of Social Movements in West Germany: Between Activism and Social Science." In *Research on Social Movements: The State of the Art in Western Europe and the USA,* ed. Dieter Rucht. Frankfurt: Campus.

Rucht, Dieter. 1993a. "Eine Institutionalisierte Bewegung." *Politische Ökologie* 11, no. 31 (May/June): 36–43.

Rucht, Dieter. 1993b. "Parteien, Verbände und Bewegungen als Systeme Politischer Interessenvermittlung." In *Stand und Perspektiven der Parteienforschung in Deutschland,* ed. Oskar Niedermayer and Richard Stoss. Opladen: Westdeutscher Verlag. Pp. 251–275.

Rucht, Dieter. 1994. *Modernisierung und Neue Soziale Bewegungen: Deutschland, Frankreich und USA im Vergleich.* Frankfurt: Campus.

Rucht, Dieter. 1995a. "Think Globally: Act Locally? Needs, Forms, and Problems of Cross-National Environmental Groups." In *European Integration and Environmental Policy,* ed. C. D. Liefferink, P. D. Lowe, and Arthur P. J. Moll. New York: Wiley. Pp. 75–96.

Rucht, Dieter. 1995b. "Ecological Protest as Calculated Lawbreaking: Greenpeace and Earth First! in Comparative Perspective." In *Green Politics Three,* ed. Wolfgang Rüdig. Edinburgh: Edinburgh University Press. Pp. 66–89.

Rucht, Dieter. 1996. "The Impact of National Contexts on Social Movement Structures: A Cross-Movement and Cross-National Comparison." In *Comparative Perspectives on Social Movements,* ed. John D. McCarthy and Mayer N. Zald. New York: Cambridge University Press. Pp. 186–204.

Rucht, Dieter. 1997. "Limits to Mobilization: Environmental Policy for the European Union." In *Transnational Social Movements and Global Politics,* ed. Jackie Smith, Charles Chatfield, and Ron Pagnucco. Syracuse, NY: Syracuse University Press. Pp. 195–213.

Rucht, Dieter. 1999a. "The Transnationalization of Social Movements: Trends, Causes, and Problems." In *Social Movements in a Globalizing World,* ed. Donatella della Porta, Hanspeter Kriesi, and Dieter Rucht. New York: St. Martin's Press. Pp. 206–222.

Rucht, Dieter. 1999b. "Michels' Iron Law of Oligarchy Reconsidered." *Mobilization: An International Journal* 4, no. 2(fall): 151–169.

Rucht, Dieter. 2001. "Lobbying or Protest? Strategies to Influence EU Environmental Policies." In *Contentious Europeans: Protest and Politics in an Emerging Polity,* ed. Doug Imig and Sidney Tarrow. Lanham, MD: Rowman and Littlefield. Pp. 125–142.

Rucht, Dieter, Barbara Blattert, and Dieter Rink. 1997. *Soziale Bewegungen auf dem Weg zur Institutionalisierung.* Frankfurt: Campus.

Rucht, Dieter, and Jochen Roose. 1999. "The German Environmental Movement at the Crossroads." In *Environmental Movements: Local, National, and Global,* ed. Christopher Rootes. London: Frank Cass. Pp. 59–80.

Rucht, Dieter, and Jochen Roose. 2001a. "Neither Decline nor Sclerosis: The Organizational Structure of the German Environmental Movement." *West European Politics* 24, no. 4 (October): 55–83.

Rucht, Dieter, and Jochen Roose. 2001b. "Zur Institutionalisierung von Bewegungen: Umweltverbände und Umweltprotest in der Bundesrepublik." In *Verbände und Demokratie in Deutschland,* ed. Annette Zimmer and Bernhard Weßels. Opladen: Leske + Budrich. Pp. 261–290.

Rucht, Dieter, and Jochen Roose. 2001c. "Von der Platzbesetzung zum Verhandlungstisch? Wandel von Aktionen und Struktur der Ökologiebewegung." In *Protest in der Bundesrepublik: Strukturen und Entwicklungen,* ed. Dieter Rucht. Frankfurt: Campus. Pp. 173–210.

Rüddenklau, Wolfgang. 1992. *Störenfried: DDR-Opposition 1986–1989, mit Texten aus den "Umweltblättern."* Berlin: BasisDruck.

Rüdig, Wolfgang. 1980 "Bürgerinitiativen im Umweltschutz: Eine Bestandsaufnahme der Empirischen Befunde." In *Bürgerinitiativen in der Gesellschaft: Politische Dimensionen und Reaktionen,* ed. Otthein Rammstedt. Villingen: Neckar-Verlag. Pp. 341–350.

Rüdig, Wolfgang. 2002. "Germany." *Environmental Politics* 11, no. 1 (spring): 78–111.

Rutschke, Erich. 2001. "Ornithologie in der DDR: Ein Rückblick." In *Naturschutz in den Neuen Bundesländern: Ein Rückblick,* ed. Institut für Umweltgeschichte und Regionalentwicklung. Berlin: Verlag für Wissenschaft und Forschung. Pp. 109–133.

Sabatier, Paul A. 1992. "Interest Group Membership and Organization: Multiple Theories." In *The Politics of Interests,* ed. Mark A. Petracca. Boulder, CO: Westview. Pp. 99–129.

Sachs, Wolfgang. 1996. "Liegt Greenpeace Vorn im 21. Jahrhundert?" In *Das Greenpeace Buch: Reflexionen und Aktionen,* ed. Greenpeace Deutschland. Munich: C.H. Beck. Pp. 291–295.

Salamon, Lester M., and Helmut K. Anheier. 1998. "The Third Route: Government-Nonprofit Collaboration in Germany and the United States." In *Private Action and the Public Good,* ed. Walter W. Powell and Elisabeth S. Clemens. New Haven, CT: Yale University Press. Pp. 151–162.

Sale, Kirkpatrick. 1993. *The Green Revolution.* New York: Hill and Wang.

Salisbury, Robert H. 1969. "An Exchange Theory of Interest Groups." *Midwest Journal of Political Science* 13, no. 1 (February): 1–32.

Sander, Reinhard. 1992a. "Die Neue Kooperation der Umweltverbände und Gewerkschaften." *Forschungsjournal Neue Soziale Bewegungen* 5, no. 3 (September): 18–23.

Sander, Reinhard. 1992b. "Umweltverbände und Gewerkschaften." In *Ökologische Reform der Wirtschaft,* ed. Karin Roth. Cologne: BUND Verlag. Pp. 211–227.

Sbragia, Alberta M. 2000. "Environmental Policy: Economic Constraints and External Pressures." In *Policy Making in the European Union,* ed. Helen Wallace and William Wallace. New York: Oxford University Press. Pp. 293–316.

Schaeffer, Martin B. 1995. "The Internal Dynamics of Environmental Organizations: Movement Interest Groups, Communal Advocacy Groups, and the Policy Process." *Policy Studies* 14 (summer): 183–194.

Schenkluhn, Brigitte. 1990. "Umweltverbände und Umweltpolitik." In *Im Dienst der Umwelt und der Politik: Zur Kritik der Arbeit des Sachverständigenrates für Umweltfragen,* ed. Helmut Schreiber and Gerhardt Timm. Berlin: Analytika. Pp. 129–157.

Scheuch, Fritz. 1999. "Marketing für NPOs." In *Handbuch der Nonprofit Organisation: Strukturen und Management,* ed. Christoph Badelt. Stuttgart: Schaffer-Poeschel. Pp. 241–256.

Schiller, Theo. 1984. "Bürgerinitiativen und die Funktionskrise der Volksparteien." In *Bürgerinitiativen und Repräsentatives System,* ed. Bernd Guggenberger and Udo Kempf. Opladen: Westdeutscher Verlag. Pp. 188–212.

Schlipköter, Hans-Werner, and Gerhard Winneke. 2000. "Gesundheitsschutz durch Umweltschutz: Umwelthygiene Gestern, Heute und Morgen." In *Umweltpolitik mit Augenmaß,* ed. Henning von Köller. Berlin: Erich Schmidt Verlag. Pp. 133–148.

Schmid, Carol. 1987. "The Green Movement in West Germany." *Journal of Political and Military Sociology* 15, no. 1 (spring): 33–46.

Schmitt-Beck, Rüdiger, and Cornelia Weins. 1997. "Gone with the Wind (of Change): Neue Soziale Bewegungen und Politischer Protest im Osten Deutschlands." In *Politische Orientierungen und Verhaltensweisen im Vereinigten Deutschland,* ed. Oscar W. Gabriel. Opladen: Leske + Budrich.

Schmitter, Philippe C. 1979 . "Still the Century of Corporatism?" In *Trends toward Corporat-*

ist Intermediation, ed. Philippe C. Schmitter and Gerhard Lehmbruch. Beverly Hills, CA: Sage. Pp. 7–52.

Schmitter, Philippe C. 1982. "Reflections on Where the Theory of Neo-Corporatism Has Gone and Where the Praxis of Neo-Corporatism May Be Going." In *Patterns of Corporatist Policy Making,* ed. Gerhard Lehmbruch and Phillippe C. Schmitter. Beverly Hills, CA: Sage. Pp. 260–279.

Schmitz, Hans-Peter. 1984. "Naturschutz - Landschaftsschutz - Umweltschutz." In *Mit Uns Zieht die Neue Zeit: Die Naturfreunde,* ed. Jochen Zimmer. Cologne: Pahl-Rugenstein. Pp. 194–204.

Schmoll, Friedemann. 2003. "Paul Schultze-Naumburg: Von der Ästhetischen Reform zur Völkischen Ideologie." In *Naturschutz Hat Geschichte,* ed. Stiftung Naturschutzgeschichte. Essen: Klartext. Pp. 100–112.

Schnaiberg, Allan, and Kenneth A. Gould. 1994. *Environment and Society: The Enduring Conflict.* New York: St. Martin's Press.

Schnurrbusch, Gottfried. 1979. "Die Flurneugestaltung: Ein Beitrag zur Intensivierung der Landwirtschaft." In *Natur und Umwelt: Beiträge zur Sozialistischen Landeskultur,* ed. Fachausschuß Landeskultur und Naturschutz, Kulturband der Deutschen Demokratischen Republik. Berlin: Fachausschuß Landeskultur und Naturschutz, Kulturband der Deutschen Demokratischen Republik. Pp. 34–38.

Schoenichen, Walther. 1930. "Aus der Entwicklung der Naturdenkmalpflege." In *Der Deutsche Heimatschutz: Ein Rückblick und Ausblick,* ed. Gesellschaft der Freunde des Deutschen Heimatschutzes. Munich: Kastner & Callwey. Pp. 222–227.

Schoenichen, Walther. 1954. *Naturschutz, Heimatschutz: Ihre Begründung durch Ernst Rudorff, Hugo Conwentz und Ihre Vorläufer.* Stuttgart: Wissenschaftliche Verlagsgesellschaft.

Scholz, Reiner. 1989. *Betrifft Robin Wood: Sanfte Rebellen gegen Naturzerstörung.* Munich: Beck'sche Reihe.

Schönborn, Gregor. 1995. "Offensive Umweltkommunikation von Unternehmen." In *Umweltbewußtsein und Massenmedien,* ed. Gerhard de Haan. Berlin: Akademie Verlag. Pp. 149–155.

Schreiner, Johann, and Peter Wörnle. 1994. "Ökosponsoring: Eine Brücke zwischen Naturschutzverbänden und Unternehmen Wird Geschlagen." In *Ökosponsoring: Werbestrategie oder Selbstverpflichtung,* ed. Bayrische Akademie für Naturschutz und Landschaftspflege und Norddeutsche Naturschutzakademie. Schneeverdingen: Norddeutsche Naturschutzakademie. Pp. 5–6.

Schreuers, Miranda A. 2002. *Environmental Politics in Japan, Germany, and the United States.* Cambridge, MA: Cambridge University Press.

Schumacher, Ulrike. 2003a. "Ressourcen im Wandel: Der Strategische Umgang von Umweltorganisationen mit der Diversifizierung Ehrenamtlichen Engagements." In *Mission Impossible? Strategien in Nonprofitorganisationen,* ed. Arbeitskreis Nonprofit-Organisationen. Frankfurt: Eigenverlag des Deutschen Vereins für Öffentliche und Private Fürsorge. Pp. 96–122.

Schumacher, Ulrike. 2003b. *Lohn und Sinn: Individuelle Kombinationen von Erwerbsarbeit und freiwilligem Engagement.* Opladen: Leske + Budrich.

Schuster, Kai. 2003. "Image und Akzeptanz von Naturschutz in der Gesellschaft." In *Naturschutz in Deutschland: Eine Erfolgsstory,* ed. Deutscher Rat für Landespflege. Bonn: Deutscher Rat für Landschaftspflege. Pp. 80–89.

Schutzgemeinschaft Deutscher Wald. 2003. "Ziele und Aufgaben." Downloaded 30 January 2007. http://www.sdw.de/

Scott, W. Richard, and Gerald F. Davis. 2007. *Organizations and Organizing: Rational, Natural, and Open System Perspectives.* Upper Saddle River, NJ: Pearson Prentice-Hall.

Selznik, Philip. 1957. *Leadership in Administration: A Sociological Interpretation.* Evanston, IL: Row Peterson.

Shaiko, Ronald G. 1999. *Voices and Echoes for the Environment.* New York: Columbia University Press.

Sieferle, Rolf Peter. 1984. *Fortschrittsfeinde: Opposition gegen Technik und Industrie Von der Romantik bis zur Gegenwart.* Munich: C.H. Beck.

Skocpol, Theda. 2003. *Diminished Democracy: From Membership to Management in American Civil Life.* Norman: University of Oklahoma Press.

Smith, David Horton. 1994. "Determinants of Voluntary Association Participation and Volunteering: A Literature Review." *Nonprofit and Voluntary Sector Quarterly* 23, no. 3 (fall): 243–263.

Smith, Gordon. 1989. "Structures of Government." In *Developments in West German Politics,* ed. Gordon Smith, William E. Paterson, and Peter H. Merkl. Durham, NC: Duke University Press. Pp. 24–39.

Smith, Jackie. 1997. "Characteristics of the Transnational Social Movement Sector." In *Transnational Social Movements and Global Politics: Solidarity beyond the State,* ed. Jackie Smith, Charles Chatfield, and Ron Pagnucco. Syracuse, NY: Syracuse University Press.

Smith, Jackie. 1998. "Global Civil Society." *American Behavioral Scientist* 42, no. 1: 93–107.

Smith, Jackie. 2005. "Globalization and Transnational Social Movement Organizations." In *Social Movements and Organization Theory,* ed. Gerald F. Davis. Cambridge, MA: Cambridge University Press.

Snow, D. A., et al. 1986. "Frame Alignment Processes, Micromobilization, and Movement Participation." *American Sociological Review* 51, no. 4 (August): 464–481.

Snow, David, and Robert Benford. 1992. "Master Frames and Cycles of Protest" In *Frontiers in Social Movement Theory,* ed. Aldon D. Morris and Carol McClurg Mueller. New Haven, CT: Yale University Press.

SPD. 2007. "Umwelt und Nachhaltigkeit." Downloaded 27 January 2007. http://www.spd.de/menu/1687443/

Speitkamp, Winfried. 1988. "Denkmalpflege und Heimatschutz in Deutschland zwischen Kulturkritik und Nationalsozialismus." *Archiv für Kulturgeschichte* 70, no. 1: 149–193.

Spengler, Oswald. 1926. *The Decline of the West.* New York: Knopf.

Spielmann, Michael. 1997. "Eine Chance um Überhaupt zu Überleben." *Politische Ökologie* 15, no. 52 (July/August): 6.

Staggenborg, Suzanne. 1988. "The Consequences of Professionalization and Formalization in the Pro-Choice Movement." *American Sociological Review* 53 (August): 585–606.

Stark, Carsten. 2001. "Germany: Rule by Virtue of Knowledge." In *Democracy at Work: A Comparative Sociology of Environmental Regulation in the United Kingdom, France, Germany, and the United States,* ed. Richard Münch et al. Westport, CT: Praeger. Pp. 103–128.

Statham, Alison. 1997. "Ecology and the German Right." In *Green Thought in German Culture,* ed. Colin Riordan. Cardiff: University of Wales Press. Pp. 125–138.

Staudenmaier, Peter. 1995. "Fascist Ideology: The 'Green Wing' of the Nazi Party and Its Historical Antecendents." In *Ecofacism: Lessons from the German Experience,* ed. Janet Biehl and Peter Staudenmaier. Edinburgh: AK Press. Pp. 5–31.

Steenbock, Kristina. 1996a. "Greenpeace in der Politischen Arena." In *Das Greenpeace Buch: Reflexionen und Aktionen,* ed. Greenpeace Deutschland. Munich: C.H. Beck. Pp. 21–25.

Steenbock, Kristina. 1996b. "Ohne Macht Geht Gar Nichts: Vom Sinn und Nutzen der Lobbyarbeit." In *Das Greenpeace Buch: Reflexionen und Aktionen,* ed. Greenpeace Deutschland. Munich: C.H. Beck. Pp. 171–175.

Stein, Tine. 2003. *Interessenvertretung der Natur in den USA: Mit Vergleichendem Blick auf die Deutsche Rechtslage.* Baden-Baden: Nomos.

Sternstein, Wolfgang. 1980. "Bürgerinitiativen als vierte Gewalt: Wie Bürgerinitiativen sich selbst Verstehen." In *Bürgerinitiativen in der Gesellschaft: Politische Dimensionen und Reaktionen,* ed. Otthein Rammstedt. Villingen: Neckar-Verlag. Pp. 319–340.

Sternstein, Wolfgang. 1981. *Willensbildungs- und Entscheidungsprozesse in der Ökologiebewegung.* Hanover: Institut für Angewandte Systemforschung und Prognose.

Stiftung Europäisches Naturerbe (Euronatur). 2006. *Geschäftsbericht 2005.* Radolfzell: Stiftung Europäisches Naturerbe (Euronatur).

Stoß, Richard. 1991. "Parteien und Soziale Bewegungen." In *Neue Soziale Bewegungen in der Bundesrepublik Deutschland,* ed. Roland Roth and Dieter Rucht. Bonn: Bundeszentrale für Politische Bildung. Pp. 392–414.

Strachwitz, Rupert Graf, ed. 1998. *Dritter Sektor - Dritte Kraft: Versuch einer Standortbestimmung.* Stuttgart: RAABE.

Streek, Wolfgang, and Philippe C. Schmitter. 1985. "Community, Market, State - and Associations? The Prospective Contribution of Interest Governance to Social Order." In *Private Interest Government: Beyond Market and State,* ed. Wolfgang Streek and Philippe C. Schmitter. London: Sage. Pp. 1–29.

Streich, Jürgen. 1986. *Betrifft: Greenpeace.* Munich: Verlag C.H. Beck.

Stuttgarter Nachrichten. 1998. "Käpt'n Iglo um Fanggründe." Downloaded 26 June 1998. http://www.stuttgarter-nachrichten.de/dc1/html/news-stn/19980626wirt0021.htm

Tarrow, Sidney. 1991. *Struggle, Politics, and Reform: Collective Action, Social Movements, and Cycles of Protest.* Ithaca, NY: Center for International Studies, Cornell University.

Tarrow, Sidney. 1996. "States and Opportunities: The Political Structuring of Social Movements." In *Comparative Perspectives on Social Movements,* ed. Doug McAdam, John D. McCarthy, and Mayer N. Zald. Cambridge: Cambridge University Press. Pp. 41–61.

taz. 2004. "Wir Duzen uns Nicht!" *taz,* no volume, no. 7539 (12 December: 4–5.

Teichert, Volker. 1992. "Gewerkschaften und Ökologiebewegung: Die Annährung der Königskinder." *Forschungsjournal Neue Soziale Bewegungen* 5, no. 3 (September): 14–16.

Thaysen, Uwe. 1984. "Bürgerinitiativen - Grüne Alternative - Parlamente und Parteien in der Bundesrepublik." In *Bürgerinitiativen und Repräsentatives System,* ed. Bernd Guggenberger and Udo Kempf. Opladen: Westdeutscher Verlag. Pp. 124–155.

Thompson, James D. 1967. *Organizations in Action.* New York: McGraw-Hill.

Timm, Gerhard. 1985. "Die Offizielle Ökologiedebatte in der DDR." In *Umweltprobleme und Umweltbewußtsein in der DDR,* ed. Redaktion Deutschland Archiv. Cologne: Verlag Wissenschaft und Politik. Pp. 117–147.

Tnsemnid. 2007. "Charts zum Spendenmonitor 2004." Downloaded 31 January 2007. http://www.tns-emnid.com/pdf/presse-presseinformationen/2004/TNS_Emnid_10Jahre_Spendenmonitor.ppt

Touraine, Alain. 1985. "An Introduction to the Study of Social Movements." *Social Research* 52, no. 4 (winter): 749–786.

Trautmann, Günter. 1984. "Alte, Neue oder eine Andere Politik? Die Alternativbewegung gegen den Nuklearkonsens im Parteien- und Verbandssystem der Bundesrepublik." In *Bürgerinitiativen und Repräsentatives System,* ed. Bernd Guggenberger and Udo Kempf. Opladen: Westdeutscher Verlag. Pp. 333–360.

Trice, Harrison M., and Janice M. Beyer. 1993. *The Cultures of Work Organizations.* Englewood Cliffs, NJ: Prentice-Hall.

Turner, Henry Ashby. 1991. *Germany from Partition to Reunification.* New Haven, CT: Yale University Press.

Turner, Ralph H. 1970. "Determinants of Social Movement Strategies." In *Human Nature and Collective Behavior,* ed. Tamotsu Shibutani. Englewood Cliffs, NJ: Prentice-Hall. Pp. 145–164.

Uekötter, Frank. 2003a. "Umweltbewegung zwischen dem Ende der Nationalsozialistischen Herrschaft und der 'Ökologischen Wende': Ein Literaturbericht." *Historical Social Research* 28, no. 1/2: 270–289.

Uekötter, Frank. 2003b. "Sieger der Geschichte: Überlegungen zum Merkwürdigen Verhältnis des Naturschutzes zu Seinem Eigenen Erfolg." In *Naturschutz in Deutschland: Eine Erfolgsstory,* ed. Deutscher Rat für Landespflege. Bonn: Deutscher Rat für Landespflege. Pp. 34–38.

Uekötter, Frank. 2003c. *Von der Rauchpflege zur Ökologischen Revolution: Eine Geschichte der Luftverschmutzung in Deutschland und den USA 1880–1970.* Essen: Klartext.

Uekötter, Frank. 2005. "Erfolglosigkeit als Dogma? Revisionistische Bemerkungen zum Umweltschutz zwischen dem Ende des Zweiten Weltkriegs und der 'Ökologischen Wende.'" In *Natur und Umweltschutz nach 1945: Konzepte, Konflikte, Kompetenzen,* ed. Franz-Josef Brüggemeier and Jens Ivo Engels. Frankfurt: Campus. Pp. 105–123.

Uhdehn, Lars. 1993. "Twenty-five Years of the Logic of Collective Action." *Acta Sociologica* 36, no. 3: 239–261.

Ulbricht, Justus H. 1993. "Grün als Brücke zu Braun." *Politische Ökologie Special "Grün Heil"* 11, no. 34 (November/December): 7–12.

Umkehr e.V. 2007. "UMKEHR e.V. - Eine Kurze Vorstellung." Downloaded 28 January 2007. http://www.umkehr.de/

Umwelt: Kommunale Ökologische Briefe. 2004. "Im Großen und Ganzen Okay." *Umwelt: Kommunale Ökologische Briefe* 9, no. 11 (26 May): 3.

Upmann, Augustin, and Uwe Rennspieß. 1984. "Organisationsgeschichte der Deutschen Naturfreundebewegung bis 1933." In *Mit Uns Zieht die Neue Zeit: Die Naturfreunde,* ed. Jochen Zimmer. Cologne: Pahl-Rugenstein. Pp. 66–111.

Urselmann, Michael. 1998. *Fundraising: Erfolgreiche Strategien Führender Nonprofit-Organisationen.* Bern: Verlag Paul Haupt.

Vandamme, Ralf. 1996. "Über den Wandel von Öffentlichkeit als Ort des Politischen: Von der Gegenöffentlichen zur Kampagnenöffentlichkeit von Greenpeace." *Forschungsjournal Neue Soziale Bewegungen* 9, no. 1 (January): 101–103.

Vandamme, Ralf. 2000. *Basisdemokratie als Zivile Intervention.* Opladen: Leske + Budrich.

van der Heijden, Hein-Anton. 1997. "Political Opportunity Structure and the Institutionalization of the Environmental Movement." *Environmental Politics* 6, no. 4 (winter): 22–50.

van der Heijden, Hein-Anton. 2002. "Political Parties and NGOs in Global Environmental Politics." *International Political Science Review* 23, no. 2 (April): 187–201.

van der Heijden, Hein-Anton. 2005. "European Politics of Nature: Actors, Discourses and Actors." Paper presented at the Annual Meeting of the European Consortium for Political Research. Budapest.

van der Heijden, Hein-Anton. 2006. "Globalization, Environmental Movements, and International Political Opportunity Structures." *Organization and Environment* 19, no. 1 (March): 28–45.

van der Heijden, Hein-Anton, Ruud Koopmans, and Marco G. Giugini. 1992. "The West European Environmental Movement." In *Research in Social Movements, Conflict and Change, Supplement 2,* ed. Matthias Finger. Greenwood, CT: JAI Press. Pp. 1–40.

van Kersbergen, Kees, and Frans van Waarden. 2004. "Governance as a Bridge between Disciplines." *European Journal of Political Research* 43, no. 2 (March): 143–171.

van Koppen, C.S.A. (Kris), and William T. Markham. 2007. "Nature Protection in Western Environmentalism: A Comparative Analysis." In *Protecting Nature: Organizations and Networks in Europe and the USA,* ed. C.S.A. (Kris) van Koppen and William T. Markham. Cheltenham: Edward Elgar.

van Tatenhove, Jan P. M. 2002. "Policy Making in an Institutional Void: Multilevel Governance and Innovative Environmental Policy Arrangements in the European Union." Paper Presented at the World Congress of Sociology, International Sociological Association. Brisbane, Australia.

van Tatenhove, Jan P. M., and Pieter Leroy. 2003. "Environment and Participation in a Context of Modernization." *Environmental Values* 12, no. 2 (May): 155–174.

van Til, Jon. 1988. *Mapping the Third Sector: Voluntarism in a Changing Social Economy.* New York: Foundation Center.

Visser, Jelle. 2006. "Union Membership Statistics in 24 Countries." *Monthly Labor Review* 129, no. 1 (January): 39–49.

von Alemann, Ulrich. 1989. *Organisierte Interessen in der Bundesrepublik.* Opladen: Leske + Budrich.

von Kodolitsch, Paul. 1975. "Gemeindeverwaltungen und Bürgerinitiativen: Ergebnisse einer Umfrage." *Archiv für Kommunalwissenschaften* 14, no. 2: 264–278.

von Reuter, Eduard. 1926. "Organisation der Naturschutzbewegung in Bayern." *Blätter für Naturschutz und Naturpflege* 9, no. 3 (December): 83–86.

von Treuenfels, Carl-Albrecht. 2003. "40 Jahre WWF Deutschland: Prägende Persönlichkeiten, Ideen, Organisationsentwicklung." In *Das Große Buch des WWF. 40 Jahre Naturschutz für und mit den Menschen,* ed. Klaus-Henning Groth. Steinfurt: Tecklenborg Verlag. Pp. 14–23.

Voss, Gerhard. 1990. *Die Veröffentlichte Umweltpolitik: Ein Sozio-ökologisches Lehrstück.* Cologne: Kölner Universitätsverlag.

Voss, Gerhard. 1995. "Umweltschutz in den Printmedien." In *Umweltbewußtsein und Massenmedien: Perspektiven Ökologischer Kommunikation,* ed. Gerhard de Haan. Berlin: Akademie Verlag. Pp. 123–129.

Walker, Jack L. 1991. *Mobilizing Interest Groups in America.* Ann Arbor: University of Michigan Press.

Wallmeyer, Gerhard. 1996. "Der Scheck als Stimmzettel: Ziele und Methoden des Fundraising." In *Das Greenpeace Buch: Reflexionen und Aktionen,* ed. Greenpeace Deutschland. Munich: C.H. Beck. Pp. 93–101

Walzer, Michael. 2004. *Politics and Passion: Toward a More Egalitarian Liberalism.* New Haven, CT: Yale University Press.

Wapner, Paul. 1996. *Environmental Activism and World Civic Politics.* Albany: SUNY Press.

Warren, Mark E. 2001. *Democracy and Association.* Princeton, NJ: Princeton University Press.

Webster, Ruth. 1998. "Environmental Collective Action: Stable Patterns of Cooperation and Issue Alliances at the European Level." In *Collective Action in the European Union,* ed. Justin Greenwood and Mark Aspenwall. London: Routledge. Pp. 176–196.

Wegener, Uwe. 2001. "Ohne Sie Hätte Sich Nichts Bewegt: Zur Arbeit der Ehrenamtlichen Naturschutzhelfer und -Helferinnen." In *Naturschutz in den Neuen Bundesländern: Ein Rückblick,* ed. Institut für Umweltgeschichte und Regionalentwicklung e.V. Berlin: Verlag für Wissenschaft und Forschung. Pp. 89–107.

Weidner, Helmut. 1991a. "Reagieren statt Agieren: Entwicklungslinien in der Bundesrepublik Deutschland." *Politische Ökologie* 9, no. 23 (August/September): 14–22.

Weidner, Helmut. 1991b. "Umweltpolitik: Auf dem Alten Weg zu einer Internationalen Spitzenstellung." In *Die Bundesrepublik in den Achtziger Jahren: Innenpolitik, Politische Kultur, Außenpolitik,* ed. Werner Süß. Opladen: Leske + Budrich. Pp. 137–151.

Weigand, Petra. 1995. "Öffentlichkeitsarbeit von Umweltorganisationen." *Forschungsjournal Neue Soziale Bewegungen* 8, no. 2 (June): 102–107.

Weiger, Hubert. 1987a. "Leistungsvermögen und Leistungsversäumnisse der Naturschutzverbände." In *Jahrbuch für Naturschutz und Landschaftspflege 39,* ed. Bundesverband

Beruflicher Naturschutz e.V. Bonn: Bundesverband Beruflicher Naturschutz e.V. Pp. 154–162.

Weiger, Hubert. 1987b. "Wie Können Naturschutzverbände Naturschutzpolitik Gestalten?" In *Strategien einer Erfolgreichen Naturschutzpolitik - Laufener Seminarbeiträge 2/87,* ed. Akademie für Naturschutz und Landschaftspflege. Laufen: Akademie für Naturschutz und Landschaftspflege. Pp. 59–67.

Weiger, Hubert. 1997. "Ablaßhandel Zerstört die Glaubwürdigkeit." *Politische Ökologie* 15, no. 52 (July/August): 8.

Weinbach, Kerstin. 1998. *Umweltpolitisches Engagement im Vereinigungsprozeß.* Koblenz: Fölbach.

Weinzierl, Hubert. 1991. "Naturschutzverbände als Lobby der Umwelt." *Berichte der Akademie für Naturschutz 15,* ed. Akademie für Naturschutz und Landschaftspflege. Laufen: Akademie für Naturschutz und Landschaftspflege. Pp. 5–13.

Weinzierl, Hubert et al. 1990. "...Ende der Kompromißbereitschaft: Reden Anläßlich der Delegiertenversammlung in Ludwigshafen am 11. Juni 1989" (booklet). Bonn: BUND.

Weisbrod, Burton A. 1988. *The Nonprofit Economy.* Cambridge, MA: Harvard University Press.

Welz, Wolfgang. 1984. "Literaturüberblick - Literaturverzeichnis." In *Bürgerinitiativen und Repräsentatives System,* ed. Bernd Guggenberger and Udo Kempf. Opladen: Westdeutscher Verlag. Pp. 436–471.

Wensierski, Peter. 1985. "Die Gesellschaft für Natur und Umwelt: Kleine Innovation in der Politischen Kultur der DDR." In *Umweltprobleme und Umweltbewußtsein in der DDR,* ed. Redaktion Deutschland Archiv. Cologne: Verlag Wissenschaft und Politik. Pp. 151–168.

Weßels, Bernhard. 1989. "Politik, Industrie und Umweltschutz in der Bundesrepublik: Konsens und Konflikt in einem Politikfeld 1960–1986." In *Konfliktpotentiale und Konsensstrategien: Beiträge zur Politischen Soziologie der Bundesrepublik,* ed. Dietrich Herzog and Bernhard Weßels. Opladen: Westdeutscher Verlag. Pp. 269–306.

Wesser. 2001. "Wesser Job-Haus." Downloaded 15 May 2001. http://www.ferienjob-center.de/

Wettengel, Michael. 1993. "Staat und Naturschutz 1906–1945: Zur Geschichte der Staatlichen Stelle für Naturdenkmalpflege in Preußen und der Reichsstelle für Naturschutz." *Historische Zeitschrift* 257, no. 2 (October): 355–399.

Wey, Klaus-Georg. 1982. *Umweltpolitik in Deutschland: Kurze Geschichte des Umweltschutzes in Deutschland seit 1900.* Opladen: Westdeutscher Verlag.

Wickert Institut. 1989. "Bekanntheitsgrad und Image Bund für Umwelt und Naturschutz Deutschland e.V." (booklet). Bonn: Wickert Institut.

Wiesenthal, Helmut. 1993. "Akteurkompetenz im Organisationsdilemma: Grundprobleme Strategisch Ambitionierter Mitgliederverbände und Zwei Techniken Ihrer Überwindung." *Berliner Journal für Soziologie* 3, no. 1 (March): 3–18.

Wiesenthal, Helmut. 2004. "From a Nest of Rivals to Germany's Agents of Change: Remarks on 'Values and Conflicts' with Regard to the German Greens in the 1980s and 1990s." Paper presented at the workshop The Origins of Green Parties in Global Perspective at the German Historical Institute, Washington, D.C.

Wilhelm, Sighard. 1994. *Umweltpolitik: Bilanz, Probleme, Zukunft.* Opladen: Leske + Budrich.

Williams, John Alexander. 1996. "'The Chords of the German Soul are Tuned to Nature': The Movement to Preserve the Natural Heimat from the Kaiserreich to the Third Reich." *Central European History* 29, no. 3: 339–384.

Williams, Rhys H. 2004. "The Cultural Context of Collective Action: Constraints, Opportunity, and the Symbolic Life of Social Movements." In *The Blackwell Companion to Social*

Movements, ed. David A. Snow, Sarah A. Soule, and Hanspeter Kriesi. Malden, MA: Blackwell. Pp. 91–115.

Wilson, Frank L. 1990. "Neocorporatism and the Rise of New Social Movements." In *Challenging the Political Order: New Social and Political Movements in Western Democracies,* ed. Russell J. Dalton and Manfred Keuchler. New York: Oxford University Press. Pp. 67–93.

Wilson, Graham K. 1990. *Interest Groups.* Oxford: Basil Blackwell.

Wilson, Graham K. 1992. "American Interest Groups in Comparative Perspective." In *The Politics of Interest,* ed. Mark Petracca. Boulder, CO: Westview. Pp. 80–95.

Wilson, James Q. 1995. *Political Organizations.* Princeton, NJ: Princeton University Press.

Wimmer, Frank, and Heiko Wahl. 1995. "Value Change and Environmental Awareness in Germany." In *The Green Agenda: Environmental Politics and Policy in Germany,* ed. Ingolfur Blühdorn, Frank Krause, and Thomas Scharf. Keele: Keele University Press. Pp. 25–51.

Winterer, Franz. 1955. "Die Gründung des Wiener Touristenvereins 'Die Naturfreunde.'" In *Touristenverein "Die Naturfreunde:" Denkschrift zum Sechzigjährigen Bestehen 1895–1955,* ed. Zentralausschuß der Naturfreunde-Internationale. Zurich: Zentralausschuß der Naturfreunde-Internationale. Pp. 9–26.

Wöbse, Anna-Katharina. 2003a. "Lina Hähnle und der Reichsbund für Vogelschutz: Soziale Bewegung im Gleichschritt." In *Naturschutz und Nationalsozialismus,* ed. Joachim Radkau and Frank Uekötter. Frankfurt: Campus. Pp. 309–330.

Wöbse, Anna-Katharina. 2003b. "Lina Hähnle: Eine Galionsfigur der Frühen Naturschutzbewegung." In *Naturschutz Hat Geschichte,* ed. Stiftung Naturschutzgeschichte. Essen: Klartext. Pp. 113–130.

Wolf, Angelika. 1996. *Die Analyse der Reformfähigkeit eines Umweltverbandes am Beispiel des Bund für Umwelt und Naturschutz Deutschland e.V. (BUND): Aufgaben und Struktur.* Doctoral dissertation, Fachbereich Umwelt und Gesellschaft, Berlin: Technische Universität.

Wollenweber, Marianne. 1994. "Bonns Grüne Lobbyisten." *Natur,* no volume, no. 1 (January): 13–17.

Worcester, Robert M. 1993. "Public and Elite Attitudes to Environmental Issues." *International Journal of Public Opinion* 5, no. 4 (winter): 315–334.

World Bank. 2000. *World Development Report 2000/2001.* Oxford: Oxford University Press.

Wunderer, Hartmann. 1977. "Der Touristenverein 'Die Naturfreunde': Eine Sozialdemokratische Arbeiterkulturorganisation (1895–1933)." *Internationale Wissenschaftliche Korrespondenz zur Geschichte der Deutschen Arbeiterbewegung* 13, no. 4 (December): 506–520.

Wunderer, Hartmann. 1980. *Arbeitervereine und Arbeiterparteien: Kultur und Massenorganisationen in der Arbeiterbewegung.* Frankfurt: Campus.

Wünschmann, Arnd. 2003. "Naturschutz für den Menschen und mit den Menschen: Der Weg des WWF Deutschland." In *Das Große Buch des WWF. 40 Jahre Naturschutz für und mit den Menschen,* ed. Klaus-Henning Groth. Steinfurt: Tecklenborg Verlag. Pp. 26–39.

Wünschmann, Arnd et al. 2003. "40 Jahre Umweltstiftung WWF Deutschland: Eine Chronik." In *Das Große Buch des WWF: 40 Jahre Naturschutz für und mit den Menschen,* ed. Klaus-Henning Groth. Steinfurt: Tecklenborg Verlag. Pp. 271–290.

Würth, Gerhard. 1985. *Umweltschutz und Umweltzerstörung in der DDR.* Frankfurt: Peter Lang.

Wurzel, Rüdiger K. W. 2002. *Environmental Policy-Making in Britain, Germany and the European Union: The Europeanization of Air and Water Pollution Control.* Manchester: Manchester University Press.

Wüst, Jürgen. 1993. *Konservatismus und Ökologiebewegung: Eine Untersuchung im Spannungsfeld von Partei, Bewegung und Ideologie am Beispiel der Ökologisch-demokratischen Partei (ÖDP).* Frankfurt: IKO-Verlag für Interkulturelle Kommunikation.

Wüstenhagen, Hans-Helmuth. 1975. "Erfahrungen mit Bürgerinitiativen für Umweltschutz." *Blätter für Deutsche und Internationale Politik* 20, no. 10 (October): 1105–1115.

WWF-Deutschland. 1999. *Jahresbericht 1998.* Frankfurt: WWF-Deutschland.

WWF-Deutschland. 2000. *Tätigkeitsbericht 1999.* Frankfurt: WWF-Deutschland.

WWF-Deutschland. 2001a. "Young Panda." Downloaded 20 May 2001. http://www.wwf.de/c_youngpanda/c_youngp_index.html

WWF-Deutschland. 2001b. *Jahresbericht 2000.* Frankfurt: WWF-Deutschland.

WWF-Deutschland. 2002. *Jahresbericht 2001.* Frankfurt: WWF-Deutschland.

WWF-Deutschland. 2003. *Jahresbericht 2002.* Frankfurt: WWF-Deutschland.

WWF-Deutschland. 2004. *Jahresbericht 2003.* Frankfurt: WWF-Deutschland.

WWF-Deutschland. 2005. *Jahresbericht 2004.* Frankfurt: WWF-Deutschland.

WWF-Deutschland. 2007. "Unsere Themen." Downloaded 10 February 2007. http://www.wwf.de/unsere-themen/

WWF-Deutschland. No date. "WWF 2000: Ziele, Auftrag, Programm und Positionen der Umweltstiftung WWF Deutschland für die 90er Jahre" (booklet). Frankfurt: WWF-Deutschland.

WWF-International. 2006a. *Working Together: WWF Annual Review 2006.* Gland, Switzerland: WWF-International.

WWF-International. 2006b. "WWF - Who We Are and How We Came About." Downloaded 8 June 2006. http://www.panda.org

WWF-Journal. 1998a. "Neu: Der Panda Renditenfonds DWS." *WWF-Journal,* no volume, no. 1 (January): 11–14.

WWF-Journal. 1998b. "Klimaschutz Duldet Keinen Aufschub." *WWF-Journal,* no volume, no. 2 (April): 12.

WWF-Journal. 1998c. "Partner für die Natur." *WWF-Journal,* no volume, no. 2 (April): 46.

WWF-Journal. 1998d. "'Eines der Erfolgreichsten Jahre des WWF.'" *WWF-Journal,* no volume, no. 3 (July): 19.

WWF-Journal. 1998e. "Partner für die Natur." *WWF-Journal,* no volume, no. 3 (July): 46.

WWF-Journal. 1998f. "Young Panda Erfolgreich Gestartet." *WWF-Journal,* no volume, no. 3 (July): 49.

WWF-Journal. 1999a. "Starke Partner aus der Wirtschaft: Für Naturschutz Unverzichtbar." *WWF-Journal,* no volume, no. 1 (January): 6–7.

WWF-Journal. 1999b. "Die Öko-Manager des Jahres 1998." *WWF-Journal,* no volume, no. 1 (January): 30.

WWF-Journal. 1999c. "Ein Testament für die Natur." *WWF-Journal,* no volume, no. 1 (January): 45.

WWF-Journal. 1999d. "Die Vereinigten Nationen des Naturschutzes." *WWF-Journal,* no volume, no. 4 (October): Part I, 4–5.

WWF-Journal. 1999e. "Panda Versand Bleibt Öko-Vorreiter." *WWF-Journal,* no volume, no. 2 (April): 11.

WWF-Journal. 1999f. "Unterstützen Sie Uns mit Realistischen Vorschlägen." *WWF-Journal,* no volume, no. 4 (October): Part II, 22–23.

WWF-Journal. 1999g. "Mehr Wildnis für Europa." *WWF-Journal,* no volume, no. 3 (July): 30–31.

WWF-Journal. 2000a. "Global 2000: Die Arche Noah der Lebensräume." *WWF-Journal,* no volume, no. 1 (January): 4–5.

WWF-Journal. 2000b. "Die Ökomanager des Jahres 1999." *WWF-Journal,* no volume, no. 1 (January): 7.

WWF-Journal. 2000c. "Ems-Sperrwerk: Baustopp Aufgehoben." *WWF-Journal,* no volume, no. 1 (January): 8–9.

WWF-Journal. 2000d. "Großes Votum für Nationalpark Kellerwald." *WWF-Journal,* no volume, no. 1 (January): 11.

WWF-Journal. 2000e. "Global Player in Naturschutz." *WWF-Journal,* no volume, no. 2 (April): 4.

WWF-Journal. 2000f. "PEFC Waldsiegel Dokumentiert Ökologischen Stillstand." *WWF-Journal,* no volume, no. 3 (July): 9.

WWF-Journal. 2000g. "WWF Geschäftsbericht, 1999." *WWF-Journal,* no volume, no. 3 (July): 14–15.

WWF-Journal. 2000h. "Die Retter der Auen: Das WWF-Auen-Institut in Rastatt Feiert seinen 15. Geburtstag." *WWF-Journal,* no volume, no. 4 (October): 10.

WWF-Journal. 2000i. "Hamburger WWF-Gruppe im Dschungelfieber." *WWF-Journal,* no volume, no. 4 (October): 40.

WWF-Journal. 2001. "Ohne die Natur Wird der Mensch Nicht Überleben." *WWF-Journal,* no volume, no. 4 (October): 35–39.

WWF-Journal. 2002a. "Helfen Sie dem WWF, noch Größer und Stärker zu Werden." *WWF-Journal,* no volume, no. 2 (April): 43.

WWF-Journal. 2002b. "Werden Sie Internet-Aktivist gegen den Fischerei-Wahnsinn!" *WWF-Journal,* no volume, no. 3 (July): 9.

WWF-Journal. 2002c. "Die Letzten 450 Sibirischen Tiger: Paten Gesucht!" *WWF-Journal,* no volume, no. 3 (July): 21.

WWF-Journal. 2002d. "Das Krombacher Regenwald Projekt: Ein Überragender Erfolg." *WWF-Journal,* no volume, no. 3 (July): 45.

WWF-Journal. 2002e. "In Kleinen Schritten Großes Leisten." *WWF-Journal,* no volume, no. 4 (October): 4.

WWF-Journal. 2002f. "Hüter des Bedrohten Schatzes." *WWF-Journal,* no volume, no. 4 (October): 6–11.

WWF-Magazin. 2003a. "Power Switch! Umschalten auf Saubere Energien." *WWF-Magazin,* no volume, no. 3 (July): 4.

WWF-Magazin. 2003b. "WWF in Berlin." *WWF-Magazin,* no volume, no. 3 (July): 5.

WWF-Magazin. 2003c. "Krombacher: Genießen für den Regenwald." *WWF-Magazin,* no volume, no. 3 (July): 20–21.

WWF-Magazin. 2004. "Rendite für Ihr Geld: Gewinn für die Natur." *WWF-Magazin,* no volume, no. 4 (October): 23.

WWF-Magazin. 2005. "Menschen Überzeugen." *WWF-Magazin,* no volume, no. 3 (July): 18–19.

Zahrnt, Angelika. 1991. "Marketing Strategie und Moralischer Diskurs." *Ökologische Briefe,* no volume, no. 51/52 (18 December): 18–19.

Zahrnt, Angelika. 1993. "Die Enttäuschung ist Da." *Politische Ökologie* 11, no. 31 (May/June): 66–67.

Zahrnt, Angelika. 2002. "Der Kapitalismus- und Wachstum Ansatz." In *Die Zukunft der Umweltbewegung: Analysen und Strategien in 10 Interviews,* ed. Nick Reimer. Leipzig: Baerens & Fuss. Pp. 11–15.

Zald, Mayer N., and Roberta Ash. 1966. "Social Movement Organizations: Growth, Decay, and Change." *Social Forces* 44, no. 3 (March): 327–341.

Zeller, Thomas. 2003. "'Ganz Deutschland Sein Garten': Alwin Seifert und die Landschaft des Nationalsozialismus." In *Naturschutz und Nationalsozialismus,* ed. Joachim Radkau and Frank Uekötter. Frankfurt: Campus. Pp. 273–308.

Zentrale Kommission der Natur- und Heimatschutzfreunde. 1956. *...und der Zukunft Zugewandt: Die Arbeit der Zentralen Kommission Natur- und Heimatschutz von 1950 bis 1956.* Berlin: Zentrale Kommission Natur- und Heimatschutzfreunde im Kulturbund zur Demokratischen Erneuerung Deutschlands.

Zeuner, Bodo. 1991. "Die Partei der Grünen: Zwischen Bewegung und Staat." In *Die Bundesrepublik in den Achtziger Jahren: Innenpolitik, Politische Kultur, Außenpolitik,* ed. Werner Süß. Opladen: Leske + Budrich. Pp. 53–68.

Zidek, Werner. 2003. "Der Gläserne WWF: Organisationsstruktur, Verwendung der Spendengelder, MitarbeiterInnen-Profil." In *Das Große Buch des WWF. 40 Jahre Naturschutz für und mit den Menschen,* ed. Klaus-Henning Groth. Steinfurt: Tecklenborg Verlag. Pp. 223–229.

Zillessen, Renate, and Dieter Rahmel. 1991. *Umweltsponsoring: Erfahrungsberichte von Unternehmern und Verbänden.* Wiesbaden: Betriebswirtschaftlicher Verlag Dr. Th, Gabker.

Zimmer, Annette. 1996. *Vereine – Basiselement der Demokratie.* Opladen: Leske + Budrich.

Zimmer, Jochen. 1984. "Mit Uns Zieht die Neue Zeit." In *Mit Uns Zieht die Neue Zeit: Die Naturfreunde,* ed. Jochen Zimmer. Cologne: Pahl-Rugenstein. Pp. 12–30.

Zimmer, Jochen. 1989. "Soziales Wandern: Zur Proletarischen Naturaneignung." In *Besiegte Natur: Geschichte der Umwelt im 19. und 20. Jahrhundert,* ed. Franz-Josef Brüggemeier and Thomas Rommelspacher. Munich: C.H. Beck. Pp. 158–167.

Zucker, Lynne G. 1983. "Organizations as Institutions." In *Research in the Sociology of Organizations 2,* ed. Samuel B. Bacharach. Greenwich, CT: JAI Press. Pp. 1–48.

Zucker, Lynne G. 1987. "Institutional Theories of Organizations." *Annual Review of Sociology* 13: 443–464.

Zuhorn, Karl. 1954. *50 Jahre Deutscher Bund Heimatbund.* Neuss am Rhein: Deutscher Heimatbund.

INDEX

Index note: page references with an *f* or *t* indicate a figure or table.

M

N